城市设计篇（第二版）

变城记

City Changing

中国建筑工业出版社

图书在版编目(CIP)数据

变城记：城市设计篇/伍新凤编著. —2版
北京：中国建筑工业出版社，2011.10

ISBN 978-7-112-13659-9

Ⅰ. ①变… Ⅱ. ①伍… Ⅲ. ①城市规划—建筑设计
Ⅳ. ①TU984

中国版本图书馆CIP数据核字(2011)第201161号

责任编辑：常 燕 付 娇

变城记：城市设计篇（第二版）
伍新凤 编著

*

中国建筑工业出版社出版、发行(北京西郊百万庄)
各地新华书店、建筑书店经销
贵州天海规划设计有限公司制版
广州市一丰印刷有限公司印制

*

开本：546×886毫米 1/12 印张：31¼ 字数:941千字
2011年10月第二版 2011年10月第二次印刷
定价：250.00元
ISBN 978-7-112-13659-9
(21433)

编委 Compiling Members

作者简介——伍新凤

1958年生于湖南凤凰，毕业于贵州大学艺术学院美术系
贵州天海规划设计有限公司法人代表、董事长、总经理、首席策划、规划、设计师
贵州民族学院建筑工程学院副教授
中国景观设计专业委员会副主任
国际ICAD注册A级职业景观设计师
贵州省美术家协会理事
贵州省旅游规划评审专家组成员
贵州省设计艺术委员会副主任

经历风雨，刚毅顽强，立志高远，他做过知青，当过矿工，办过工厂，在深圳打过工，曾在贵州民族学院美术系任教，现任教于贵州民族学院建筑工程学院，2002 年创建了贵州天海规划设计有限公司……这些承载了他 50 载的经历。

十多年前，当刚接触到规划设计行业时，他突然感觉到这就是其人生理想之所在。天行健，君子以自强不息，地势坤，君子以厚德载物，在他的脑海里，世间有一种比山峦更高更远的风景，那就是天空；世间有一种比江河更宽更深的景象，那就是海洋，普天之下，四海之内，万物囊于其中。于是 2002 年他注册了贵州天海规划设计有限公司。他立志于通过天海规划这样一个平台在体现个人价值的同时，为贵州乃至全国的规划设计行业贡献自己的一份力量，为他的学生带去更鲜活更具竞争力的理念。

博观约取，厚积薄发，天海自成立以来正在以惊人的速度蓬勃发展，给予世人不断的震撼。

每个企业都需要有个领军人物，带来好的思路，先进的管理方法，与时俱进的开拓精神，引导这个企业的健康发展，显然，他便是这个赋予天海灵魂的人。做人当心志要苦，意趣要乐；气度要宏，言动要谨；修学不以诚，则学杂；行事不以诚，则事改；自谋不以诚，则是欺自心而自弃其忠；与人不以诚，则是丧其德而增人之怨，这便是他的人生信条。正是这样一个人，以他丰富的人生阅历，深刻的人生感悟，独到的眼光，诚信刚毅的品质，演绎着天海的完美绝伦，让天海之船驶向辉煌。

Introduction of the Author —Wu Xinfeng

He was born in Fenghuang Hunan 1958. As a design major, he graduated from the Art Institute of Guizhou University.
Wu Xinfeng—Legal representor, general manager, prime planner, designer of Guizhou Tianhai Planning and Design Co., Ltd.

A professor of Guizhou National University
Vice Director of Professional Committee of Landscape Design of China
Class A Professional Landscape Designer of International ICAD
Executive Member of the Artists Association of Guizhou Province
One of the Expert Group of Tourism Planning of Guizhou Province
Vice Director of Design and Art Committee of Guizhou Province

With massive life experience of working in the countryside as educated youth, in the mines as miner, in some corporations of Shenzhen, and running his our plants, he has become stronger in mind and more promising,And now he is a professor in Architecture Institute of Guizhou National Collage, he set up Tianhai Planning and Design Company in 2002, which is on the basis of his fifty years life experience.

Once he contacted the field of planning and design more than ten years ago, he was aware that it is the very point of his life expectation. Thus he registered the foundation of his company with the name of Tianhai (天海) Planning and Design Company with the hope to achieve his individual value, contribute to the planning and design profession of Guizhou province and even China, and bring his students more vivid and competitive concepts. The Chinese characters "天海" mean the universe and oceans. While the universe keeps rolling on orbit, a superior man makes untiring endeavour for advancement In Wu xinfeng's mind, universe is the highest and most distant landscape, while ocean the broadest and deepest scenery, of the two which everything is contained.

Tianhai has been developing at the incredible speed since its foundation, which recuires massive learning and accumulation, and of which people are shocked continually.

A leader, with great guidance, advanced managing method and pioneer spirit of keeping pace with the times, is needed in an enterprise. Obviously, he is the very person who endows Tianhai with soul. With his life credo of bearing hardship, pleasing mind, and extensive tolerance and cautious behavior, he believe that sincerity is the basic principle of doing everything and the guarantee of successful enterprise, without sincerity, one may fail in learning and lose his friends and even himself. It is he, with his rich life experience, profound life understanding, unique viewpoint and faithful and stouthearted quality, leads Tianhai going towards brilliance.

“一个行业游戏规则六十年未变，城市却已经变化巨大、沧海桑田，由此而带来灾难性的后果，那就是所有中国城市同质化现象严重，变得千城一面。”

千城一面何时止?!
——对话伍新凤
A THOUSAND OF THE SAME CITY

- 中国 30 年的变化让世界感到震撼，特别是北京、上海这样的大城市，现代化程度可以与纽约媲美。
- 中国一直在宣传自己五千年的历史文化，长期引以为豪。遗憾的是，在中国的城市里很难看到五千年历史文化的辉煌。
- 我走在中国的大城市，如果没有中文，街上行人不是中国人，我根本不知道自己身处于哪个国家。

这三句话，是一位美国专家访问中国，被问及关于中国城市的观感时所言。这三句话一直令伍新凤耿耿于怀。作为一个城市规划的圈内人，也许他更加有切肤之痛。

伍新凤是较早认识到这个问题严重性的业内人士之一，记得很多年前采访他时，提起这个问题，焦急、愤怒、心有不甘便溢于言表。不幸的是，几年过去了，“千城一面”的现象非但没有得到有效遏制，反而有愈演愈烈之势。有人戏称，中国 200 个城市一奶同胞。英国《卫报》说：“中国是一个由一千座雷同的城市构成的国家。”不知道是调侃，还是“千城一面”的英文直译。作为每一个生活在其中的人，没有理由不感到压抑和悲哀。“高楼、高密度、大广场”，“呆板、毫无生气的、火柴盒般的钢筋水泥森林”——这便是当今中国大多数城市的真实写照。举个例子，中国县级城市的外环线无入口，无一例外地有一个圆形大转盘，盘中立着一个宏伟的不锈钢雕塑，多半为球形。全国城雕艺术委员会主任、中国雕塑院院长吴为山披露，近 30 年来，全国各地“托起的球”不下 100 万尊。许多城市沦为“无特点、无文化、无历史”的三无城市。

These three sentences, said by an American expert when he came over to China and was asked about his impression of Chinese cities, has been lingering over Wu Xinfeng. As a person in the occupation of urban design, he may be more suffered.

Wu Xinfeng is one of the professional insiders who is aware of the seriousness of this problem. I remember many years ago I interviewed him, he was worried, angry and enterprising when talking about this issue. Unfortunately, in stead of being effectively confined, the phenomenon of "singularity of thousands of cities" has become more and more intense. It is quipped that more than 200 cities of China are twins. The UK Guardian said:"China is a country contributed with one thousand similar cities". Is it ridicule, or the literal translation of the "singularity of thousands of cities"? Nobody has no reason to be pressed and sad living in such cities. "high buildings, high density, large squares ", "stiffness, no activeness, and the

"The rules hadn't been changed for sixty years , but the cities have transformed completely, which causes badly similarity of Chinese cities"

What time is the end of "singularity of all cities" An interview with Wu Xinfeng

A THOUSAND OF THE SAME CITY

- The world is shocked by the transformation of China in 30 years, especially the metropolises such as Beijing and Shanghai, of which the modernization can compete with New York.
- China has publicized and been proud of its five thousand years of historical culture. However, it is hardly to see the brilliance of the five thousand years of historical culture.
- When I walk through the metropolises of China, I even can't know which country I am in if without Chinese characters and pedestrians

中国城市的“千城一面”广受诟病非今日始。英国皇家城市规划学会主席帕白森曾痛心疾首：“中国历史和文化的传统太珍贵了！不能允许它们被西方来的这些虚假的、肤浅的、标准化概念的洪水所淹没。我确信，你们遭到了这种威胁。”其实，这已经不是威胁，而是活生生的现实。上海世博会的总设计师，法国人马丁·罗班也说中国城市普遍缺乏特点。从他住的酒店望出去，城市风貌都差不多。著名建筑师库哈斯则这样调侃他的中国同行：“中国建筑师的数量是美国建筑师的十分之一，在五分之一的时间内设计了五倍数量的建筑。这就是说，中国建筑师的效率是美国同行的 2500 倍。”

在国内，城市的“千城一面”则更是受到不断质疑。作家冯骥才说：“近年来，我国的城市有非常迅猛的发展，但是也有不可挽回的损失，‘千城一面’失去了文化个性，失去了自己独有的文化灵魂，破坏了文化的多样性。”上海大学教授林少雄给博士生放幻灯片，讲述在国内不同城市的游历观感，放着放着就串了，切换了好多张，好像还没有走出一座城市，其实已经是好几座城市了。

matchbox-shape concrete forest"— this is the real image of most of the present Chinese city. For example, the outer races of county-level cities in China are without entrances, in stead, a large round platform with a stainless steel sculpture, usually in a ball shape which, revealed by Wu Weishan, director of national urban sculpture art committee and the president of sculpture institute of China, were no less than one million in China. Many cities turn down to be without specialty, culture and history.

In China, the "singularity of all cities" has drawn much attention. "In recent years, although the cities have developed rapidly in China, we got irretrievable loss. Singularity of thousands of cities causes the loss of culture and the unique cultural soul, and the damage of cultural diversity", an author Feng Jicai said. "It seems to be in one city, but already run through many cities", displayed by Professor Lin Shaoxiong of Shanghai University to his doctoral students.

林少雄说："土耳其诗人希格梅有句名言，'人一生中有两样东西是永远不能忘却的，这就是母亲的面孔和城市的面貌。'可我们的城市只有同一副面孔。几乎百分之百的中国城市互相抄袭，三线看二线、二线看一线，只有楼房高矮、汽车多少等量的区别，没有质的区别，这就叫'同质化'。"许多历史、人文学者也表达了同样的观点。《中国青年报》甚至评论说："没有思想和个性的城市犹如没有灵魂的人。对于具有几千年城市史的中国来说。如果城市市区特点，将历史文化的底蕴破坏殆尽，'千城一面'都像一个模子里刻出来的，那么，不仅是中国建筑和中国城市的悲哀，也是对中国文明的糟蹋。"

悲哀，莫大的悲哀！

罗素先生说："须知参差多态，乃是幸福的本源。"亿万中国人生活在一模一样，毫无差别的城市里，生活的幸福感受，不说毁之殆尽，至少也大打折扣，这也难怪为什么官方统计的生活幸福指数，总和普罗大众的切身感受有着相当大的出入。

要知道，个人是无法脱离群体和环境生存的。人们之所以选择进入城市生活，就是要享受城市生活的品质。因为只有城市这种集聚形式，才集中了现代文明和先进科技的生活方式，为生活提供了优越的条件。但是，假如一个城市以建设为名，将历史的传承印记从城市上抹去，将优秀的文化遗产破坏殆尽，将值得继承和发扬的传统资源弃若敝屣的时候，城市无所凭借，变成无根之城；人心失去依托，灵魂漂浮半空，于是物欲横流、人情冷漠、道德沦丧。看上去毁掉的是城市，实质上泯灭的是世道人心。这也是为什么"千城一面"并不仅仅是一个城市发展路径的选择问题，也是一个很严峻的社会现实问题的重要原因。

可是，为什么"千城一面"饱受质疑与诟病并非今日始，却依然愈演愈烈？为什么这个长期为社会瞩目的问题，会迟迟没有办法解决，并逐步沦为茶余饭后的谈资？

2010年末，新华网刊发了一篇题为《拒绝"千城一面"：中国西部给城市穿上"民族服装"》的报道，给解决城市的同质化现象带来了一丝新鲜气息。文章对民族地区运用来自生活的当地民族元素改造城市给予积极评价，其中特别提到了凯里，认为凯里的公共汽车站使用坡屋面和垂瓜，体育馆吸取了苗族的吊脚楼、侗族的鼓楼等特点，并使用了许多苗侗服饰上的元素，使得凯里的城市特点突出，给人印象深刻。同济大学规划六所所长，上海世博会规划顾问委员会成员，城市规划博士，《城市中国》执行副主编苏运升在文

Lin Shaoxiong said, "A turkey poet said 'In one's life,there are two things that are unforgettable for ever;one is his mother's face,the other is his town's image.' However, our cities are in the same image, almost all the Chinese cities copy each other. Small cities imitate medium cities, while medium cities imitate big city. The difference is not essentially, but the different level of constructions and different number of cars. This is so-called 'homogenization'". Many historical and human learners also express the same idea. The Youth Daily even presents comment that cities without thought and uniqueness are like people without soul. As a country with thousands of years of urban history, China will be sorrowful of its constructions and cities, and ravaged of its civilization, if the cities lost their features and were damaged totally on their historical culture.

Sorrow, badly sorrow!

Mr. Luosu says: "we should recognize that the diversity is the origin of happiness. " However, millions of Chinese are living in the cities which are in the same image. No wonder the happiness index by authority statistics is fair different from the real public experience.

We should know that an individual can not exist without the groups, so that people choose to live in the cities to enjoy the living quality of cities. Only this collective living style can provide the superior conditions with modern civilization and advanced technology. Nevertheless, if a city were constructed by erasing the historical inheritance, destroying the cultural heritage and deleting the valuable traditional resources, it will become a city without root, and which is the cause of shallow relationships and moral bankruptcy. In all, "singularity of thousands of cities" is not only an issue of the developing method of a city, but also a severe realistic social problem.

Why "singularity of thousands of cities" has become more and more intense since the query is not from today? Why such a hot spot hasn't been solved, but turned off to be conversaticn resources?

At the end of 2010, "Xinhua Net" issued a report named "Reject 'Singularity of Thousands of Cities': The West of China Put Ethnic Costume on the Cities" which introduced a fresh idea to figure out the problems of urban homogenization. The paper presented positive comments on the urban transformation using local ethnic elements of the ethnic regions. Kaili City was especially mentioned in the paper as a city with prominent features and will leave people deep impression

2011年5月，一个阳光明媚的早晨，带着对中国城市"千城一面"的关注与困惑，我们造访了位于贵阳华坤大厦16楼的天海规划设计有限公司，在伍新凤的办公室进行了一场对话。

对公共空间可能提供的美、安慰和关怀，人们总是寄希望于规划。然而现实给人们的一个嘲讽是，大部分所谓的规划正在成为一种破坏。举目四望，"规划"所到之处，城市原有的魅力尽失。"规划正在变成城市魅力的杀手。"

中评价说："比如凯里的体育场，风雨廊和墙体的结合从场内和场外看都和水平线是平衡的，看起来很舒服。"

凯里的城市规划设计者，正是伍新凤。

多年来，伍新凤一直致力于解决"千城一面"的思路与模式的探索，并在九年前创办天海规划作为实践。天海成立九年来，坚持"民族现代化、现代民族化"的城市规划新理念，坚持城市发展的"可延续性"和挖掘城市的"核心价值"，在此基础上打造特色城市。

目前，经过九年探索，天海规划设计已经形成了一套规划设计方法和模式：在项目规划满足功能需求的同时更强调对规划地核心价值的深入挖掘，对规划地民族性、文化性和本土性的全力彰显并最终使之体现于项目中；整合策划、规划、建筑、创意、效果表现、动画等与规划设计密切相关的各个行业于一体，使得规划设计从资源评估、背景分析到战略定位、规划布局、建筑创意、效果表现一气呵成，最大限度地克服了传统规划设计模式中各个程序脱节的弊端。

多年来，伍新凤和天海规划以独特的规划设计理念、颠覆性的操作模式以及极富想象力的创意，承担和完成了60多项大中型规划设计项目，包括云南省丘北县县城整体设计及普者黑景区规划设计、贵州省凯里市城市整体形象设计、贵州省铜仁市整体设计、贵州省三都县县城整体设计、贵州省黔西县县城整体设计、贵州省威宁县县城整体设计、山东黄金主题公园规划设计、华藏世界多国佛教园规划设计、贵州省东风湖百里画廊休闲度假区规划设计、贵州省香纸沟温泉度假区规划设计、贵州省博物馆创意设计等。这些项目的实践，都是伍新凤为改变"千城一面"现状而作的可贵尝试。

2010年8月，伍新凤根据多年思考和实践经历编著的《变城记——城市规划改造篇》、《筑魂记——原创建筑篇》、《梦游记——旅游规划篇》、《造景记——原创景观篇》四部专著由中国建筑工业出版社出版，这四部专著总结了伍新凤八年来从事规划设计行业的经验及独到见解，同时将天海的项目操作流程及规划设计理念向读者展示，试图对规划设计行业引入新的操作模式和表现手法，并为这行业带来一些新鲜观点供有关人士借鉴与参考。

as its bus stations were designed adopting pitched roof and gourd and the gymnasium was designed absorbing the features of the Diaojiao pavilion of Miao and the Drum pavilion of Dong and some ethnic elements of the costumes of Miao and Dong.

It is Wu Xinfeng who designed Kaili City.

Wu Xinfeng has devoted himself to the exploration of tackling "singularity of thousands of cities" and organized Tianhai Planning and Design to make practices for nine years. Tianhai has been focusing on the new concept of urban design "Modernization of Nationality, Nationalization of Modernity" and "sustainability" of city development and digging the "core value" of cities, so as to create special cities.

At present, Tianhai Planning and Design has concluded a certain method and mode of planning and design after nine years of exploration: In addition to meeting the needs of functions, the projects are stressed more on excavating the core value and highlighting the nationalities, culture and localities of the planning land, so as to express them in the design; The involved professions such as scheming, planning, architecture, design, 3D display and animation display are integrated with each other to complete a project from the resources assessment and background analysis to strategic orientation, planning layout, architectural design and image display, so as to avoid the processes apart.

During these years, Tianhai Planning and Design has undertaken and completed more than 60 large and medium projects with its unique planning and design ideas, subversive operation mode and innovative creation. The works, such as the project in Qiubei City of Yunnan, Kaili City, Tongren City, and Sandu City of Guizhou, and the Golden Theme Park of Shandong, the design of Sino Tibetan World of Qingzhen, the creative design of Guizhou Museum, planning and design of Xiangzhigou Resort of Guiyang, are the valuable practice that Wu Xinfeng engaged in to change the the current situation of "singularity of thousands of cities".

In August 2008, Wu Xinfeng completed his four monographs named "City Changing—Urban Design", "Spirit Shaping—Original Architecture", "Dreamlike Traveling—Tourism Planning", "Landscape Building—Original Landscape" which are issued by China Architecture and Industry Press, and conclude Wu Xinfeng's experience and unique ideas after eight years working on this profession.

A sunny morning in May, 2011, we visited Tianhai Planning in the 16th floor of Huakun Building and had a dialogue with Wu Xinfeng in his office.

People always expect that planning can make the public space beautiful, comfortable and agreeable. However, the reality is so ridiculous that most of the so called "planning" is turning out to be ruin. "Planning is becoming the assassin of urban charm"

对话 // Dialogue

Newtarget：中国城市“千城一面”的现象已被诟病多时，但到底是什么造成这种现象的原因众说纷纭，作为一个长期从事规划设计并在这个问题上有所建树的专家，您是怎么看待这个问题的?

伍新凤：造成中国城市“千城一面”的原因有三。

第一是城市规划设计行业的操作模式和游戏规则六十年来未变。六十年不变的城市规划行业，怎么能适应和跟上城市翻天覆地的变化?

第二是体制性的问题，行政主导学术，个别政府官员将城市建设作为升迁的政绩，忽略了这个专业的特殊性，不重视专家的意见，搞“政绩工程”，因此作出的规划往往不是出自科学和文化论断，而是出自个人判断和喜好。

第三是行业管理上的问题。由于行业管理上多年不变，思想僵化、缺乏创新，造成了对决策部门的误导。

正是这三个原因，造成了中国城市的“千城一面”。其中，游戏规则和操作模式六十年未变，是最核心和最根本的原因。

Newtarget:那么，应对这样的问题，有没有什么行之有效的解决办法?

伍新凤：一个城市规划项目，同时要面对六七个不同的学科和专业，现行体制下，涉及城市规划的各个领域之间互不相关，甚至相互拆台，这样的现实，怎么能做好城市规划?

改变“千城一面”的现实，关键在于发现潜入城市肌理的文脉。当一种资源成为一个城市的特色，不仅为世人所知，而且被挖掘出来，通过一定的手段展现，为这个城市打上深深的烙印，那么这种资源就构成了这个城市的核心价值。所以，在城市规划之前要首先对城市进行资源评估和定位，再围绕核心价值和定位进行功能布局，创造方便舒适内容丰富的新生活。最后，通过艺术创作手段，使城市建筑或景观的现实性升华为公众所能感受到的具象。

规划要在设计之前，城市规划师一定要具备人文素质和城市战略眼光，而不是一个只会规划设计的工匠。中国建筑大师梁思成，是梁启超先生的大公子，从小就攻读《史记》、《左传》，有很深的国学功底，念书时爱好美术和音乐，是校刊的美术编辑。在宾夕法尼亚大学学习建筑期间，自学中国古代历史。正是这样的历史积淀，才使得他的眼光超越了时代，对当时的北京城市建设提出许多真知灼见，当时没有被采纳。直到 1993 年，北京的城市规划方案，最终还是回到了梁思成的思路上。

Newtarget：In China, the phenomenon of "singularity of thousands of cities" has existed for a long time, and there are different versions of the cause of this phenomenon. As an expert working on planning and design and making contributions to this field, what do you think about this problem ?

WU：There are three causes of "singularity of thousands of cities" in China. First of all, the operation mode and occupation rules never changes for sixty years. But the cities has experienced earth-shaking changes during the sixty years, so how can the old urban design concept adapt to the changing cities? Secondly, it is a systematic issue, the administration confines the academy. Some of the governors treat urban construction as his merit for his promotion, but ignore the specialty of this occupation and the experts' suggestions, so that the planning is generally not according to scientific and cultural conclusion, but the personal judgement and favor. Thirdly, it is the management.

These three points are exactly the cause of "singularity of thousands of cities", and the core and essential reason is the unchanged operational mode and rules for sixty years.

Newtarget：Then, do you have any methods to figure out these problems effectively ?

WU：There are six or seven different objects and specialties involved when carry out a project of planning and design of a city, unfortunately, the involved departments don't cooperate with each other or even discriminate against each other due to the present system. Under this circumstance, how to make a good urban design.

The key to change the reality of "singularity of thousands of cities" is to discover the cultural context which permeates through the city. If a kind of resource is one of the specialties of a city, it will be known and displayed by certain methods to brand the city deeply, and it, therefore, will contribute to the core value of this city. Thus, the process of urban design is, firstly, assessing resources and making orientation, then working out the functional layout according to the core values and orientation to create convenient and comfortable new life, and lastly, transforming the urban architectures and landscape to be real constructions by art methods, so that the public can enjoy them.

Planning should be prior to design. Urban planners can't be a workman of planning and design, but should be of humanistic quality and urban strategy. The Chinese architectural master Liang Sicheng, the first son of Liang Qichao, accumulated wide Chinese ancient civilization as he had study "Historical Records" and "Zuo Zhuan" since his childhood, served as the art editor of the school journal due to his favor of art and music when he was in the school, and learned by himself the ancient Chinese history during his study in University of Pennsylvania. All the experiences endowed him with foresight, so

《中国新闻周刊》和新浪网曾经进行一项名为“你的城市理想”的网上调查，16877 人中，有 29.46% 的人们最希望自己所居住的城市建筑具有的品质是“外观很美，有艺术感的”。在“你希望居住的城市的历史年龄”一项中，38.38% 的人希望居住城市的历史年龄在 200 以上。一旦离开城市，多数人表示，除去经济因素，最受不了的一点是“生活缺少变化”。根据第十五次中国互联网报告发布的数据，69% 网民的年龄在 35 岁以下，可见，丰富多变、有历史内涵和审美情趣的城市“魅力”，已经成为当下年轻人城市理想的价值偏好。

我们渴望生活的富足与美好，但更渴望精神的充实与尊严，我们有权利希望更好的城市文明。从这一点上，我们对伍新凤与他的实践充满期待。

Newtarget：天海也是正在做着这方面的努力？

伍新凤：是的，我们认为城市创意设计是主题性、原创性和震撼性的综合体——以体现城市定位为核心和主题，抓住城市文脉特征，从人文背景中挖掘创意设计元素，从公众的视觉、听觉感受出发，以新颖独特为生命进行城市建筑和景观的创意设计。

同时，在进行创意设计时还要注重现代与民族融合。民族与时尚并不矛盾，民族并不代表落后，也不仅存在于历史中，只要合理挖掘，城市规划完全可以将民族的变成时尚的主体，带领人们穿越时空，去感受民族与现代的交融。

在操作层面，天海整合策划、规划、建筑、创意、效果表现、动画等与规划设计密切相关的行业于一体，使得规划设计从资源评估、背景分析到战略定位、规划布局、建筑创意、效果表现一气呵成，力图以全新理念与操作模式来改变当前行业的弊端。

Newtarget：在您眼里，一个美好的城市应该是什么样子？

伍新凤：马丁·罗班说："一百年前说到城市，我们讲得最多的是经济增长、工业发展，很少有人站在人的角度去评价城市。如今我们认识到，城市是为人而建，人才是城市里最重要的元素。衡量一个城市是否有魅力，最重要的标准是，当地人在当地文化历史背景下，对空间是否感觉舒适惬意。"

维也纳与音乐、鹿特丹与航海、哥本哈根与童话、威尼斯与水……一个成功的城市应该是有性格、有魅力的个性城市，是建基于自己文化传统、社会习惯和民风民俗基础上的集合。我到法国考察，法国议会大厦浮雕上的灰尘鸟粪很厚，据说一两百年没打扫，这是一种对自身文化的敬畏与自豪，这种态度很值得今天的城市规划者们学习。

一个美好的城市，不仅是 GDP 增长快速的城市，而应该是性格鲜明、文化积累深厚、历史传统清晰、生活方便舒适，并能有长时段可延续发展的城市。而绝不是高楼大厦密集，但却让人困惑、受到伤害、感到失望甚至绝望的城市。

that he presented many valuable suggestions to urban construction of Beijing City. However, his suggestions were not adopted until 1993 when the proposal of Beijing urban design came back to Liang Sicheng's ideas.

Newtarget : Is Tianhai making effort on this aspect ?

WU : Yes, we believe that the creative urban design should be thematic, original and shocked. The proposal should be of the core and theme of describing urban orientation, and innovation and uniqueness of urban architectures and landscape with creative elements from cultural background and considering public visual and auditory sense.

At the same time, the integration of fashion and ethnicity should be focused on while make the design. Ethnicity is neither contradictory against fashion, nor the symbol of backwardness. Ethnicity can be absolutely transformed into fashionable mainstay if developed reasonably to lead people to experience the integration of fashion and ethnicity through time and space.

Tianhai unites the specialties of scheming, planning, architecture, design, 3D display and animation to complete a project entirely from the resources assessment, background analysis to strategic orientation, layout, architecture and animation display, with the hope to avoid the disadvantages of present profession with a brand-new operation concept and mode.

Newtarget : What is a nice city like in your mind ?

WU : "Martin Robain said:"We talked more about economic increase, industry development, but seldom gave comment from the views of human when it comes to cites hundred of years ago. Now we recognize that cities are constructed for human who is the most important elements of the cities. A city is charming or not mainly depending on if local people feel agreeable and comfortable "

Vienna and music, Rotterdam and voyage, Copenhagen and fairy tales, and Venice and water– a successful city should be of its character and charm, and the union on the basis of its own cultural tradition, social custom and folk styles. It is said that the sculptures of French Parliament building carry a lot of dust and bird droppingshas as they have not been cleaned for one or two hundred years, which I learned when I traveled in France. It is a kind of respect and pride of their own culture, and worth the planners learning.

A nice city should not be of rapid GDP increase only, but of striking characters, accumulated culture, clear historical tradition, convenient and comfortable enviroment, and sustainable development. It is absolutely not a nice city which is filled of confusion, damage, disappointment and desperation even with dense tall buildings.

The "China Newsweek" and Sina net have ever made a investigation named "Your Ideals of the City" on internet, among 16877 informants, 29.46% of them hope their cities with "beautiful looks and art sense", and 38.38% of the informants hope that the cities they lived are with more than 200 years history when they are asked "how old do you hope your city is". Most of them said except the economic factor, what they can't stand is "life is lack of alternation" once they leave the cities. 69% netizen are no more than 35 years old according to the statistics issued by the Fifteenth China Internet Report, thus, the urban "charm" of diversity, historical connotation and aesthetic taste has become the modern youth's preference of value. We are eager for rich and beautiful life, but what we aspire more is substantial and dignity. We have the right to expect a better urban civilization, so that we are looking forward to Wu Xinfeng's practices.

自序 PREFACE

“中国的城市化与美国的高科技发展将是深刻影响 21 世纪人类发展的两大主题。”

诺贝尔经济学奖获得者——斯蒂格利茨

"China's urbanization and America's development of high-tech will be the two great themes to impact profoundly human development in 21st century".

The winner of the Nobel Prize on economy—Stiglitz

进入 21 世纪以来，中国经济持续快速发展，城市化进程不断加快，成为贯穿本世纪的主题。在此背景下，如何确保城市战略发展方向的准确性，如何实现城市的可持续发展，如何最大限度地构建城市核心竞争力成为城市可持续发展的核心。因此，在这个时代，规划设计行业肩负起此项重任。毫不夸张地说，当下是属于规划设计行业的，是规划设计行业展示自我的大好时机——在此过程中，责任与压力并存。

路漫漫其修远兮，吾将上下而求索。回首改革开放以来的 30 年，我国的城市规划虽说取得了很大进展，但是却陷入了“千城一面”的泥潭。城市是人类文明的主要组成部分，城市也是伴随人类文明与进步发展起来的。随着工商业的发展，城市还成为了传播文明的场所。遗憾的是，当下的中国城市，文化缺失、内涵不足、特色不明、协调不力，整体上略显浮躁，没有成为传播当地文明的有效载体——中国城市处于病态中，就像每个人从出生到成长都会不同程度地经历病痛一样，这是当今中国社会形态下所必经的阶段。缓解病情、治愈疾病，提高免疫力，规划设计师责无旁贷。

没有一味地批判，没有一味地指责，仅仅是作为一个规划人对我国规划设计现状进行反思。

反思一：规划设计行业体制

在计划经济体制下规划设计工作主要是政府行为，规划设计行业统一在事业编制下，规划设计任务由上级主管部门统一下发，行业竞争及压力有限。进入市场经济体制后，这一特殊行业现象在很大程度上延续下来，与此共同沿袭下来的还有几十年来一成不变的规划设计模式，缺乏与时俱进的创新思维，结果便是规划设计中规中矩，却缺乏特色，缺乏灵气，缺乏创意，更缺乏社会责任感。

反思二：规划设计模式

一个项目需要多个专业之间的配合才能完成。然而这些相关专业都是各自独立的，专业之间各自为政，本位主义较强。一个项目仅凭一个专业解决问题是远远不够的。现状是：从策划、规划到设计缺乏协调，严重脱节，导致规划逐步细化后与最初构想相去甚远。

By 21st century, with the rapid and continual development of economy in China, urbanization has speeded up and become the theme throughout this century. Under this circumstance, the core of sustainable development of cities has depended on how to make sure the right direction of strategic development of cities, accomplish sustainable development and form urban core competitions as strong as possible. And this is the responsibility that the planning and design profession should take. Thus, we dare say that this is the era of planning and design field and the best opportunity for the planning and design unities to express themselves with responsibility and stress.

The way stretches endless ahead, I shall search. What a long journey ahead!

Although urban planning in our country has made great progress during the thirty years since reform and opening up, it has been put in an awkward situation of "one thousand cities, one appearance". As the main components of human civilization, cities have developed along with the human civilization and progress. With the development of industry and business, cities have become the places spreading local civilization, too. Nevertheless, the current cities, with deficient culture and connotation, unobvious characteristics, and weak coordination abilities, are not successful to disseminate local civilization effectively. Chinese cities are in sick! This is the inevitable period for cities under the current social formation in China just like every people should go through lots of pain in their growth. Releasing sickness and enhancing immunity are the responsibilities of the planners and the planning and design profession.

As there is no absolute criticism, there is no absolute blame. Just as a planning person, I've made reassessment of the current status of the planning and design of our country.

First, system of planning and design

Planning and design is the governmental work during the planned economy that the field of planning and design is controlled by the governmental departments and the tasks of planning and design are arranged by the state, which caused little competition and press in this field. This phenomenon has been lasted after entering market economy, as well as the planning mode which has not changed for many decades and has been lack of innovative ideas. Thus, the planning works are regular and common, without characteristics and soul.

Second, mode of planning and design

Completing a project needs the cooperation of several separate specialties which, however, are leaded by different departments, and it's hard for them to reach a consensus on the project, as a result, the processes of scheme, planning and design are divorced seriously from each other and the initial image is different a lot from real effect, also working efficiency is not in control.

反思三：规划设计教育

规划设计师的职责是宏观地制定区域发展路线，定格区域发展基调，合理地进行区域功能布局。这需要宏观经济学、社会学、历史学等大量学科的知识积累，只有这样才能站在正确的战略高度上去计划区域未来的发展。在现今的教育制度下，更多的规划设计专业将大量精力放在了诸如行业规划标准，管网道路铺设，制图技术等微观战术层面上，对宏观战略的理解把握严重缺乏。除此之外，规划设计行业还需要艺术思维、创新思维的体现，但是现今“正统的”规划设计行业对审美、造型艺术等方面的教育缺乏甚至是空白。很难想象，一个缺乏大局观、缺乏审美眼光的规划设计师如何能使城市摆脱“千城一面”的窘境。

八年前，带着对行业的反思，怀揣着自己的信念，我创办了天海规划设计有限公司。从公司创立之初，我就不断吸取内外部经验教训，进行剖析比对，敢冒“天下之大不韪”，推行了一系列的改革措施：

规划理念方面。在项目规划时最大限度地满足功能需求的同时更强调对规划的核心价值的深入挖掘，对规划的民族性、文化性和本土性的全力彰显并最终使之体现于项目中。

规划模式方面。整合策划、规划、建筑、创意、效果表现、动画等与规划设计密切相关的各个行业于一体，使得规划设计从资源评估、背景分析到战略定位、规划布局、建筑创意、效果表现一气呵成，形成流水线规划模式，最大限度地克服了传统规划设计模式中各个程序脱节的弊端。

八年来，经过60多个项目的成功实践，在市场上赢得许多开发商和政府部门的认同和赞誉。事实证明我的改革与创新是有价值的。现在，我把近十年来的改革创新的规划设计成果分为城市设计、建筑设计、景观设计以及旅游规划四部分集结成册分别命名为“变城记”、“筑魂记”、“梦游记”、“造景记”，用我的挚热和真诚与大家共同交流探讨，同时更希望能与同行们携手为推动全国规划设计行业的改革与发展，贡献绵薄之力。

Third, education of planning and design

Urban planners, who are requested to be of large accumulation of knowledge in economics, sociology, and history and so on, take the responsibilities of working out the developing proposal of urban regions macroscopically, determining basic tone of cities and reasonably arranging urban functional districts. Only the ones who are competence enough can design the future development of the cities on a correct strategic view. Under the current educational system, many urban planners only know some about urban functional layout and distribution, various planning norms and setting of pipes and roads, but they are badly lack of macroscopic innovative consciousness. Besides, artistic and innovative thought are needed to be demonstrated in the planning and design works. But the education of aesthetic and art etc. are lost in the existing "orthodox" planning and design profession. It's hard to image how can a planner without overall sense and macro view help cities escape from the awkward situation of "one thousand cities, one appearance".

Eight years ago, I found Tianhai Planning and Design Company with rethinking about this sector and my strong conviction. We have totally discarded traditional operation mode, which was regarded "fighting a rearguard action against the mass of public opinion" at that time.

In the aspect of planning idea, besides meeting the needs of functions, our projects are stressed on excavating deeply the core value and highlighting the nationalities, culture and localities of the planned place so as to show them in the project eventually.

In the aspect of planning mode, the closely relative professions of scheme, planning, architecture, creativity, effect display, and animation etc. have been integrated in a unity, so that the planning and design work can be accomplished smoothly and completely from assessing resources and background analysis to strategic orientation, planning and layout, architectural innovation and displaying effects, so as to avoid maximally the disconnection of each step in the traditional planning and design mode.

More than 60 successful items within 8 years has won accept of many investors and relevant governments and proved that our reformation and innovation are significant. Now, we divide our works based on reform and innovation into four parts: urban design, architectural design, landscape design and tourism planning and named respectively "City Changing", "Spirit Shaping", "Dreamlike Traveling" and "Landscape Building". We hope to communicate with you sincerely and passionately and also make great efforts and contribution along with the people in the same field to the reformation and development of the planning and design field in our country.

INTRODUCTION
导言

“什么时候蛙鸣蝉声成了记忆，什么时候家乡变得如此拥挤，高楼大厦到处耸立，七彩霓虹把夜空染得如此俗气……”一首《一样的月光》唱出了人们对都市巨变的怅然若失与对精神家园的怀想。

然而，变化是世界永恒的主题。改革开放以后，中国进入了一个大变革的时代，最明显的变化就是城市化进程不断加快。随着中国经济的高速发展，城镇人口剧增，城市负担不断加大，旧城区的老化日益严重，城市原有基础设施越来越无法满足经济快速发展的要求。在这样的情况下，城市设计在很多城市如火如荼地开展起来。大规模的旧城更新和重建在短时间内大大改善了城市的物质环境，在城市内部实现了空间的拓展，但也随之出现了许多问题。例如盛行于大江南北各大城市的“城市化妆”运动，缺乏人文关怀和环境意识的大型公共建筑，目标单一的城市建设等，导致南方北方一个样，大城小城一个样，城市历史文脉被割裂，大城市面貌雷同，空间形态呆板，缺乏鲜明的特色和个性，小城市具有民族风格和地域特色的城市风貌也正在消失，千城一面的现象日趋严重。

当人们欣喜于人类改造自然的能力如此巨大之时，当人们沉醉于所谓的现代文明中时，是否曾想过本土文化的根基正在渐渐瓦解，我们的精神家园

"When have the songs of frogs and cicadas become memory? When have our hometowns become so crowd, blocks of tall buildings are standing everywhere, colorful light is rendering the night sky so garish"… a song of《the Same Moonlight》speaks out the people's emotional sense of the great urban changes and the miss of their spiritual hometown.

However, change is the eternal theme in the world. China has come into a time of change after reform and opening-up, the most obvious change is the accelerating process of urbanization. With the highly economy development in China and the rapid population increase in the cities, the old infrastructure of cities are more and more unable to meet the requirements of rapid economic development. In this circumstance, urban design has become more and more popular in many cities. Large scale urban renewal and reconstruction has greatly improved the physical environment, and realized the internal space expansion of the cities, but also brought a lot of problems. For instance, the campaign of "urban make-up" prevailed in major cities throughout our country, the large public buildings lacking humanistic care and environmental awareness, and urban buildings with signal goal, and so on, which causes the historical context of cities have been cut that the big cities are lack of characteristics and personality, and the urban style of ethnic and geographical features of small cities are disappearing. As a result, the cities, no matter in south or north, big or small, are in the similar appearance, and this problem has become more and more serious.

正在离我们远去。探究原因我们看到，我国目前的城市设计研究和实践没有跟上时代的发展和环境的变迁，许多城市的规划建设缺乏系统科学的指导，对城市空间环境没有进行有效的引导和控制，不能站在区域发展和文化传承的高度来研究城市设计方法，更新改造常常是头痛医头，脚痛医脚，城市自发无序的原始发展模式没有得到根本改变。

改变城市形象、美化城市生活是每一个城市规划师的最终目标，对于天海规划来说更是我们这个团队所有成员的光荣使命。为此，天海规划一直都在积极地探索城市设计之路，始终秉承“民族现代化，现代民族化”的理念，不断开拓创新，经过八年孜孜不倦的研究和实践，形成了自己的一套城市设计方法和模式。在这里，我们将这些城市设计的心得和经验结集成册，并取名为《变城记》，希望我们的变化之道能够为中国的城市建设理论和实践添砖加瓦，并与广大读者朋友共同探讨、一起分享。由于水平和时间有限，若有未能尽善尽美之处，请各位朋友批评指正！

When people are delighted in their strong power of transforming nature, and indulged in the so-called modern civilization, have they ever thought that the foundation of native culture is being gradually collapse, and their spiritual home is being away from them? Exploring the reasons, we have seen the main reason of this phenomenon is the current urban design research and practice have not kept pace with the development of times and change of environment that people couldn't study the methods of urban design at a high level of regional development and cultural heritage, and the planning and construction of many cities are in need of a systematic scientific guidance, which makes the disorder and spontaneous mode of original development of cites haven't been changed.

For every planner, it is his ultimate goal to change the urban image and improve urban life, which, moreover, is the glorious mission for all members of our team, Tianhai. Therefore, Guizhou Tianhai Planning and Design Co., Ltd. have been exploring methods of urban design innovatively and creatively. After seven years of study and practice, we've formed our own method and mode of urban design. We would like to transform our studies and experience to be a book and named "A Tale of Changing Cities" with the hope that our methods of change can make some contribution to the theory and practice of urban construction of China, and also to discuss and share with friends and readers. Also, we welcome criticism and advice if you find any deficiency or unsatisfactory due to our limitation time and knowledge.

城市设计篇（第二版）

变城记

City Changing

Contents 目录

1

从城市形象说起

Speaking from Urban Image

Urban competition is not only the competition of scale and economy in the modern time, but the competition of the taste, culture, living environment and quality of the cities. Essentially, it is the competition of urban image. Therefore, the construction of urban image is a higher stage of urban development following the construction of production and facilities in the process of urban modernization. It is in the modern social situation of paying more attention to image that urban image has become more important and practical.

今天的城市竞争，再也不仅仅是规模的竞争、经济的竞争，而更表现出城市文化、城市品位和城市生活环境与生活质量的竞争，实质就是城市形象的竞争。因此，城市形象建设是城市现代化过程中继生产建设、公共设施建设之后迎来的城市发展的更高阶段。尤其是在当代这样一个比过去更加关注形象、更加依赖形象而生存的社会环境中，城市形象变得比以往任何时候都更加重要、更有实际的意义。

1.1 城市形象
Urban Image

每一个城市由于地域环境、人文环境的差异，都已经有它给予大众独特的形象认知。例如谈到国内外的一些知名城市，我们都会有许多联想，如巴黎是浪漫之都、新加坡是狮城、伦敦是雾都、威尼斯是水城，北京是帝王之都……我们通过城市独有的景观或文化特征，对这些城市形成了一些固有的形象认知，说明这些城市已经给大众留下了其独有的印象。

对于什么是城市形象，美国学者凯文·林奇在 1960 年发表的著作《城市意象》中对人的"城市印象"给出了他的理解。他用一般市民认同的五个元素来构成城市形象。①路径 (path)——是通道；②边界 (Edge)——是界线、轮廓、断面等；③节点 (node)——是战略性地点、广场、交点等；④地区 (district)——是功能性、活动（购物、娱乐、文化）的集中地等；⑤地标 (landmark)——是外向性的标志。这五个方面其实就是今天所说的城市道路、城市边界、城市区域、城市节点、城市标示物。

进入 21 世纪，城市形象的概念已经得到了新的诠释——它是指人们对城市价值评判标准中各类要素如景观、自然、文化、经济等形成的综合性的总体印象和特定共识，是能够激发人的思想感情活动的城市形态。其中，城市景观是指建筑、街道（步行街）、广场、滨江滨湖滨海地带、公园和绿地，自然环境如城市的气候、地质、地貌、水文等特征；城市人文因素如：民族、民俗、政治、宗教、文化、军事等；城市经济因素如：人口规模、产业构成、经济布局等。所有这些因素的突出点均可能成为激发人的思想感情活动的因素，从而成为影响城市总体形象的关键所在。但我们也看到，尽管城市形象的概念被注入了自然、人文、经济等新的要素，但城市形象主要是由城市景观构成的这一概念并没有改变。总而言之，城市形象代表着城市的身份和个性，它反映着城市自然地理形态、历史文化的"文脉延伸"、产业结构特点、城市功能和整体视觉的特色。

We often have some association of cities when it comes to the famous cities of the foreign countries, for example, Paris is a romantic city, Singapore a lion city, London a flog city, Venice a water city, and Beijing an emperor city and so on. Then what brings us this association? The answer is the urban image. We get some image cognition through the unique landscape and cultural characteristics of cities, which proves that the cities have left people their characteristics, respectively.

What is urban image? Kevin • Lynch, who had cared about urban image earlier, had given his explanation of "urban image" of people in his book "Urban Image", published in 1960. The urban image combined with five factors in his idea: ① Path–the passage; ② Edge–the borderline, outline or cross section; ③ Node–the strategic location, square, or cross point and so on; ④ District–the functional places for activities(shopping, entertainment, culture); ⑤ Landmark– the external symbol. In fact, these five aspects are the urban street, urban boundaries, urban area, urban nodes and urban marker.

After entering 21st century, the "urban image" has a new concept: it is the complex overall impression and specific consensus formed with the factors in people's standard evaluation on urban value, such as landscape, nature, culture and economy and so on. It is also the urban formation which can inspire people's mental activities. Among them, urban landscape: buildings, streets (pedestrian streets), squares, riverside and costal zones, parks and lawn; natural environment: the features of urban climate, geology, geomorphology, hydrology and so on; human factors of cities: nation, folk, politics, religion, culture, military and so on; economic factors of cities: population, industrial composition, economic layout and so on. Any distinguishing features of all these factors are likely to be the factor to stimulate people's activities of thinking and feeling, and to be the key factor to affect the overall urban image. The urban image, however, was added with the new factors of nature, culture and economy, the concept of the urban image is mainly composed of the urban landscape has not been changed. In all, urban image stands for the identity and personality of the cities, and reflects the patterns of natural geography, the "extended context" of historical culture, the characteristics of industrial structure, urban function and features of the overall view.

1.2 城市形象的品牌效应
The Brand Effect of Urban Image

一个环境优美、适于人居，并且有自己独特气质、鲜明的城市标志和城市理念的城市更具魅力。因为，今天城市之间的人才、资金、市场的竞争，说到底，已经演变为城市形象和城市文化的竞争，城市形象的优劣对于城市的发展有着至关重要的作用。良好的城市形象是一个城市宝贵的物质和精神财富，是一个城市综合竞争实力的标志。它凝结着城市精神，浓缩着城市文化，张扬着城市个性，使城市树立起自己的品牌。而优秀的城市品牌可以使城市拥有更大的功能效应和环境效益，吸引投资者、人才、旅游者以至中央政府的关注，对招商引资、吸引人才、发展旅游及经济结构调整都将起到重要作用，反过来这些因素又促进城市地位的提升。优秀的城市品牌还可以唤起市民的自信心和凝聚力，有利于市民的整体素质的提高。

通过改变城市形象，树立城市品牌而提升地位的城市屡见不鲜，澳大利亚的悉尼就是最好的例子。1955 年，悉尼市通过一次国际建筑设计竞赛选中了一个奇形怪状的建筑设计方案，而后花了近二十年的时间和几乎相当于预算二十倍的费用才建成了这座建筑。这就是如今大名鼎鼎的悉尼歌剧院。近几十年来，悉尼的地位越来越高，使得堪培拉、墨尔本等城市都感到嫉妒。而悉尼地位的上升与这座歌剧院有着不可分割的联系。有人说，悉尼歌剧院每年在无形中为悉尼做了上百亿美元的广告。无法断定这种说法是否有点夸大其词，但有一点是可以肯定的，那就是悉尼歌剧院大大提升了悉尼的城市形象，而这的确为悉尼带来了巨大的文化和经济效益。时至今日，人们要提到悉尼，首先会想到的就是那座歌剧院。悉尼的城市形象就是以这座歌剧院为标志的，它给人留下了难以磨灭的印象。所以，城市形象的品牌效应是一个城市的“人气”、“财气”之源，是城市发展的基本动力。

Actually, it is the same in the modern society. A city, with beautiful environment, comfortable to live, unique feature and obvious urban mark and idea, will be more charming. Because the competition of talents, finance and market among the cities has become the one of urban image and culture essentially that the urban image has crucial effect of the development of cities. Good urban image is a material and spiritual wealth of a city, and also the symbol of competitiveness of comprehensive strength. It concentrates the spirit, culture and personality of the cities, and is to create brand for the cities. While the outstanding urban brand will endow the cities with more functional effects and environmental benefits, as well as attract the attentions of investors, tourists and even the Central Government, which will play an important role in investment, attracting talent, developing tourism and restructuring economy; the outstanding urban brand can also promote public awareness of self-confidence and cohesion, which is good to improve the overall quality of the citizens.

It is not innovative to improve the cities by changing urban image and creating urban brand. Australia is a good example. In 1955, Sydney City chose a strange design proposal through an International Architectural Design Award, and spent around twenty years and almost twenty times of the budget to finish this building, which is the well-known Sydney Opera House. In the recent decades, the position of Sydney has become higher and higher, which made Canberra, Melbourne and other cities be jealous. Obviously, the promotion of Sydney City is mostly because of this opera. It is said that Sydney Opera House contributes to Sydney city with about one billion dollars advertisement every year. We can not determine whether this statement is exaggerated, but we can be sure that Sydney Opera House has greatly promote the urban image of Sydney, and this promoted urban image surely brought Sydney huge cultural and economic benefit. Up to now, when it comes to Sydney, people will think about firstly that opera which has become the symbol of urban image of Sydney and left people an unforgettable impression. Therefore, the brand effect of urban image is the resource of "human spirit" and "finance" of a city, and the basic driving forcing for city development.

1.3 城市形象的景观识别

Landscape Identification of Urban Image

Landscape identity of urban image means that the urban landscape should be rationally adopted and created in order to convey ideas correctly, which will be good for producing and improving urban image. Specifically, it is to choose key factors from the above factors (nature, humanity, economy) of urban image, and illustrate it with words schemas, and then arrange the different landscape section orderly in the urban design.

城市形象的景观识别是指通过城市景观的合理组织和营造来传辞达意，为树立或改善城市形象服务。具体地说，就是在上述城市形象的自然、人文、经济等若干要素中提取关键因素，并用建筑语言进行表述，然后在城市设计中针对各种景观构成要素进行统筹的安排，主要处理城市的公共界面，如建筑、广场、街道（步行街）、滨江滨湖滨海地带、公园和绿地等城市景观。

这些城市景观是如何实现城市形象的塑造的呢?

1. 建筑：作为构成城市最小的单体、最基本的细胞，一个建筑单体的存在应该立足于承载城市的理念，城市根源，城市文化、城市记忆的思路，它的存在不单是告诉了这个城市的地理特征，还是在述说着这个城市的历史。街道两侧矗立着的高低错落、鳞次栉比的建筑，具有城市细胞中“最能代表城市格调，最能展示城市意韵，最能彰显城市根源”的特性。

2. 广场：广场类型多种多样，一般有疏散、休闲、购物、文化、集会、纪念等功能，多数广场均是上述功能的复合。 广场在城市中有特殊地位，而城市中心广场又被喻为城市的客厅，是一个城市对外展示的窗口，因而也是城市形象表达的关键。广场的形象意义已不在于广场本身，而是城市整体形象及面貌的客观反映。

3. 街道（步行街）：步行街商业繁荣，人流密集，具有广袤的吸纳能力，是城市中的磁性点；也是行为主体和行为对象之间的一个极佳交互场所，是城市标志性的空间，具有形象的传播意义。步行街的设计不仅要在缤纷繁杂的商业气氛中寻找城市的整体统一，而且还要弘扬城市的地域文化特点，宣示出城市的时代特征。

4. 滨江、滨湖、滨海地段：这些地段均为城市的重要展示界面，大多被开发为休闲、旅游的主要区域。江河湖海为城市提供了一个全景的空间场所，城市的生活气氛、城市的人文环境在此得到充分的展现。

5. 城市软硬质景观：软质景观为城市植被。花草树木的季节变化色彩纷呈，中国古代文人墨客历来喜好以植物作为颂咏的题材，久而久之使很多植物均具有人文色彩及人格化意义，并且植物的芬芳对人的感官有直接作用，所有这些均是城市形象的景观资源。硬质景观指道路铺装、围墙、栏杆、标牌、电话亭等城市组件。该部分内容与人关系最为亲和，是人可触摸范围，也是视觉可精细辨别的领域。

How to produce urban image with these urban landscape?

1. Building: As the smallest monomers and the basic elements, buildings keep record of the concept, resource, culture and memory of a city. The existence of buildings not only shows the geographic feature of a city, but also tells the history of a city. The buildings, stand on both sides of the streets, are different in size and large number in quantity. They are the best of marking urban style, demonstrating urban charm, and memorizing urban resource.

2. Square: The square types are various. Generally, the functions of the squares are recreation, shopping, expressing culture, assembly, and memorizing ceremonies and so on; most of them are with all the above functions. Squares have special position in the cities, and the central squares are named urban living room. They are the key expression of urban image. The significance of appearance of square is not on the square itself, but the objectives reflect of the overall image and appearance of the city.

3. Street (Pedestrian Street): With flourishing business and dense crowd, pedestrian street has a vast capacity of absorbing. It is the magnetic point in the city, and also the best interaction place between the behavior agent and object, so it is a symbol place in the city. The design of pedestrian street should not only seek overall unity with the whole city in a complex commercial atmosphere, but also promote the geographical and cultural characteristics of a city, as well as express the characteristics of times.

4. Riverbank, Lakeside, Coastal Area: These lots are important showcases of cities. Urban rivers and lakes provide a panoramic view of the space place. The musical outline, shining atmosphere and mysterious space of the city can be demonstrated thoroughly in this way.

5. Soft and Hard Landscape of Cities: The soft landscape is urban vegetarian. Seasonal flowers and trees change colors of scene, which has attracted the Chinese ancient poets and calligraphers to make works to praise them. With time passing by, many plants have been with significance of humanities and personification; besides, the fragrance of plants has direct effect to people's senses, all of which are the landscape resources of urban image. Hard landscape refers to the road pavements, fences, railings, signs, telephone booths and other urban components. They are most close with people in the city as they can be touched and visual identified by people.

2

认识城市设计

Outline of Urban Design

In the process of urbanization, the cities of our country are in an important step of transforming from the traditional urban model to the modern one, an unpredicted climax of design and construction of cities is flourishing and expanding in our country, which makes the cities face with a big change and reconstruction of material space and humanistic space. The expansion and connotation of Chinese cities have changed obviously and profoundly from improving infrastructure and pursuing rural modernization to emphasizing on urban image and creating the best residential environment. Therefore, as the important method of recreating urban image, the urban design has become the same subject in such change that many cities of our country have faced in the process of development.

在日益推进的城市化进程中，我国的城市正处于一个从传统城市模式向现代城市模式转型的重要阶段，一场前所未有的城市设计高潮正在全国蓬勃兴起，城市正面临和进行着物质空间和人文空间的大变动和重新建构。从强调改善基础设施和追求粗放型的城市化到关注城市形象和创建一流的人居环境，中国城市建设的外延与内涵都在发生着明显又深刻的变化。城市设计作为重塑城市形象的重要手段，成为我国众多城市在发展过程中所面临的一个共同课题。

2.1 城市现状与城市设计
Current Status of Cities and Urban Design

2.1.1 城市现状

当下的城市现状主要体现在旧城之中，一般人都会将“旧”理解为陈旧、过时、年代久远，与“新”相对。过去人们对旧城的理解也是一片破败的城市建成区。但天海理解的旧城却不仅仅如此。其实，一切总体形象与城市发展不协调的城市地段都可以称为旧城，除了破败的建筑区外，还应包括建筑功能和质量很好，但外观设计没有本土特色，缺乏整体形象的地段，或有本土特色但没有被开发展示出来的地段，以及濒临损毁的历史遗存。需要说明的是，所谓的“旧城”是一个具有一定时间性和相对性的概念，随着人类社会历史车轮的滚滚前行，部分新城区也会进入旧城区的行列，这是历史发展的必然，所以城市更新改造是城市发展不断要面对的问题。

2.1.1 The Current Status of Cities

The current status is expressed mainly in the old cities. Generally, in people's concept, the word "old" means existed for a long time, or belong to past times; it is opposite to "new". People understood the "old city" as the general ruined area, but in our idea it isn't enough. Instead, any place with the overall image cannot match with the development of the city. Besides the general ruined area, the lots with good architectural function and quality, but lack of local characteristics design and overall image, or the local characteristics haven't been developed and demonstrated, and the historic remains faced ruin. What we should point is that the so called "old city" is a concept along with time and relative. As time flying, some of the new cities will enter the rank of the old city, this is a historical necessity, so the renewal of old cities is the continuous subject in the development of cities.

而旧城之所以需要改造，就是因为其具备一些与城市发展不协调的特征：

The reason that the old cities need to be renewed is that they are with some features not coordinate with the development of cities.

(1) 物质老化或缺少本土特色

这些区域，或者功能上已经不能有效满足使用者实际生活需求，在结构上日益危旧的各类建筑，或者建筑功能和质量很好，但外观设计没有本土特色，缺乏整体形象，或者外观设计和功能设计本身是合理的、成套的、完善的，但由于历史变迁和住房严重短缺，而造成现状使用上的不合理。

(1) Physical Aging or Lack of Local Characteristics

These areas, some of which can not meet the real life need of the users, or are with good architectural function and quality, but lack of local characteristics in the whole appearance, or the appearance and function design is rational and complete, but irrational to use caused by the historical changes and the acute shortage of housing.

(2) 环境质量差

旧城区一般人口和建筑密度都很高，卫生条件差，污水横流，气味难闻，而且基础设施超负荷运行，供水管道排污等市政管网设施落后。这些与现代发展背道而驰的环境质量问题，也是旧城区更新改造中的重点。

(2) Interior Environmental Quality

Generally, the environmental problems which hinder the cities development such as the poor sanitation due to high density of both population and buildings, overloading infrastructure use and interior municipal pipes facilities, are the main parts in process of cities renewal.

(3) 交通状况有待提高

地段内机动车、行人混杂在狭窄的街道内，安全性较差，废气、噪声影响到环境的舒适感；有些地段交通与城市交通之间缺少良好的衔接，便利性、可达性差；有些地段缺少停车场，车辆乱停乱放，影响空间的观赏性等。

(3) Needed Improved of Traffic Status

Lots of motor vehicles and pedestrians are crowding in the narrow streets, which declines the environmental comfort caused by poor security, emissions and noise. There lack of good convergence of traffic between some places and cities; and the parking lots are in need in some places, so that the vehicles are everywhere, which declines the space view.

(4) 缺少公共空间

由于历史原因，城市人口的不断膨胀，许多街道空间被居民占用，各种临时建筑在蚕食着地段内的公共空间。各部分公共空间之间又缺乏有机的联系，不能构成一个有机的整体，空间的表现力有待深入挖掘。

(5) 城市配套设施不足

一些城区功能设计本身就存在欠缺，缺乏城市应有的基本功能，导致居住者基本生活需求得不到满足。如缺少路灯、公共厕所、指示牌等配套设施，给当地市民和旅游者造成诸多不便。20 万城镇人口的城市没有公共体育场馆，运动会都只能借用一些学校的场地。医院、学校、幼儿园的规模等跟不上城市的发展与城镇人口的增长。

2.1.2 城市设计

正是因为旧城的上述种种问题影响破坏了城市形象，使城市的社会经济发展受到制约，所以要通过城市设计来解决这些问题。城市设计的目的就是城市外部形态结构的优化，将与城市发展不协调的城市地段，进行有计划的重建、改建或维护措施，重塑城市形象，提高城市居住环境质量，保护和延续城市历史文脉，使城市旧区适应城市发展的需要，重焕生机和活力。具体来说，它有以下涵义：

城市设计是对旧城物质环境的提升。城市的物质环境就是人的物质环境，而人是城市的主体，是城市文明的创造者、发展者，又是城市文明的体现者。城市设计是以改善人的居住条件和环境，充分发挥改造地段的经济效益和社会、环境效益为目标，以实现城市现代化为目的。它最终的受益者是居住于城市中的人。

城市设计是改善人居环境、实现土地利用集约化的重要途径，是激发城市生机与活力的重要手段。科学合理的城市规划，是确定城市发展方向和发展目标，高效利用城市各种资源，促进城市合理布局和有序建设的战略性、先导性和基础性工作。

(4) Lack of Public Space

Because of history and ever-expanding urban population, many street spaces have been occupied by residents. Besides, various temporary buildings and adding buildings are occupying the public space. The organic links are in need between the different parts of public spaces.

(5) Deficient Urban Facilities

The facilities planning are inadequate and the basic functions of a city are not included in some designs of urban function that the need of the residents' basic life can't be met. For instance, people living in a town without a public gymnasium can only hold the sports meeting in the school. The development of the basic facilities, such as hospitals, schools and kindergartens, can not keep pace with the increasing of the population.

2.1.2 Urban Design

The above problems of old cities hinder the appearance of the style of historic location and destroy the urban image, which restricted the social and economic development of cites. So we should solve these problems by renewing cities. Renewal of old cities, which refers to optimizing the external structure of cities, means that reconstruct, renew or maintain the urban areas which are incompatible with urban development to recreate the urban image and enhance the residential environmental quality of cities, as well as conserve and last historic context of cities, which makes the old urban areas adapt to urban development. Specifically, it has the connotation as following:

Renewal of old cities is to enhance the material environment of old cities. Physical environment of the cities is the one of human being, and human is the master of the city, and also the creator and constructor of urban civilization. Renewal of old cities is aimed to improve the residential conditions and environment, make good use of the economic, social and environmental effects of the renewal areas. Human being is the ultimate beneficiaries. Therefore, we should put emphasis on the master of cities if we want to renew the cities, which is good to embody the concept of people-oriented.

The renewal of old cities is the important way to improve living environment, realize the intensive land use and stimulate the vitality and vigor of the cities. The scientific and reasonable urban planning is to determine the direction and goal of development of cities, make good use of city resources, and promote rational distribution and orderly construction. We must put emphasis on the planning in the renewal of old cities as it is the important part of the urban construction.

2.2 城市设计之痛
Problems in Urban Design

现代科技的广泛应用，创造了大量的建筑作品和城市空间艺术。然而，由此产生的对自然资源和文化遗产的破坏，也危及到人类自身的生存和发展，由于人们更多地强化物质文明的建设，由此带来了现代人与传统文化的分离，城市文化多样性正在消失；商品化使城市特色走向淡化，丰富多彩的城市空间艺术正在变得越来越单调和乏味。甚至可以说，城市空间艺术乃至优秀的传统文化艺术都在走向枯萎。这种破坏带来的危机已经成为全世界瞩目的问题，因此，我们必须认真总结昨天，才能更好地把握今天，积极探索明天，才能创造一个适合人类生存的、具有较高艺术价值和美学意义的、可持续发展的城市空间。

2.2.1 城市定位，目标偏差

城市定位是打造城市品牌的首要问题，是城市核心价值的集中体现。在城市设计中，旧区的准确定位更加重要。因为旧城代表着一个城市的昨天，承载着一个城市的历史，它不同于新区建设的定位，需要规划者和建设者对它去粗存精，那种完全不顾历史推倒重建的做法将会使历史文脉消耗殆尽。如果定位恰当，将城市各个方面的资源合理配置到旧区当中去，配合城市空间环境正确的表达，不仅会带动城市更快地发展，也能使旧区复活，真正体现出城市的个性和气质。但如果定位不当，不仅不能改变旧城面貌，而且还可能对旧城造成更严重的伤害，不仅浪费了城市资源，也导致城市发展方向出现偏差。目前我国许多城市在自身定位上都存在一些问题：

(1) 脱离现实，目标超前——有些城市不顾自己经济发展所处的阶段，盲目追求过高城市环境标准，例如在建设中追求过宽的马路、过于豪华的灯光夜景和绿地草坪等。有些城市不问区位，不计各种资源，在不具备条件的地方规划“大城市”，城市规模、空间体量、人均用地面积、广场都向国际化城市看齐。

(2) 刻舟求剑，固步自封——城市定位根据城市竞争力的变化而变化。目前，许多城市在其城市竞争力发生变化的情况下，其城市定位没有进行相应变化，结果不但原来的地位和城市形象没有保住，新的机会也没有抓住。例如，一个城市曾一度为某个省区的经济、金融中心，但随着国内外区域经济格局的变化，区域经济金融中心的地位已不复存在，但其继续坚持这一定位，最终该市不仅没有成为区域经济金融中心，反而错失了建设其他中心，实现成功发展的机会。

The extensive application of the modern technology has created a large number of architectural works and urban space art which, however, caused damages of natural resource and cultural heritage, and also threatened human survival and development of their own. Because people strengthen more the material civilization, the modern human beings has been far away from the traditional culture, and the variety of urban culture has been disappearing. The commercialization has diluted the urban characteristics, and rich and colorful urban space art has been becoming more and more monotonous and boring. One could even say that the urban space art, even the splendid traditional cultural art are too wither. The crisis caused by this damage has become the problem that the world pays attention to. Therefore, we must conscientiously learn from yesterday in order to grasp today better, and actively explore tomorrow to create an urban space fit for human surviving, and with high artistic value and aesthetic significance and sustainable development.

2.2.1 Urban Orientation, Target Deviation

City orientation is the first step of city construction, and the concentrated image of the core value of cities. In the renewal of old cities, the correct orientation of the old urban areas is more important. Since the old city represents the past of and carries the history of a city. It is different from the orientation of the new districts that delete the rough and keep the fine by the planners and builders, which is the way without considering history will exhaust the historical context. If the orientation is proper, the resources of different aspects in the city can be rationally allocated in the old city, along with the correct expression of the urban space environment. In this circumstance, not only the city can be accelerated to develop more rapidly, but also the old district revived, finally the urban personality and charm can be embodied truly. However, if the orientation is not proper, the appearance of the old cities won't be changed, even more they may be destroyed seriously. Not only the urban resources will be wasted, but also the direction of the urban development will caused to be and lost the good opportunity to develop. At present, some problems exist in the orientation of many cities in our country.

(1) Divorce from the reality, ahead of the goal—Some of the cities pursue blindly a higher standard of urban environment without considering the stage of economic development. i.e. building too wide road, too luxurious night lighting and green lawn and so on. Some cities are planned with the international standard without considering location, resources and the other conditions.

(2) Take measures without attention to change in circumstances—The orientation is according to the different urban competitiveness. Nowadays, many cities didn't make appropriate changes of their orientation as their urban competitiveness has changed. As a result, neither the original status and urban image nor the new opportunities left. For example, a city has been the economic and financial center of a province. However, with the domestic and international changes in the pattern of the regional economy, this position of regional economic and financial center of this city no longer exists, but it continued this orientation, finally, this city never kept its regional economic and financial center, and lost the opportunities to develop successfully.

(3) 避重就轻，以点代面——城市的定位取决于多种因素，而且这些因素的地位和作用十分不同。但目前不少城市，仅仅依据一个或几个不重要的因素，就轻率地进行定位。例如，有的城市因自然环境优越，就将其定位为人居中心；有的城市因有港口，就将其定位为物流中心。由于并不具备其功能形成等其他关键因素，定位落空，城市失去最佳发展时机。

(4) 目标模糊，定位不准——现在我们的很多城市，在功能定位上相互矛盾。例如珠江三角洲某城市，在规划一片新城区时，方圆 50 多平方公里，既要成为生态旅游休闲区，又要摆上炼钢厂、化工厂。这样做想面面俱到，反而顾此失彼。

(5) 盲目跟风，功能趋同——这一现象目前在我国十分普遍。例如旅游城市如雨后春笋般崛起，有些城市的强项明明是制造业与工业，其两项产值远远高于其旅游业产值，但由于看到周边城市纷纷兴起旅游热潮，并且取得了巨大的成效，也不顾自身旅游资源是否丰富，是否能在旅游市场中赢得一席之地，就如“赶时髦”一般开始大力发展旅游业，有些省份的旅游城市占了城市数量的 80%。

拥有一个准确的城市定位非常重要。如果城市定位不准确，这样做的结果往往是顾此失彼，投入的资金又不能收回，给国家造成极大的损失。在中国社科院财贸经济研究所年度课题《2008 中国城市竞争力报告》成果发布会上，课题组负责人倪鹏飞就指出，“中国传统城市定位的误区，极大地摧毁了传统城市的协调、持续发展”。

(3) Avoid importance and dwell on trivial, fan out from point to area—City orientation is determined by many factors, and the status and roles of these factors are quite important. But the orientations of a plenty number of cities was just according to a few of unimportant factors. For example, some cities were oriented as the residential center just because of the superior natural environment; some cities the logistics center because of a port. As they don't have the other key factors, the orientations were not correct that the cities lost the best opportunities to develop.

(4) Fuzzy objective, inaccurate orientation—The function orientations of many of the current cities are contradictory with each other. For example, people have made a planning for a new district of a city of the Pearl River Delta, which was the area of eco-tourism, but the steel plant and chemical plant were included. Finally, they wanted to strike a balance, but they were failed.

(5) Blindly follow the trend without considering the advantage of the city—This phenomenon is popular in our country at present. For instance, the number of the tourism cities has been increasing rapidly. Some cities, however, are good at manufacturing and processing industry that their industrial output is much higher than the value of its tourism industry, but they still develop the tourism just because their neighbor cities did and succeed, so they followed this trend blindly without considering if their tourism resource is enough or not. In some provinces, the tourism cities accounted for 80% of their total cities.

If the orientation is not appropriate, the result is often not ideal. The capital can not recover, which caused a great deal of loss. On the outcome conference of "2008 China Urban Competitiveness Report" of the Institute of Finance and Trade Economics of Chinese Academy of Social Sciences, Ni Pengfei, the group principal pointed out that the errors of the traditional urban orientation of China has greatly ruined the coordination and sustainable development of the traditional cities.

2.2.2 城市布局，功能缺失

城市设计，老百姓最关心的是什么？无疑是与他们生活密切相关的交通、基础设施等经过改造以后是否变得更加便利，环境是否更加舒适，是否有足够的场所可以满足各类人群的不同需要。然而，现在很多城市规划者都有这样的误区，随着人的感觉特征、行为心理、社会文化心理、生活水平的提高，对于精神生活追求更为渴望，在规划中不能真正从人的行为方式和当地人的生活习惯出发，致使改造后生活在城市中的人们还是感觉不到生活的便利和舒适。这主要表现在改造后城市基础设施人均拥有量仍然不足，功能设计不够人性化，而且有的布局不合理，再次造成城市功能缺失。

2.2.2 Urban Layout, Function Missing

What do the people care about the renewal of cities? Undoubtedly, there are the traffic, infrastructure and environment that close to people's life. However, many of the planners consider neither enough the human sensory characteristics, behavioral psychology, social and cultural psychology and the higher standard of the spiritual life as the living level raised, nor people's behavior and the living habits of the locals, so that people living in cities can't feel convenient and comfortable in their life after the renewal of the cities, which mainly manifested that the urban infrastructure is still not enough in average after the renewal, the function design not humanism enough, and some of the layout not rational that cause the loss of urban function again.

(1) 配套设施人均拥有量仍然不足

公共文娱等配套设施的建设其实是衡量一个城市文明程度的标准。从某种意义上来说，所谓文明就是人与社会的和谐，让社会中的每一个人自由地享用公共资源，快乐地生活在自己的城市中，是一个城市所能达到的最高文明。然而，现在中国的城市特别是小城镇，基础设施拥有量严重不足。最近，一家报纸报道了珠海一名喜欢运动却找不到一个合适场所的年轻人致电该报社，发出“好想有个打篮球的地方”的感叹，期盼有更多的体育设施能够对外开放。公共活动设施不足反映出城市规划建设以及更新改造中对群众文娱体育设施考虑欠缺的不足。许多城市设计在对城市重新规划的过程中，更多地追求城市的空间美学意义，一切为城市美化服务，而没有真切地考虑到改造对城市生活的再造和重组。城市里高楼大厦越建越多，住宅小区越来越高档，可是社会生活空间却越来越狭小，而普通市民能享用的免费公共文娱设施也成了稀缺品，想要在小区内找个打球健身的地方更是困难——物质生活水平上来了，文化生活水平却没有得到应有的改善。

(1) The facilities are still inadequate in average

The construction and opening of public recreational facilities is the standard of judging civilization of a city: In a certain sense, the so-called civilization is the harmony of man and society, so that everyone in society can freely enjoy the public resources that he is entitled to, and live happily in his own city, which is the highest civilization that a city should achieve. However, the possession of infrastructure is serious short. Recently, a newspaper reported a young man, who likes sporting very much, but can not find a suitable place to do sport, sighed: "if only we had a place to play basketball" with the hope that more gym facilities for the public. The inadequate public facilities reflect the deficient consideration of recreations and athletic facilities for public in the process of urban planning and construction. There are more and more tall building in the cities, and the residential quarters are more and more high-grade, but the social living space becomes smaller and smaller, and the free public recreational facilities that open to the public become scare. It is hard to find a place to do sports in the residential area.

(2) 功能设计不够人性化

由于旧城是在以往的城市规划理念上进行建设的，许多功能已经远远不能满足今天社会发展和居民生活的需要。比如，以前规划的城市马路往往比人行道宽很多，缺少合理的健康的城市公共步行系统，道路、广场等公共场所没有为残疾人设计的无障碍设施以及特别为母婴、老人、儿童设计的户外设施。这些本该是我们在城市设计过程中应当特别注意改善的地方，然而在现代主义设计思想指导下，我国很多城市存在着刻板的功能分区、非人性的大尺度、对弱势人群的需求欠缺考虑等设计缺陷，对公共空间的酝酿、策划、营造、设计、使用、维护和改造缺乏以人为本的理性思考，造成了城市功能和形象特征的弱化、宜人尺度的丧失等城市景观环境问题。

(2) Functional design is not humanization enough

As the design of the old cities was according to the former ideas of urban planning, many functions haven't met the need of the social development and residents' life. For instance, the urban road in the former planning is generally much wider than the sidewalks, which is lack of rational and health public walking system, roads and square and so on. Also there is no planning of barrier-free facilities for the disabilities and the special outdoor facilities for the mothers and babies, the elderly, and children. All of these are the places that people should pay more attention to in their planning of urban reconstruction. But the defective design exists in many cities of our country, and the lack of rational people-oriented thinking for the planning, creating, designing, using, maintaining and renewing to the public space under the guidance of modern design ideas, which causes the problems of urban landscape and environment, such as the weakening of urban function and features, the loss of comfort, and so on.

(3) 布局不合理造成功能缺失

还有一些城市功能的缺失，其实是因为规划布局不当造成的缺失，2005 年竣工的四川绵阳九州体育馆就是最好的例子。为了满足某一次国际赛事的需要，相关部门投资 1.5 亿元在绵阳市高新技术开发区内建造了九州体育馆，与绵阳射击场和奥林匹克体校配套使用。然而，尽管馆内保安、保洁、值班等人员配备齐全，使用者却寥寥无几。一处大型的体育馆建在离市区十几公里的地方，不通公交车，乘出租车一次得花费二三十元，来往非常不便，因此普通市民一般不到这里来运动。体育馆建成后，除开展过几次较大活动外，主要是一些单位在这里举办系统内赛事，由此被长期闲置。仅为比赛而建，不考虑体育场馆后期使用和市民使用的便利是导致体育馆闲置的主要原因。一方面绵阳城区需要一个大型的体育场馆，而另一方面这个大型体育馆又被闲置，造成城区基本功能的缺失。

(3) Irrational patterns cause the loss of function

Besides, the loss of function in some cities is because of irrational patterns actually. The Jiuzhou Gymnasium of Mianyang, Sichuan is the best example. In order to meet the requirement of an international competition, the relevant department invested 1.15 billion to build the Jiuzhou Gymnasium in the innovative and high technology industrial development zones of Mianyang City, planning to use along with the Mianyang Shooting Range and Olympics Gym School. However, although the securities and services are available in the gymnasium, the users are few. So it is very inconvenient that a big gymnasium is very far away from the center of cities. On the one hand, Mianyang city needs a large-scale gymnasium, and on the other hand, this large gymnasium was idle, which causes the loss of the urban function.

2.2.3 Urban Landscape, Similar in Different Cities

Convergence is a sad cultural trend of urban construction in current China. The variety of landscape for thousands of years in different cities hasn't existed, instead of the similar ones after the reconstruction. People have made great effort to renew almost all domestic cities, which make the similar cities no matter in the north or south, small or big, abroad or foreign. The so-called planning is generally to redivide the city according to most popular and simple method. i.e. to build several business streets or pedestrian streets in the center of cities, of course including something like residential areas and cultural gallery with slender, cylindrical, and sharp shape, golden color, pure glass material or revolving restaurant, the higher the more awe-inspiring, grouping to be an area is even more awe-inspiring. Later, with the unknown inspiration, people began to be keen to replace the wall with railings, named permeable wall. The white tiles, more strange, are stuck on the outside wall of the houses almost all over the country. It was dubbed the "toilet blocks", although this is indeed harsh, it is true that beauty is in need.

2.2.3 城市景观，千城一面

趋同化是当前中国城市建设的一个可悲的文化走向。千百年来风情各异的景象不复存在，经过再造的城市全都似曾相识，甚至千城一面。几乎是所有的中国城市都在城市设计上不遗余力，北方模仿南方，小城模仿大城，国内模仿国外。而所谓的规划，一般都是按照最流行、最简单的方法重新分割城市。比如在市中心造几条商业区或者步行街，当然还有住宅区和文化长廊这一类的东西。细长的、圆柱式的、尖的、金的、纯玻璃的、带旋转餐厅的，而且愈高愈威风，凑成一片更觉得威风。后来也不知是受了什么启发，人们开始热衷于把实墙拆掉，换上栏杆，名曰通透墙。更奇怪的是白瓷砖，全国各地的房屋差不多全用瓷砖把外墙贴满，其中大部分多为白色。有人戏称为“厕所砖”，虽然这种说法苛刻了些，但缺乏美感却是事实。

还有可笑的水泥树。城市里有树和人造景观当然好，但当树变成水泥树的时候就变得有些诡异了。想想看，北方飘雪时节，绿意泛滥的水泥树是多么扎眼。 而且似乎没有地域差异，一色的椰子树，常年硕果累累。于是我们感到自己生活的城市愈来愈陌生，难怪有人对中国城市的发展现状用“一千个雷同城市构成的国家”来形容。这不禁让人联想起了计划经济时代，所有物品，走遍全国都是一个样子的景象。

虽然我们看到，趋同化有一个国际大环境，不单是中国独有的现象，但当中国城市变成千城一面的时候，我们不妨把问题再深入一些，去想想生活方式的趋同由谁主导，以及谁会有权自由选择不同的生活方式。中国城市经济学会秘书长刘维新先生就在很多场合下强调，城市设计千万不要把人气、文气改没了，不可一蹴而就。

Besides, there are the ridiculous cement trees. It is surely good that there are trees and man-made landscape in the cities, but it will be bizarre when the trees are replaced with cement trees. Just thinking, what a loud color of the cement trees when it snows in north! Moreover, it seems to be no geographical differences with the same color of coconut trees and perennial fruit in them. Even the skyline of the cities can be so similar. No wonder people describe the development status of Chinese cities with "a country constituted one thousand similar cities", which makes people think of the planned economy era when all items are in the same face throughout the country.

We know that convergence is the international trend, but not the unique phenomenon in China. However, why not we think deeper that what led the trend of our lifestyle and who have the right to choose freely the different way of life. Mr. Liu Weixin, the general secretary of China Society of Urban Economy, stressed in many occasions that never lose the human spirit and cultural heritage in the renewal of old cities.

2.2.4 Urban Buildings, Removing and Reconstruction

Many people, who left their hometown for many years, found that their hometown had changed thoroughly when they came back, they even couldn't know it anymore. The streets, old trees and memorial archways in their childhood could be only seen in the pictures. Thus people often describe that their hometown has undergone enormous changes. This phenomenon reflects the problems that many cities have faced in their renewal for a long time in our country, which is removing and building thoroughly. This simplest and roughest method of renewal can make nothing, but the damage of old cities and buildings that never can be recovered.

2.2.4 城市建筑，推倒重建

一些阔别家乡多年的人常常有这样的感受，当他们回到故里的时候，发现家乡完全变了，甚至变得自己都不认识了。儿时玩伴们嬉戏过的街巷，打酱油常常路过的古树和牌坊，都变成了只能在照片中回味的影像。因此，人们常常会用“翻天覆地”来形容家乡快速发展后的景象。这一现象反映了我国长期以来许多城市在更新改造中面临的问题，那就是大拆大建。这种最简单、最粗野的改造方式，可能对于城市的重新划分有所用处，但对古城和古建筑造成了永难修复的破坏。

英国文物建筑学会指出，20 世纪 70 年代发展的旧区改造所破坏的具有文物性质的建筑竟比第二次世界大战中被炮火摧毁的还要多。我国文物保护界也有类似说法，即中国改革开放 30 年来以建设的名义对旧城的破坏超过了以往 100 年。1982 年始，我国先后公布了 114 个国家级历史文化名城，但在古建筑保护与城市发展的冲突中，牺牲的往往是前者。1999 年 11 月 11 日，国家历史文化名城襄樊千年古城墙一夜惨遭摧毁，郑孝燮、罗哲文等专家称之为 20 世纪末恶劣破坏历史文化名城的事件。之前，还有福州三坊七巷的建设性破坏、贵州遵义和浙江舟山市定海的老街区被拆。1992 年 7 月 1 日，矗立了 80 多年的济南标志性建筑——具有典型日耳曼风格、可与近代欧洲火车站媲美的济南老火车站被拆除，起因是有官员说“它是殖民主义的象征，看到它就想起中国人民受欺压的岁月，那钟楼的绿顶子（穹隆顶）像是希特勒军队的钢盔有什么好看的”？照此逻辑，号称万国建筑博览会的上海外滩建筑群是否理应夷为平地?

The Architectural Heritage Society of the United Kingdom pointed out that the heritage ruined in the renewal of old districts were more than the ones that destroyed by the fire in the World War II. The Heritage Conservation sector had a similar claim that the old cities damaged in the 20 years during the reform and opening had been more than the ones destroyed over 100 years before. From 1982, China has announced 99 state-level historical and cultural cities, but in the conflict of preservation of ancient architecture and urban development, the former is often sacrificed. In November 11,1999, the millennium ancient city wall of Xiangfan, the national historical and cultural city, was ruined over one night. Zheng Xiaoye, Luo Zhewen and other experts called it as the worst event of destruction of historical and cultural cities in the end of the 20th century. Before, there were the constructive destruction of the Sanfang-Qixiang of Fuzhou, the demolishment of old districts of Zunyi, Guizhou and Zhoushan, Zhejiang. In July 1, 1992, the old station of Jinan, which was the landmark of Jinan and with the a typical German style, also can comparable to the modern Europe railway station, was demolished with more than 80 years since it stood. The cause was some officials said that it was a symbol of colonialism, whenever saw it, the green dome of the clock tower would remind them of helmets of Hitler's army in the years of bullying. By analogy, the buildings along the Bund of Shanghai, known as the Great Exhibition of Architecture, should be razed to the ground?

旧城的破坏已成为上个世纪中国城市建设者们最短见的城市行为。然而，历史的载体不是在教科书中，而是在有质感、有形体、有生命痕迹的城市里，口说无凭，只有古道巷陌和文物才令历史文明活色生香。旧城被完全拆除重建之后，似乎达到了破旧立新的目的，但是宝贵的文明传统被割断，新城市对未来表现得无所适从，只好一味地抄袭，拆了又建，建了又拆，无休止地折腾。这种方式对现状采取了完全否定的态度，忽略甚至摧毁了城市历史环境中的诸多有价值的东西，导致了城市宜人环境的丧失。 可见保护性的更新改造对城市文脉的保存是多么重要，试想没有对文物建筑科学的复原重修，我们今天就不可能看到完整的北京明十三陵的昭陵，颐和园的景明楼、澹宁堂，居庸关的城墙城楼、寺观衙署，广州光孝寺的钟楼，苏州的瑞光塔，辽宁朝阳的北塔，河北赵州桥等古迹。

The destruction of old cities has become the most undesirable city act of Chinese urban constructors in the last century. However, the historical carriers are not in the textbooks, but in the cities with texture, physique and traces of life. Only the ancient streets and heritage can make the historical civilization alive. After the old cities demolished and reconstructed, it seemed the objective of destroying the old and establishing the new achieved, but the valuable civilization and tradition was cut, which made the new cities to be at a loss for the future, only to be torn down and rebuilt. This method took a completely negative attitude; ignored, and even destroyed plenty of valuable things in the historical environment of cities, which causes the loss of the pleasant environment of the cities. Therefore, it is obvious that the conservative renewal is so important to the conservation of the urban context. If we haven't restored and recovered scientifically the heritage buildings, we could not enjoy the scenes of Ming Tombs of Beijing, the Summer Palace, the clock towel of Guangxiao Temple, the Ruiguang Towel of Suzhou, and so on.

对待古物尚且如此，对于一般性旧建筑的再利用，就更加缺乏重视了。由于我国对建筑再循环的理论研究刚刚起步，尚未形成一套切实可行而又行之有效的理论体系，且大多数人，包括政府和开发商对一般性旧建筑的再利用没有引起足够的重视和形成充分的共识，再加上国内也缺乏具有普遍指导意义的优秀的改造建筑实例。因此，很多旧建筑再利用是在朴素的经济观念驱使下完成的，缺少明确的设计指导思想及对原有建筑内涵的充分挖掘，没有在与新功能相配合的功能空间、外部特征和环境的协调与重组等方面进行完整的综合设计工作，而是仅仅停留在简单的室内外装修和恢复阶段，焦点大多集中在物质建设层面上，缺少从社会、经济和环境的角度的系统分析研究，表现出了一定的盲目性和片面性，普遍忽视建筑重建所带来的其他社会生活问题。

The theory of architectural recycling in our country has been just in the primary step that it hasn't formed a practicable and effective theory system, and the majority of people, including the government and the developers haven't paid enough attention and formed fully consensus to the recycling of the general old construction, also the outstanding examples of renewal and construction with universal guiding significance have been inadequate in our country. Therefore, the recycling of many old buildings was fulfilled with the simple economic concept, so that it was lack of clear guiding ideology and full exploration of the connotation of the original buildings, and haven't completely and comprehensively design the factors of functional space along with the new function, the external features and the coordination and recombination of environment, but just stayed at the step of simply indoor and outdoor decoration and restoration, and focused on the material construction, which was lack of systematical analysis and study from the aspects of society, economy, and environment, and generally ignored the other social problems caused by the reconstruction.

3

变城记

City Changing

如何营造魅力之城

How to Create Charm for a City

介绍天海城市设计步骤和方法

Introduce the steps and methods of urban design of Tianhai

The problems came out in the practice of urban design in our country have provided valuable examples for the way of urban design of Tianhai. We always keep the lessons in mind, strive to explore, innovate and create, and always adhere to the concept "Modernization of nation, while nationalization of modernity". After seven years tireless research and practice, we have formed our methods and models of designing cities and counties. Any design must be based on the full understanding of the design object, so we believe that the careful investigation and research, accurate knowing the urban resources and grasping the is the fist step in the renewal of cities, so we believe that we should firstly investigate and research carefully, accurately understand the city resources, and grasp the city context, and then determine the developing goal and core idea of the cities, which is called creating city soul. On this basis, we will fully tap the urban design elements, and finally combine these design elements and landscape to create an urban space which is fit for human living, and with high artistic value and sustainable development.

我国城市设计实践中出现的弊端为天海的城市设计之路提供了难得的借鉴。我们谨记教训，努力探索，始终秉承“民族现代化，现代民族化”的理念，不断开拓创新，经过八年孜孜不倦的研究和实践，形成了自己的一套城市设计方法和模式。任何设计都要建立在对设计对象充分认识了解的基础上，所以我们认为城市设计首先要通过细致的调查研究，准确认识城市资源、把握城市文脉，然后确定城市发展目标和核心理念，即塑造城市灵魂；在此基础上，充分挖掘城市设计元素；最后将这些设计元素与景观塑造相结合，创造出一个适合人类生存、具有较高艺术价值的、可持续发展的城市空间。

3.1 调查——准确把握城市文脉

Investigation—to Master Urban Cultral Context Accurately

城市文脉是一个城市的灵魂所在，是城市赖以存在的背景，包括自然条件、文化氛围及社会经济背景等要素。文脉的识别是建立在现有资源的基础上的，其显性形态包括人、建筑、景观及环境中的各种可见要素；隐性形态指那些对城市形成与发展有潜在深刻影响的因素，如政治、经济、文化、历史和习俗。在提倡“人文精神”的今天，准确地认识和评价城市资源，把握城市文脉，通过塑造人性化的城市公共空间环境来体现设计者对使用者的关怀，是天海规划设计有限公司的潜心追求。

The cultural context of a city is the soul and the background of the city, including natural conditions, cultural atmosphere and social and economic background and other factors. The identity of cultural context is based on the current resources, of which the dominant form including human, buildings, landscapes, and the various visible elements of environment; recessive form refers to these elements that have profound effects to the forming and development of the cities, such as politics, economy, culture, history and custom. Today, with the advocation of "human spirit", Tianhai Planning and Design Co., Ltd. is making their great effort to precisely understand and evaluate the urban resources, grasp urban cultural context, and show the designers concern for users by the way of create humanized public space and environment.

例如凯里和铜仁，这两个地理位置相距不足 160km 的城市，在做城市设计时，我们对它们各自文脉的把握和认识就有所区别。

For example, Kaili City, and Tongren City, these are two cities with the distance no more than 160km, but we have different understanding and grasping of their cultural context respectively when we renew them.

凯里是一个典型的、具有浓郁民族特色的城市，它的文化以苗族侗族为主体。同时，凯里是“芦笙的故乡”、“歌舞的海洋”，素有“百节之乡”的美称，具有得天独厚的自然景观、古朴浓郁的民族风情和丰富独特的民族文化。而铜仁虽然也有苗、侗等少数民族文化，但多民族的兼容性更强，其城市形象更多依托的是“山水文化”。“山”是梵净山，“水”是锦江水，自古以来是铜仁城市的灵魂与命脉，是铜仁连通中原的纽带，所以铜仁的“文化”更多的是杂合型的，融合了弥勒文化、江浙与川渝商贾文化和铜仁本土少数民族文化，这可以说是铜仁城市的精神底蕴所在，其中佛文化是其魂、商贾文化是其体、少数民族文化是其肢，这与凯里的文脉是有所区别的。所以，在这两个城市的设计过程中，对民族元素的运用既有相似处，又有不同点，关键是要突出它们的各自独特文脉特征，要充分调查研究，抓住最核心的文化脉络，才能为尔后的城市定位、城市设计、景观塑造打下良好的基础。

Kaili, with the Miao and Dong culture as its main culture, is a typical city with thick ethnic characteristics. Meanwhile, Kaili is the "Hometown of Lusheng", "Sea of songs and dances"; it has enjoyed the fame of "the County of Hundred of Festivals", and is endowed with outstanding natural landscapes, simple and thick ethnic custom, as well as rich and unique ethnic culture. Tongren, although also has the Miao and Dong ethnic culture, compatibility of multiple ethnics is stronger. And its urban image mainly depends on "mountain, water, and culture". "Mountain" is the Fanjing Mountain, "water" is the water of Jin River which have been the soul and vital of Tongren City, and the bridge that linked Tongren with the Central Plains, thus the "culture" of Tongren is hybrid, which was a blend of Buddhism culture, the culture of Jiangsu, Zhejiang, Sichuan and Chongqing and the native ethnic culture of Tongren. Among them, Buddhism culture is its soul, the culture of business its body, and the ethnic culture its limbs, which is different from the cultural context of Kaili. Therefore, in the process of reforming the two cities, there is both the same and difference using the ethnic elements. The key point is to highlight their unique cultural context respectively, and investigate carefully to grasp the core cultural context, so that make good conditions for the following urban orientation, urban transformation and landscape producing.

城市设计中表达城市文脉还要抓住城市的关键节点，例如进入城市的主干道、城市标志性的地段及其建筑，对它们进行尽显民族文化意蕴的包装和渲染，通过规范得体的、恰到好处的“穿衣戴帽”和功能划分，将城市文化在建筑外形上符号化和在人文气息上意识化（如建筑风格的民族化、街道名称的民族化、城市标识的民族化等），赋予城市原本就应该拥有的极具冲击力的民族文化风韵，让人们无论从外在形式上还是深层内涵上感受到强烈的地域民族特色。

In addition, to express the cultural contest of cities in the process of urban transformation, we should seize the key node of the cities. i.e. to decorate and render the main urban street and the symbolic urban lots and buildings with ethnic cultural implication; endow the cities with ethnic cultural charm that they should have, so that people can feel the strong regional ethnical features no matter from the outside shape or the inside connotation.

3.2 定位——确定城市发展目标

Orientation—to Determine the Objection of Urban Development

城市定位就是对城市未来发展方向，发展目标的拟定。它是城市总体发展战略的核心部分，体现了城市的文脉特征、建设重点和形象定势。城市要想在区域发展中避免被边缘化，占有优势地位，就必须有一个明确的定位。它是建立在对城市文脉作准确的概括和提炼的基础上的，它要从城市自然、人文、经济等错综复杂的对象中抽取其中的要点并概念化，应能综合反映出城市的地域性、文化性、时代性特征，充分表达城市文脉，体现城市鲜明的个性。

同样以凯里和铜仁为例，为了突出凯里以苗、侗为主体的少数民族文脉特征，我们在城市设计中将其定位为“贵州少数民族都城”，目的是将凯里打造成中国乃至世界级的极具民族生态意韵的人文旅游胜地，使之成为中国民族文化旅游城市示范地。而在铜仁多元文化融合的背景下，铜仁旅游城市的总体形象定位为“梵天净土，桃源铜仁；娱情山水，一度悠然；十里锦江，不夜之城”。从这一定位中，我们可以看到铜仁城市“山”、“水”这两大核心资源的体现，即佛教圣地梵净山和穿城而过的锦江，而有意识地对铜仁的少数民族特征予以弱化，体现出铜仁有别于周边少数民族城市的独特形象。

城市总体定位是城市规划设计的核心步骤，城市各个地段的改造都是为体现城市定位服务的，它们的发展目标是在总体定位这个总体目标下分解的若干分支目标，它们的改造、创意以及每一阶段的规划设计工作都要围绕总体目标这个核心来开展。

以铜仁旅游城市设计为例，在“梵天净土，桃源铜仁；娱情山水，一度悠然；十里锦江，不夜之城”这个铜仁城市总体形象定位下，我们对迎宾大道（梵净山大道）、金鳞大道、东太大道、锦江大道、清水大道等地段的改造，都是围绕这个目标定位来开展的。迎宾大道是从南面进入铜仁市的主干道，是城市设计的重点，要将它打造为铜仁市的“门户”，入口牌坊融合了徽派、苗、侗等多民族的设计元素，让游人一进入铜仁就感受到这座城市多元文化融合的气息，“铜仁欢迎您”的字样也使迎宾大道在整个城市设计中的“门户”作用体现得淋漓尽致，迎宾大道两侧分段设置街心花园，所有街心花园都营造了喜庆浪漫的情怀，同“桃源铜仁”的定位相契合；锦江大道是沿江而行的一条干道，东侧临江，我们考虑到该路段近水楼台之势、丰富的河滨自然景观、文物古迹，充分利用建筑表现、广场设计等手段打造滨江风光带，尽展江水风光，并配合开发锦江夜景旅游的构想，充分体现了总体目标中“十里锦江，不夜之城”的定位。

Urban orientation is to determine the development of direction and goal of a city. It is the core part of the overall developing strategy, which reflects the features of cultural context of a city, the constructional focus and the image status. If a city wants to avoid to be ignored, but take an advantage, it must have an excellent orientation, which is based on accurately grasping and abstracting the cultural context of the city, and also can comprehensively reflect the urban feature of region, culture and time, express completely the urban cultural context and demonstrate urban personality.

Also, we take Kaili City and Tongren City as example. In order to highlight the features of cultural context of the Miao and Dong ethnics as the main culture of Kaili, we gave it the orientation of "the Ethnic Capital of Guizhou" in the process of transformation with the goal to create Kaili to be a tourism resort with extreme ethnic eco-charm in China and even in the world, so as to make it to be the example of tourism city of ethnic culture of China. "Buddha Pure Land; Peach Orchard Tongren; Enjoying Yourselves in Mountains and Rivers Leisurely; Ten Miles of Jinjiang River, The Sleepless City", in this orientation, we emphasized on the two core resources of "mountain" and "water" of Tongren City. There are the Fanjing Mountain which is the Holy Land of Buddhism and the Jinjiang River which is through the city. Whereas we decline the features of the ethnics of Tongren on purpose, so as to show the unique image that Tongren is different from the nearby ethnic cities.

The overall orientation is the core step of urban planning and design. The renewal of every urban lot is to embody the urban orientation, of which the developing goals are the branches of the overall goal, the renewal, creation and the planning and design of each step have to fulfill according to the overall goal.

Take the creation of Tongren City as example, with the orientation of "a city locates under the mountains and with Jinjiang River, flowers, and Buddhism culture", we reformed the Guest Avenue, Jinlin Avenue, Dongtai Avenue, Jinjiang River Avenue, Qingshui Avenue and so on. The Guest Avenue, the main way to Tongren City from the south, is the emphasis in the transformation. We planned to produce it as the "window" of Tongren City. The entrance memory arch is integrated with many ethnic elements of Hui, Miao and Dong, which will make the tourists feel the atmosphere of multiply cultures of this city as soon as they enter the city. The Chinese characters of " 铜仁欢迎您 " (Welcome to Tongren) also strengthens the function of "window" for the Guest Avenue. On the two sides of the avenue there are gardens in sections, and all the gardens make a festive and romantic atmosphere, which fit the orientation of "a city with flower". Jinjiang Avenue is a trunk line along the river with the east side near the river. Considering the advantages of rich natural riverside landscape and heritage, we make good use of ways of architectural manifestation and square design to create scenic spots along the river to show completely the river scenery, and along with the planning of developing Jinjiang River tourism in the night to achieve the orientation of the overall goal "ten miles Jinjiang River, a city without night".

3.3 规划——安排城市功能布局

Layout—to Arrange Urban Functional Layout

如果说城市定位是城市设计的目标的话，那么城市的功能布局就是实现这目标的重要手段。我们认为，设计城市就是设计生活，就是设计人与城市的关系，因为人是城市的主角。良好的功能布局能够带动人们改变传统落后的生活方式，而创造方便舒适、内容丰富的新生活，能让人感到亲切，甚至产生眷恋感，从而提升城市的吸引力。一个富有活力的城市首先应该是一个具有健康生活的城市，各种市政设施应较为齐备，能满足市民政治、经济、文化、休闲娱乐等生活需求，并能通过各种文体场馆的建设，促进市民业余文化生活品位的提高。

If the urban orientation is the goal of urban transformation, then the functional layout of a city is the important way to realize this goal. We believe that designing cities is designing lives and the relation between people and cities, as people is the main actors of cities. A good functional layout can accelerate people to change their traditional and backward lifestyle. To create a comfortable, convenient and colorful life can make people feel warm and nostalgic, even have love to their lives, so as to enhance the urban attractiveness. A city with activity firstly should be a city with health life, and with a variety of municipal facilities that can meet the citizens' need of politics, economy, culture, and entertainment and so on; also can accelerate the citizens to enhance their taste of culture and life by construction of all various venues.

同时，在规划布局中不仅要考虑普通人群的需求，还要充分考虑细分人群（儿童、成年人、老人、母婴、残疾人）的需要，不仅要考虑本地居民的感受，也要着重关注外来旅游者活动的特点与体验，不仅要从城市景观方面展现其特色与魅力，更要从活动、氛围、设施的实用性等方面进行改造，使城市成为一个能不断满足居民和游客不同的休闲游憩需求的城市和让游客流连忘返的旅游目的地。

Meanwhile, we consider not only the needs of the common people in the planning, but also the segmenting group (children, adults, the elderly, mothers and babies, and disabled people); not only consider the feeling of the locative people, but also pay attention to the characters and experience of the foreign tourists; not only show its features and charm from the urban landscape, but also from the aspects of activity, atmosphere and equipments; so that make the city satisfy the residents and the tourists from different aspects and leave the tourists excellent impression that they don't want to leave.

针对贵州城市市政基础设施普遍不完善的特点，我们在做城市设计时，首先注重对街区基本设施的完善，如根据市民日常生活需要，按照合理、就近、方便的原则，在街道两侧添置补充公交车站台、路灯、垃圾桶、休闲长椅、公厕、电话亭、治安亭和人行天桥，并在设计中增加了无障碍设计，使城市街道形成合理、健康、适用的城市公共步行系统。其次，随着生活条件的改善，人们对公共游憩娱乐设施的需求有所增加，标准也有所提高，因此，在城市设计中我们根据用地的性质和规模以及所应配备的设施及所承担的功能，布置规划宾馆、餐饮、文化娱乐点、度假村、游乐场、游泳场、公共停车场、码头等公共服务设施，并注重广场、游憩商业区、绿地系统的布局，保证市民有足够的休闲游玩空间。再次，为了提高市民生活品位、丰富生活内容，在有条件的城市适当增加一些大型文体场馆，使人们在物质生活提高的同时，在精神生活上也得到丰富。

As the municipal infrastructure is not adequate generally in the cities of Guizhou, we firstly put emphasis on enhancing the basic equipment of the districts. For example, to meet the citizens' daily needs, we set additional bus stations, street lights, trashcans, coaches, public toilets, telephone boots, security boots, and pedestrian bridges with the principle of ration, nearby and convenience, and add no-barrier design in the planning to make the urban streets form a rational, health and practicable urban public walking system. Secondly, as the living conditions improved, people's need of public recreational equipment has increased, so has the standard. Therefore, we should make planning of the hotels, restaurants, cultural entertainment spots, resorts, recreational park, playgrounds, swimming pool, public car parks, piers and other public service facilities according to the properties and scale of the planning land as well as the equipments that should be set and their functions; meanwhile, we should pay attention to the layout of the squares, walking and business districts, and the lawn system to guarantee enough recreational space for the citizens. Last, In order to improve citizens' life taste and enrich their lives, we should add some large cultural and sport venues in the cities with enough conditions, so as to enrich people's spiritual life while their material life enhanced.

同样以铜仁市城市设计为例。铜仁市区具有三江汇流穿城而过的景观，蜿蜒的河道将市区划分出 12 个半岛，以此为基础将市区也划分为 12 个半岛区域，每个半岛都承担了不同的城市功能，（见下表）按照它们各个自的功能分区加以规划改造。铜仁市城镇人口 2006 年已突破 12 万，城镇化率也达到 35% 以上，而且逐年递增，市民的文化娱乐生活需求不断增加，现有的市政场地设施不仅数量少，而且档次低，所以我们在做城市设计时，规划了铜仁市游泳馆、铜仁地区体育场、铜仁市歌舞剧院、铜仁地区博物馆、铜仁图书馆等 5 大公共文娱场馆。

Take the transformation of Tongren City for example again. Tongren is endowed with the scenery of a river converged of three rivers running through the city, the twisted river divides the city into 12 peninsulas, which is the basis to divide the city into 12 districts, and the each has the function as following form shows. We will make planning and reconstruction according to their functions respectively. The number of population of Tongren City has been more than 120 thousand, and the rate of citizens is more than 35% and even increases year after year. The citizens' needs of entertainment increases continuously, but the current administrative equipments are not only in small quantity, but also in interior quality, so we've designed five large public entertainment stadiums of the Swimming Pool of Tongren City, the Gymnasium of Tongren Peninsula the Opera of Tongren City, the Museum of Tongren Pres, and the Tongren Library when we made the planning of urban transformation for Tongren.

铜仁城市 12 个半岛功能分区

序号	半岛名称	功能分区
1	南岳半岛	民族风情园
2	桐梓坳半岛	滨江景观区
3	金滩半岛	休闲小吃街及休闲广场
4	曹园半岛	婚庆主题时尚游览区 ——贵州民族婚庆园
5	江宗门半岛	综合旅游区
6	五显庙半岛	城市亲水旅游区
7	大江坪半岛	文化休闲游憩区
8	三江公园	城市休憩公园
9	鲤鱼嘴半岛	滨江休闲旅游地
10	渔梁滩半岛	商业娱乐区
11	水晶阁半岛	高端住宿接待区
12	黄连冲和鹭鸶岩半岛	生态湿地公园

The functional layout of the 12 peninsulas of Tongren

NO.	PENINSULA NAME	FUNCTIONAL AREA
1	Qingshuitang	Ethnic Park
2	Tongzhiao Peninsula	Riverbank Landscape
3	Jintang Peninsula	Leisure Snake Street and Leisure Square
4	Caoyuan Peninsula	Fashionable Theme Wedding Travel Area ——Guizhou Ethnic Wedding Park
5	Jiangzongmen Peninsula	Comprehensive Tourism Area
6	Wuxianmiao Peninsula	Urban Water Tourism Area
7	Dajiangping Peninisula	Culture and Leisure Walking Area
8	Sanjiang Park	Urban Rest Park
9	Liyuzui Peninsula	Riverbank Leisure Tourism Area
10	Yuliangtan Peninsula	Commercial Entertainment Area
11	Shuijingge Peninsula	High-end Lodging Reception Center
12	Huanglianchong and Lusiyan Peninsula	Ecological Wetland Park

3.4 创意——挖掘形成设计元素

Creativity—to Discover and Form Design Elements

创意，在英语中以“Creative、Creativity、Ideas”表示，是创作、创制的意思，有时也可以用“Production”表示，它是表达设计主题并最终形成美好意境的一种创造性思维活动。我国的创意产业是在20世纪90年代才逐渐兴起的。许多行业都需要创意，但在不同的领域它具有不同的涵义。在城市规划领域，城市设计创意是指设计师对设计对象（建筑、景观）所进行的创造性思维活动，是通过艺术创作手段，对城市的主题定位、功能布局、表现形式等所进行的新颖的艺术形象构思，使城市建筑或景观的潜在现实性升华为公众所能感受到的具象。

The Chinese characters " 创 意 " means "Creative、Creativity、Ideas" in English, also means "Production" sometimes, which refers to a kind of creative thinking activity of expressing design theme and finally forming good image. The creative industry of our country has gradually emerged in the 90s of the last century. The creative design is used widely in many fields, but it has different connotation in different fields. In the field of urban planning, urban design creative refers to the creative thinking activity that the designers to the design objective (architecture, landscape), and the innovative idea of artist image of theme orientation, functional layout and demonstrative form etc. by the method of artist creation, which makes the potential actuality of urban architecture and landscape can be enhanced to be the specific image that the public can feel.

天海规划经过多年的创意实践，积累了很多实用的城市设计创意经验，我们认为城市设计创意是主题性、原创性和震撼性的综合体。首先，城市设计创意要以体现城市定位为核心和主题，抓住城市文脉特征，从城市的人文背景中（如民族建筑、文字、诗歌）挖掘设计元素，获取创造的素材。其次，城市设计创意要以人为本，以情趣生动为手段，从公众的视觉、听觉感受出发，并立足现实，把身处城市的人们带入一个印象深刻、浮想联翩、妙趣横生、难以忘怀的境界中去，以引发公众的共鸣。再次，城市设计创意要以新颖独特为生命，不要随意模仿，给人雷同与平庸之感。唯有在创意上新颖独特才会在众多的城市中一枝独秀、鹤立鸡群，从而产生感召力和影响力。由此，城市设计创意巧妙地把主题性、原创性和震撼性融为一体，才能成为具有深刻感染力、经久不衰的设计作品。

After many years of creative practice, Tianhai Planning has accumulated practicable creativity experience of urban transformation. We believe that the urban design and creativity is the complex of theme, originality and impact. Firstly, the urban design and creativity should take the urban design and creativity as the core and theme, grasp the features of urban cultural context, cultivate the design elements from the human and cultural background of cities (such as the national architecture, characters, poems etc.). and obtain the materials for creativity. Secondly, the urban design and creativity should be based with the concept of people-oriented, the method of humor and vivification, and the vision and auditory of the public, as well as the reality, so that give people in the city a profound, interesting and unforgettable impression, and induce the public resonance. Thirdly, urban design and creativity should be innovative and unique; do not imitate at random, in case giving a sense of similarity and mediocrity. Only the novel and unique design can outshine others, and give appealingness and influence. As a result, urban design and creativity skillfully integrates the theme, originality and impact that can become a design work with profound impact and enduring quality.

以凯里城市设计的创意设计为例，凯里市迎宾大道的门头设计就极具创意。它融合了侗族的风雨桥、苗族的铜鼓、土家族的大刀梁等民族特色，体现出凯里“贵州少数民族都城”的鲜明形象，使进入凯里的人们一眼就感受到凯里的民族气质。另外，在公交车站、公厕、垃圾桶、路灯等公共基础设施的创意设计上，我们也十分注意民族符号的融入。例如，在铜仁锦江广场的改造中，我们运用了极具地域代表性的三圣雕塑、四面鼓牌坊、白虎图腾、梵文化主题雕塑。三圣雕塑与铜仁地名息息相关，体现了铜仁的起源、文脉；四面鼓是铜仁苗族特有的乐器，演奏时鼓声震天，气势磅礴，古时曾进京演出。以此为主题创作的四面鼓牌坊代表了独特的铜仁苗族文化；白虎是土家族图腾，以此雕塑代表土家文化；梵文化主题雕塑寓意铜仁市主题形象“梵天净土，桃园铜仁”。

Take the urban transformation of Kaili as example, the door of the Guest Avenue of Kaili City is very creative. It is blended with the Wind and Rain Bridge of Dong, the Bronze Drum of Miao and the Sword Beam of Tujia etc., which reflect the clear image of "the Minority Capital of Guizhou" of Kaili, so that people can feel the national temperament of Kaili as soon as they come into this city. In addition, we also pay great attention to the integration of national symbols in the creative design of the public infrastructure, such as bus stations, public toilets, garbage cans and street lamps etc. For instance, in the transformation of Jinjiang River Square of Tongren, we applied the very geographical representation of the Trinity sculpture, the four sided drums, the White Tiger Totem, and the Buddhism Cultural Theme Sculpture. The Trinity sculpture, which closely relate with the name of Tongren, reflects the origin and cultural context of Tongren; the Four Sided Drum is the specific instrument of Miao ethnic group of Tongren, the sound is loud and the scene is magnificent when it is played, and it has been performed in Beijing in the ancient times. So the memory archway in the shape of the four sided drum represents the unique Miao culture of Tongren; White tiger is the totem of Tujia ethnic group, so the white tiger sculpture can be on behalf of the Tujia culture.

3.5 建筑——确保技术切实可行

Architecture—to Make Sure the Feasibility of Technique

如果说创意设计是提升城市视觉效果的手段的话，那么建筑就是确保创意效果能够变成现实的重要环节，它的作用是确保城市设计方案在技术上可操作，在经济上可施行。

If we say that creative design is the means to enhance the urban visual effects, then the architecture is the important part of ensuring the creativity becomes the real construction, which plays the role of ensuring that the reforming programs are technically operable and economically implemented.

首先，要判断改造地段哪些建筑该保留，哪些建筑该拆除重建。一座城市，或多或少存在历史文化的痕迹。那些伴随我们成长的街道、院落和建筑物，尤其是那些具有历史保留价值的古城区，随着岁月的流逝，逐渐显示出优雅的气质和迷人的风貌。它们都是城市的记忆和不可再生资源，代表着城市的昨天。我们在做城市设计之前，要做好细致的调查，对街区内的建筑按其历史的保存状况来分类，保留什么、改造什么要有清晰的判断。对于一般生活区的破败建筑，如果没利用价值，应当将其拆除，以免影响城市形象；而对于有历史意义的遗迹，应当做抢救性的修复。

Firstly, we should determine what constructions should be reserved, and what should be demolished and reconstruct. There are more or less historic and cultural remains in a city. Those streets, courtyards and buildings, especially those ancient city area with historical preserved value, which accompanied with us in our growth, gradually shows elegance and charm as time going by. They are the memory of cities and non-renewable resources, and represent the past of cities. Before reforming a city, we should inquiry carefully to classify the buildings in a district according to the preserved status of historical information, so that have a clear judgments of what to remain and what to reform. The dilapidated buildings of the common living areas should be removed in case affect the urban image; while for the historic sites, we should repair them.

然后，要对创意所设计的新建建筑或改建建筑做必要的内部功能和外部立面的调整和技术评估，使其外部风貌与城市定位协调统一，内部功能合理，技术方案可施行。拆除重建的地段，要根据地段在城市中的具体位置来规划建筑的功能，小城镇临街建筑一般设计商住两用型，下层是商场，上层是住宅，与街道之间要设计绿化隔离带，起到隔声隔尘的作用。而对质量完好、使用年限不长的建筑地段，一般只作外立面改造，使其风格与街区或城市形象形成统一整体，这种改造主要考虑原有建筑的结构是否能够承受改造所带来的影响。

Then, we should adjust and technically evaluate the internal function and external facades of the new buildings and alterations which are according to the creative design, so that its external appearance coordinate with the urban orientation, internal function can be rational and the technical solutions implemented. For the demolished and reconstructed lots, we should plan their functions according to its specific locations in the cities. Generally, the buildings beside the streets in the small towns are designed to be used both in commerce and living; the lower is the mall and the upper is for residents; the green isolation belt should be designed between the streets in order to insulate the noise and dust. While for the architectural lots with good quality and short using time, we generally only make facade transformation, so that their style can be unified with the image of district or city, for which we consider mainly if the structure of the original buildings can withstand the impact brought by the transformation.

最后，要对规划设计的构想做经济可行性的评估，即投资估算。

At last, we should make an economic feasible assessment of the concept of planning and design, which is the investment estimation.

3.6 3D 表现——静态展示设计效果

3D Display—to Show Design Effect Statically

有独特的设计思想和创意，还需要通过准确、清楚的表达，才能让人们看到设计效果，理解设计师的设计意图。规划设计的最终效果的好坏除与方案本身的好坏有关外，还与后期的效果图的制作有密切的关系。从事规划设计的人都知道，3DMAX 就是制作建筑、景观效果图非常实用的三维设计软件，是静态展示设计效果的最佳手段。设计师的设计从最初的空间概念草图到二维面上的投影图的表现，再到三维的最终效果的表达，是一个从四维到二维再到三维的过程。规划设计的制图，实际也就分成了创意手稿阶段的方案设计推敲，CAD 的二维表达，再到 3D 模式下的精确制作。其中，CAD 是面向专业人员的，用于指导实际的建设工作，而 3D 是直接面向客户终端的，它的意义不只是给建筑施工人员以直观的建成印象，更重要的是它将向非专业人员的社会大众呈现建成后最直观的视觉效果，是最通俗，而又无声的解说方式，可以为前一阶段的城市定位、规划、创意等设计思想提供极好的展示手段，其效果的仿真性，与意境的营造将会增大规划地块社会正面效应。因此，设计人员除搞好设计创意外也必须精于此道，应特别关注后期的 3D 效果图的制作，才能将优秀的设计构思完美呈现。

天海规划有一支实力雄厚的 3D 制作队伍，积累了丰富的城市设计、建筑景观的后期制作经验。

Only the unique design ideas and creativity are not enough to let people see the design effect and understand the designers' design purpose, but precise and clear expression are also needed. The ultimate effect of the planning and design is good or not relaying not only on program itself, but also having close relation with the producing of the late effect pictures. Anyone who is engaged in planning and design knows that 3DMAX is the practicable three-dimensional software of making the effect pictures of architecture and landscape. It is the best method to display statically the design effect. Designers' design is from the initial concept sketches to the display of the projection picture on the two-dimensional surface, and to the expression of the ultimate effect, which is a process from four-dimension to three-dimension and then to two-dimension. Then the drawing of planning and design is actually divided into the steps of the manuscript of the creative design, the expression of the two-dimension of CAD and the accurate producing of 3D model. Among them, CAD is used to instruct the construction for the professionals; while 3D is for the customers directly, which not only gives the constructors a direct finished impression, but more important provides the public with a direct visual effect of the finished image. It is the most popular and silent explanation method, which can provide the last step of design ideas, such as urban orientation, planning and creativity etc. with the best method of demonstrations. Therefore, besides the design and creativity, the designers must master the 3D producing especially the late producing of the 3D effect pictures, so that perfectly display the excellent design ideas.

We Tianhai Planning have a powerful 3D producing team which has accumulated a wealth post-producing experience of urban design and architectural landscape.

3.7 动画——动态呈现最终愿景

Animation—to Display Final Landscape Dynamically

在计算机辅助技术广泛应用于规划设计领域的今天，人们已经不能满足于静态地观赏城市设计的设计效果，于是动画制作开始进入城市规划领域，成为动态地展现城市设计最终愿景的重要手段。这种全新的可视化技术，通过强大的三维建模技术建立逼真的三维场景，进行实时景观分析、多方案城市空间比较，对规划项目进行真实的“再现”，使规划决策者、规划设计者、城市建设管理者以及公众，在一个统一的平台上实时互动地看到规划效果，更全面地掌握城市的未来形态和理解规划师的设计意图，对城市空间同时获得视觉听觉多角度、身临其境的真实感受。这样，决策者的宏观决策将成为城市规划更有机的组成部分，公众参与也能真正得以实现。这是传统手段如平面图、效果图、沙盘等所不能达到的。

天海在动画制作方面也有十分成功的案例。

At present, with the computer technology widely used in the field of planning and design, people haven't been satisfied with the static design view of urban design effect, in this case, animation producing started entering the field of urban design, which has become the important means displaying dynamically the ultimate vision of urban design. With a powerful three-dimensional modeling techniques, this new visible technology can make a realistic three-dimensional scene, which will perform landscape analysis in time and multi-program urban space compare, and "represent" the planning project truly, so as to the decision makers, planners, designers and the public can see the planning effect timely and interactively in the same level, grasp more widely the future urban status and understand the planners'design purpose, and also get true visual and auditory feelings of the urban space. In this way, the macroscopical decision of the decision maker will become the organic integral part of the urban planning, and also the public participation can be truly realized, which the traditional means, such as floor plans, renderings, sand table and even animation, can not achieve.

Tianhai also have quite successful examples in producing animation.

4

黔途似锦

Prosperous and Splendid Cities of Guizhou

天海规划城市设计案例

Works of Urban Design by Tianhai

铜仁篇 Tongren Profile

城市设计篇（第二版）

变城记

City Changing

前期策划
Prophase Planning

铜仁市城市定位的构想

一、区位分析

铜仁市作为铜仁地区的中心，黔湘渝三地交界的区域中心枢纽城市，有着良好的基础设施和接待条件，可进入性强，是区域内具有中心接待能力和集散能力的主要城市。同时，铜仁市又拥有肩挑梵净山和凤凰古城两大旅游资源的绝佳交通区位。

可以说，铜仁市具有成为黔东旅游集散圈的交通保证、接待保证、资源保证和市场保证，既是区域旅游的起点，又是区域旅游的终点，成为整个西南地区乃至全国梵净山旅游、凤凰古城旅游必先到达的中转地，进而成为区域旅游的中心目的地城市。

二、核心形象

“锦尚铜仁，悠悠梵城”

之所以这样考虑，其一，是以“锦（江）”突出铜仁旅游城市打造的核心；其二，在于营造“悠悠”的浪漫情怀和凸显城市的核心“梵（文化）”。并在此基础上辅以“悠远”，“悠游”、“悠扬”、“悠闲”等四个次级对外宣传形象，加强主体形象“悠悠”的特质，不仅是对主体的阐述，也是对形象受众进行推介。

三、整体形象

“梵天净土，桃源铜仁；娱情山水，一度悠然；十里锦江，不夜之城”

进行如此定位是基于以下几点考虑：

第一，作为梵净山所在铜仁地区的政治、经济、文化、交通中心，其所拥有的生态环境是其他区域中心城市所不能比拟的，自古就有“武陵桃源”的美誉，自可成就“梵天净土，桃源铜仁”之说。

第二，铜仁市既然要实现跨越式的发展，那“通过发展旅游产业，构建区域旅游中心城市”，就是促成其实现这一目标的一大途径，而如今休闲度假旅游日趋火热的形式，正好为铜仁这片还未进行深度旅游开发的净土指明了方向，进行一次悠然的以山水为核心的度假，这正是“娱情山水，一度悠然”最真实的写照。

第三，铜仁市城市依托的根本在于蜿蜒穿城而过的“十里锦江”，她不仅是城市的景观，更是城市的灵魂与命脉，因此在对铜仁市城市定位的过程中，应充分的考虑将其融入，提升她、利用她、彰显她，以其塑造铜仁市城市的新形象——“十里锦江，不夜之城”。

The Conception of City Orientation for Tongren City

Ⅰ Location Analysis

As a core city in Tongren Region and central hub in the boundaries of Guizhou Hunan and Chongqing Provinces, Tongren City, which enjoys good infrastructures and reception conditions, is accessible and a major city having central reception and distributing functions. At the same time, Tongren City possesses excellent traffic location in FanJing Mountain and Ancient Phoenix City, the two major tourist resources.

We can say that, as a guarantee of traffic, reception, resources and market in East Qian Tourist Distributing Circle, Tongren City is not only the starting point but also the terminal of regional tour. It becomes a necessary conduit of Fanjing Mountain Tour and Ancient Phoenix City Tour for people from the whole southwest region and even the whole country, and further more, becomes a central destination city for regional tourism.

Ⅱ Core Image

"Tongren City on Jinjiang River; Buddha City in Leisure"

The reason for this concern is, for one thing, to give prominence to Jin River, the core of tourist city Tongren; for another, to foster "leisured" romantic feelings and to highlight "Buddha Culture", the core (culture) of this city. Based on this, Tongren City advertises its image by four sub grades "distant", "carefree", "melodious" and "leisured" to strengthen its feature "leisure", which is not only an explanation to the theme, but also a promotion to the image receivers.

Ⅲ Overall Image

"Buddha Pure Land; Peach Orchard Tongren; Enjoying Yourselves in Mountains and Rivers Leisurely; Ten Miles of Jinjiang River, The Sleepless City"

This orientation is based on the following concerns:

First, as the political, economic, cultural and transportation centre of Tongren Prefecture and the location of FanJing Mountain, it has the ecological environment that other core cities can not compared with. Since ancient times, it enjoys the reputation as "Wuling Peach Orchard", which matches the saying "Buddha Pure Land; Peach Orchard Tongren".

Second, in order to achieve leap-forward development, "establish central regional tourist city by developing tourism" is a major channel to achieve this goal. Leisure and vocation tour, the more and more popular touring method just direct for Tongren, the pure land, which has not been deeply developed yet. A tour around mountains and rivers in leisure is just a true expression of "enjoy yourselves in mountains and rivers leisurely".

Third, the essence of the base of Tongren City is "Ten Miles of Jinjiang River", which goes through the city in zigzag. She is not only the landscape but also the soul and pulse of this city. So the orientation of Tongren City should fully take into account that to integrate, improve, utilize and publicize her and create new image of Tongren through her as "ten miles of Jinjiang River, the sleepless city".

四、提升形象

锦江——中国“梦游第一江”

通过以上的分析，我们可以得出这样的结论，即铜仁市打造旅游城市的真正核心在于锦江，在于与锦江唇齿相依的12个半岛。白天的锦江自然、恬静；入夜后通过光、声、影等高科技手段，锦江呈现出令人如痴如醉的梦幻色彩，此情此景可谓“惊艳”，这不仅是带动提升铜仁城市幸福指数的一大保障，还是铜仁城市旅游的最大亮点。因而，在对铜仁打造旅游城市进行定位时，提出了“锦江——中国‘梦游之江’”这一形象，旨在营造具有梦幻色彩的锦江夜游，成全其通过旅游带动提升城市品质的最终构想。

五、形象依托

铜仁市城市形象所依托的根本在于“山水文化”

“山”是梵净山，所拥有的资源就算放在世界范围内也是绝无仅有的，极具旅游开发价值。但同时我们也应看到，作为一个拥有如此高品位的自然旅游资源，大兴土木的开发是断不可行的。因此，寻找一个城市作为它的中转，形成“山上观光，山下歇息”产品形态是必要的，也是必需的。而综观之下，铜仁市无疑是这一中转载体的最佳选择。

“水”是锦江水，她是铜仁城市的灵魂与命脉，没有她就没有铜仁城市的今天，因有她才有了铜仁城市的俊秀。因而，要想“山上观光，山下歇息”的游客在登完梵净山回到铜仁市时，能在这里住下来，就有必要做足铜仁市锦江水的文章，将游客旅游过程中的“食、住、行、游、购、娱”都移植到锦江之上，以锦江水带动铜仁市城市休闲旅游产业群的发展，形成“白天山中观景，晚上水上游戏”的有机旅游产业结构，成就铜仁市休闲度假不夜城的辉煌。

“文化”是糅合了梵净山弥勒佛文化、江浙与川渝商贾文化和铜仁本土少数民族文化的综合文化，它们是铜仁城市的精神底蕴所在，其中佛文化是其魂、少数民族文化是其体、商贾文化是其肢，对它们的挖掘、整理、提炼、展示，是对铜仁城市风貌的重塑、城市软实力的提升，可以通过具象化的建筑符号语言，实现对铜仁精神文化生活的再现、城市人文性格的梳理。

六、形象表现

依托铜仁的“山；水；文化”，紧扣铜仁的形象定位，在铜仁城市风貌的重塑上充分利用土家、徽派、苗、侗这四类建筑符号，结合现代旅游休闲城市的发展方向，将铜仁城市打造成为“既民族又现代”的旅游休闲度假城市。

同时，对铜仁市的主要公众活动场馆进行创意设计，使之成为今天引领铜仁城市风貌的地标，将来展现铜仁历史文脉的遗产。此外，与城市主干道的街景建筑相映成趣，也平添几分风情和浪漫。

Ⅳ Image Promotion

Jinjiang River–the Best River for Dreamlike Travel in China

Through analysis above, we can conclude that the core for Tongren to build a tourist city is Jinjiang River, and the 12 peninsulas attached to Jinjiang River. In day time, Jinjiang River is natural and tranquil; at night, it takes on intoxicating illusional scene, which is truly awesome and is a guarantee of Tongren happiness index as well as the biggest attraction for Tongren city tourism. Therefore, the orientation of Tongren as a tourist city raised the image that "Jinjiang River–A dreamlike travel in China" so to create illusional night tour of Jinjiang River and to achieve the final conception of improving the quality of city by tourism.

Ⅴ Image Reliance

The Image of Tongren City Essentially Relies on "Mountain, River and Culture"

"Mountain" refers to Fanjing Mountain, which owns absolutely unique resources worldwide, is worth tour and development. At the time we should see that as a natural tourist resource of such refined taste, it is impossible to go in for large-scale construction. Therefore, it is a necessity and a must to find a city as its conduit to form a tourist pattern like "view on the mountain and rest under the mountain". After comprehensive analysis, Tongren City is undoubtedly the best choice to be the conduit.

"River" refers to Jinjiang River. As the soul and pulse of Tongren City, she conceived today's Tongren City and endowed Tongren with beauty. Therefore, those tourists who want to "view on the mountain and rest under the mountain" will come back to Tongren City after climbing FanJing Mountain. To have the tourists rest here, it is necessary to take full advantage of Tongren City and Jinjiang River and transfer "dinning, boarding, transporting, touring, shopping and entertainment" in their travel to Jinjiang River, which push the development of Tongren leisure tourism so to form the organic tourism structure— "sightseeing in mountain in day time and play on river at night and to create the prosperity of Tongren, the leisured, vocational and sleepless city.

"Culture" refers to the comprehensive culture, incorporating the Buddha culture of Fanjing Mountain, business culture of Jiangzhe and Chuanyu, and local culture of ethnic minorities, all of which are spiritual essence of Tongren City. Of this, Buddha culture is its soul; ethnical minorities' culture is its body; business culture is its legs and arms. The excavation, collection, extraction, and exhibition to these cultures are a rebuild of Tongren City Landscape and a improvement of the soft power of the city so to achieve a reoccurrence of Tongren spiritual life and rearrangement of city humanitarian character by specific and visualized architectural mark language.

Ⅵ An Image Expression

Forge Tongren City into an "ethnical and modern" tourist leisure and vocation city by relying on "mountain, river and culture" of Tongren, orienting itself based on its image, taking full advantage of four architectural mark of Tujia, Hui Style, Miao, and Dong in the rebuild of city landscape of Tongren City, and combining the orientation as a modern tourist vocational city.

At the same time, we shall conduct creative design to the major venues for public activities in Tongren City so to make it into a landmark piloting landscape of Tongren City and into an expression of remains of Tongren's history and culture in the future. They correspond to the street landscape architecture of the city avenues interestingly, adding some sense of ethnical flavor and romance.

The first aspect of the key region-construction project of city landscape

The key in planning and designing of this part lies in two aspects: one is the landscape reconstruction of the two sides of the main avenue in Tongren City; the other is creative design of the public venues for public activities. Through the construction of main city landscape, the purpose is to improve humanitarian contents, increase tourist and resting scenic spots of the city, enrich cultural life of Tongren in content and pattern, form new tourist attractions, add a new advantage for constructing a tourist city in Tongren, and make the conception to construct a tourist city more feasible.

重点区域说明 Explanation of the Key Region

重点区域之一——城市风貌塑造工程

这一部分规划设计的重点主要放在两个方面，一个是铜仁市城市主干道两侧的景观化改造，另一个是主要公共活动场馆的创意设计。其目的在于通过对主体城市风貌的塑造，提升城市的人文内涵，增添城市的游憩景观，从内容到形式上丰富铜仁市的文化生活，形成新的旅游吸引物，使铜仁市打造旅游城市又新添一个有利砝码，使打造旅游城市的想法更具可行性。

一、铜仁城市主入口的设想

作为黔湘渝三地交界的中心城市，其在交通上起着区域枢纽的作用，也成就了铜仁市通达的立体式交通网。因而，对这样一个城市进行城市风貌打造，应首先考虑其城市主入口这一城市形象窗口、游客进入铜仁市的第一印象区的设置与设计。

I Conception of the Main Entrance of Tongren City

As the core city in the boundaries of Qian Xiang and Yu regions, it plays the role as a regional hub in traffic and enjoys convenient dimensional traffic net. Therefore, while constructing the city landscape in such a city, we should take the setting and design of main city entrance, the window of city image and tourists' first impression area, into consideration firstly.

二、城市主干道景观规划设计说明

铜仁城市主干道景观规划的重点在于沿街建筑的外立面改造和公共休闲设施的创意设计。通过城市建筑风貌的塑造，将文化气息之魂附于城市建筑之魄上，达到灵肉合一的效果。如此一来不仅能够提升铜仁的人文内涵，使城市具有灵气，达到更高一层的境界，同时也能成为铜仁增添新的景观，形成新的旅游吸引物，形成独具特色的同仁城市地标，使得旅游城市打造再添亮点。

II Explanation of the Plan and Design of Major Avenue Landscape in City

The transformation of outside facade of buildings along streets and the innovative design of public leisure facilities are the major part in the landscape planning of the main roads in Tongren City. The constructions of Tongren will be endowed with cultural soul and reach the effect of united spirit and body so as to enhance humanistic connotation of Tongren, and, moreover, add new landscape and form new attractions and unique urban sign in Tongren.

Ⅲ Explanation to conception and design of venues for major public city activities

As symbolic architectures of the city, the venues for major public activities have not only met the functional requirements fully, but also extended the city spirit. They are the true venues for the local residents to create collective memories and also excellent scenic spots created for nonnatives. So are the National Theatre, the Water Cube, and Bird Nest in Beijing. So this conception shall be maintained in conception and design of major venues for public activities.

三、城市主要公共活动场馆创意设计说明

城市的主要公共活动场馆，作为这个城市的代表性建筑，在充分满足功能需求的同时，也延续了这个城市精神。是为当地居民创造集体回忆的主要场所，同时也是为外地游客新建的绝佳景致。北京的国家大剧院、水立方、鸟巢，无不是如此。因此，对于铜仁市主要公共活动场馆进行创意设计时，也应持有这种思想。

1. 锦之鱼（铜仁市游泳馆）：紧扣铜仁水文化的铜仁市游泳馆，打破了传统造型，给世人预想不到的惊喜。同时，这一与功能结合紧密，具有深远寓意的建筑，无论形式还是内容，对城市内涵的提升都是大有裨益的。

2. 八龙聚首（铜仁地区体育场）：作为铜仁地区举办大型体育活动首选场所，选用松桃的滚龙作为创意设计的基本元素，八龙首首相对，分别聚首在体育场的四个主入口上，寓意着四方汇聚，八方来和，同时也象征了团结、互助的体育精神。

3. 蝉之声（铜仁市歌剧院）：以自然界中的蝉为基础，进行创意设计，使之在寓意着人文的天籁的同时，也表现了自然的天籁，两者于形意之间巧妙结合，更彰显了其深远的艺术价值。同时，又象征了有“蝉之声”美誉的侗族大歌，人文底蕴更显厚重。

4. 铜仁地区博物馆：铜仁地区作为巫傩文化的故乡之一，其地区博物馆完全应该构建以巫傩文化为主体，其他文化为补充的展示体系，以突出当地的文化主体和精华。

1. Fish in Jin River (Natatorium of Tongren City): Closely responding to River Culture of Tongren, natatorium of Tongren City breaks the traditional image, and gives unexpected surprises to people. In the mean time, this highly functional architecture has profound and far-reaching implied meanings, which is absolutely helpful to improve the internal meanings of the city in pattern or in content.

2. Eight Dragons Get Together (gymnasium in Tongren region): As the best choice to hold large scale sports activities in Tongren, it uses flying dragons as the basic element in creative design. Head to head, the eight dragons get together in four major entrances of the gymnasium separately, implying that blessings from eight directions get together in four directions, and also the sport spirits like solidarity and mutual aid.

3. Sound of Cicada (the opera house of Tongren City): The creative design is based on cicada in nature, which implies the song of people and nature. The ingenious combination of the two highlights its profound and far-reaching artistic value. At the same time, symbolizing the Great Song of Dong Nationality praised as "the sound of cicada", it shows deeper humanitarian contents.

4. Tongren Regional Museum: Tongren region is the hometown of Wu Nuo culture. Its regional museum should construct exhibition system dominated by Wu Nuo culture and complemented by other cultures so to highlight the main body and essence of local culture.

To Tongren, Jin River is not only its river of origin and of life, but also meaningful in dividing Tongren City into 12 peninsulas by zigzag watercourse gathered there rivers. Based on this, an ecological, humanitarian and visional river which perfectly incorporates "shape, spirit and meaning" and makes people lost in fantasy and gives play to their infinite creative conceptions comes into being.

重点区域之二——“黄金水脉”十里锦江（含 12 半岛）
The Second Aspect of the Key Region–"Golden River" Ten Miles of Jinjiang River (Including 12 Peninsulas)

锦江之于铜仁，不仅是铜仁的发源之河、生命之河，更大的意义是在于其三江汇流穿城而过的景观，蜿蜒的河道将铜仁市划分出 12 个半岛，并以此为基础，形成的“让人有无限遐想，能发挥无限创意空间，‘形、神、意’完美结合”的生态之江、人文之江、景观之江。在此基础上，建议对以下几个半岛进行重点打造 。

一、南岳半岛

定位：以民族风情为主体的滨江休闲游憩区。

1. 铜仁民族风情园区：由“黔东民族表演”的主体建筑、铜仁地区文化局、文工团办公楼、文工团演员宿舍、南岳半岛民族村寨几个部分构成。

2. 露天游泳场：一个环境清洁、卫生、安全的，服务完备的大型露天游泳场，满足消暑纳凉之各方面的需求。此外，冬天还作为旱冰场经营。

3. 滨河休憩景观区：该区域由飞虹伏波廊和拓展训练区两个部分组成，充分体现了其休闲游憩的主要功能。飞虹伏波廊紧邻江边，除观景功能外，还有小型茶座。

4. 酒店区：为地方民族特色浓郁的高端生态型休闲度假酒店区，包括核心酒店和产权别墅，整体设计上追求一种与自然山川相容的生态效果。

5. 旅游房产项目区：为南岳半岛发展后期的开发项目，以山水风光和民族风情为卖点，是在南岳半岛的旅游开发项目发展成熟以后而进行的二次开发，极为有利于提升南岳半岛旅游产业的价值。

I Nanyue Peninsula

Orientation: leisure, tourist and rest area along the rover dominated by ethnical flavors.

1. Tongren Park with Ethnical Flavors: It is composed of main architectures for "Qiandong ethnical performance", Tongren Regional Bureau of Culture, office building for art group, dorms of art group performers, and ethnical villages in Nanyue Peninsula.

2. Open-air Swimming Pool: It is a clean sanitary safe and well-functioned large scale swimming pool for people to remove summer heat and keep cool etc. In addition, it operates as an dry skating arena in winter.

3. Resting and Sight-seeing Area along the River: This region is composed of two parts, corridor of flying rainbow above waves and extended training area, which gives full play of its resting and sight-seeing function. The corridor of flying rainbow above waves is closely near the river, which is not only for sight-seeing and also for tea-drinking.

4. Hotel Area: It is a high-end ecological vocational hotel area with strong ethnical features, including core hotel and property right villas. The overall design pursues an ecological effect which integrates with mountains and rivers in nature.

5. Tourist Estate Project Area: It is a post development program in Nanyue Peninsula which takes the landscape of mountains and rivers as its advantage. The redevelopment after the maturity of tourist development programs in Nanyue Peninsula is absolutely beneficial to enhance the tourism value of Nanyue Peninsula.

二、江宗门半岛规划设计说明

定位：铜仁市城市起源的古城文化展示区。

项目设置：

1. 东山文化园区：以东山公园为主体，自古就是铜仁市香火鼎盛之地，现有东山寺立于其上，现在其基础上进行修缮，重修大雄宝殿、修复东山寺主殿、增设静心斋和素食斋，务求恢复其昔日鼎盛之状。

2. 中南门古城维护区：包括了古城区和立面改造区，其中还含有铜仁市博物馆，其主体铜仁中南门古城与东山寺同为国家级文物保护单位，具有极高的历史文化价值，对其的规划以维修和保护为主。

3. 黔东民族风情街区：以铜仁地区的“八县一市一特区”为基准，对街区进行划分，从建筑外形到文化内涵上，都对铜仁民族风情进行立体的展现，务求将其打造成为汇聚铜仁民族风情浓缩精华之地。

三、水晶阁及渔梁滩半岛规划设计说明

渔梁滩半岛是铜仁市“十里锦江不夜城”的核心所在，并根据其地块特性和功能配套的要求，分别突出其相应的特点，营造“玩在渔梁滩”的整体氛围。

渔梁滩“鱼梁江坊”：一条以休闲娱乐为主要功能的仿古步行街区，街区内包含多种娱乐项目于一体的综合性娱乐城、滨江酒吧街、特色商品售卖区、客栈、食肆、茶楼、洗浴中心、洗脚城等，旨在营造“铜仁的西街，不夜的江坊”，成为整条锦江乃至整个铜仁市入夜以后，最为热闹、最为灯火通明、最为不眠的夜生活休闲娱乐的区域。

依托于梵净山，将铜仁市打造成为独具民族特色和现代气息的旅游休闲度假城市，同时整合区内旅游资源、联合周边旅游景点，营造出辐射黔湘渝的区域旅游中心城市，使铜仁市形成“大铜仁”的城市经济集成圈的核心，引领并带动一方社会经济环境的持续发展。

Ⅱ Explanation of the Plan and Design of Jiangzongmen Peninsula

Orientation: ancient cultural exhibition area for the origin of Tongren City.

Project Settings:

1. Dongshan Cultural Park: dominated by Dongshan Park, it is a very popular place in Tongren City since ancient times. Now on top of the park stands Dongshan Temple, which is restored on basis of its current condition by restoration of Daxiong Grand Palace, main palace of Dongshan Temple, addition of Jinxin Room and Vegetarian Room in an effort to resume it former prosperity.

2. Zhongnanmen Ancient Town Maintenance Area: it includes ancient town and reconstruction area, in which there is Museum of Tongren City dominating Tongren Zhongnanmen Ancient City. Both Tongren Zhongnanmen Ancient City and Dongshan Temple are units that enjoy national protection for cultural relics which have extremely high historical and cultural value. Its plan should give prominence to maintenance and protection.

3. Streets with Ethnical Flavors in Qiandong: the streets are divided on basis of "eight counties one city one special region" in Tongren Region. As a dimensional expression of Tongren ethnical flavors in architectural profile and cultural contents, it shall be made into an essential miniature integrating Tongren ethnical flavors.

Ⅲ Explanation of Plan and Design of Crystal House and Yulang Beach Peninsula

Yuliang Beach Peninsula is the essence of Tongren City "a sleepless city in ten miles of Jin River". Based on its regional features and requirements in functional supporting facilities, we will highlight its related features to foster a whole atmosphere enjoy yourselves in Yuliang Beach".

"Yuliang River Lane" in Yuliang Beach: it is an archaistic pedestrian street with leisure and entertainment as its major functions. Inside the street is comprehensive entertainment city integrating many entertainment functions, Binjiang Beer Bar, selling region of special products, hotels, restaurants, tea house, bath centre, wash feet city etc. All of them are to create "Western Street of Tongren, sleepless river lane" and to make the whole Jin River even the whole Tongren City ablaze with lights and a sleepless leisure and entertainment area at night.

Relying on Fanjing Mountain, we will change Tongren City into a vocational leisured and tourist city with unique ethnic flavors and modern sense. What's more, we will integrate regional tourist resources and unify surrounding tourist scenic spot so as to create a core tourist city covering Qian Xiang Yu region and make Tongeren City the core of "Big Tongren" economic circle. In final analysis, we shall lead and push the sustainable economic social and environmental development of the area.

设计元素（徽派建筑）
Design Element

设计元素（苗族建筑）
Design Element

设计元素（土家建筑）
Design Element

全村携手走

设计元素（非物质文化）
Design Element

城市总说

Outline

Tongren, with "great mountains, beautiful water, and kind people" of the environment, is the hub of the two resources of "Natural Fanjing Mountain" and "humanistic Phoenix City", and a regional functional gateway city located in the junction of Guizhou, Hunan, Chongqing provinces. It can be said that Tongren already has the prerequisites for the development of tourism. But in fact, due to inadequate local infrastructure and the limitations of investment, its development has been repeatedly hampered and has greatly lagged behind the neighboring Phoenix City.

前言 Preface

铜仁，一个“山雄、水美、人仁”之境，一个肩挑“自然的梵净山”和“人文的凤凰城”两大资源的要冲之地，一个坐落在黔湘渝三地交界的区域功能枢纽型城市……，可以说铜仁已具备了发展旅游业的先决条件。但事实上，由于当地基础设施的不完善、可投入资金的局限，使其发展一再受阻，已大大滞后于邻近的凤凰。

如今，铜仁市旅游发展的机会来了：全球性金融危机虽然对中国旅游市场造成了一定的影响。但党中央国务院拿出了4万亿元作为拉动内需的资金，并将重点放在了完善各项基础设施上；同时国家旅游局也提出了“国民休闲计划”，以促进国内旅游市场的发展；而2009年“生态旅游年”主题口号的提出，更是为以梵净山为依托的铜仁旅游业的发展注入活力。 这一系列的措施，对于旅游发展尚处于上升期的铜仁市而言，都是利好的，它的作用在于完善铜仁市旅游基础设施的同时，鼓励当地优先考虑旅游业的发展。

就铜仁所在区域的旅游发展现状来看，虽然周边已形成了诸如梵净山、湘西凤凰古城、铜仁九龙洞、镇远古镇等旅游景区（点），但处于其交通核心位置的铜仁市并未能充分发挥其接待和集散的功能，起到连接、整合区内旅游相关产业，提升区内旅游整体品质和形象的作用。而如今进行的规划正是站在这样的高度上考虑的，通过城市风貌的改造，打造区域旅游中心城市，寻求区域旅游产业的跨越式发展。

Now, Tongren City has the development opportunities for tourism. Although the global financial crisis affected Chinese travel market on a certain extent, the CPC Central Committee and State Council took out 4 trillion yuan of funds as stimulating domestic demand, and focusing on improving the infrastructure; at the same time, the National Tourism Administration also proposed a "National Recreation Plan" to promote the domestic tourist market; the theme slogan of the 2009 "Year of Ecotourism" was proposed to booster the tourist development of Tongren backed by the Fanjing Mountain. This series of measures for Tongren City in the rising period are positive, and play the role improving the tourism infrastructure in Tongren and encouraging local tourism development with priority.

As far as the current situation of tourism development in the region where Tongren is located in is concerned, although there are tourist areas (spots) such as Fanjing Mountain, Phoenix ancient city in Western Hunan, Tongren Jiulong Cave, Zhenyuan ancient town around it, Tongren city, with the heart of traffic, is not to give full play to its reception and distribution functions, connect and integrate local tourism-related industries, and improve the overall quality and image of tourism in the region. Now, the planning is considered at such a height, through the transformation of urban style to create a regional tourist central city and develop the regional tourism industry by leaps and bounds.

第一章 基础概况 Chapter I Basic Overview

第一节 自然地理 Section I Physical Geography

一、地理位置

铜仁市位于贵州省东北部，铜仁地区东部，是铜仁地区行署和地委所在地，当地的经济、科技和文化中心。历来是渝黔湘鄂四省市边区的商品集散中心，素有“黔东门户”之称。

铜仁市地处东经 108°50′～109°28′，北纬 27°26′～27°53′，境内东西长 63km，南北宽 48km，行政区面积 1513.61km^2。东与湖南省凤凰县、麻阳苗族自治县、芷江侗族自治县毗邻，南同万山特区、玉屏侗族自治县、岑巩县接壤，西和江口县连接，北同松桃苗族自治县相依。

铜仁市到凤凰的公路行程为 63km，到麻阳 97km，到芷江 136km，到万山 41km，到玉屏 81km，到岑巩 105km，到江口 49km，到松桃 85km；到省会贵阳 420km，到湖南长沙经常德为 582km、经怀化为 640km，到重庆 738km，到张家界 311km。

I Geographical Position

Tongren City, located in the northeast of Guizhou Province and the east of Tongren Prefecture, is the seat of the administrative office and the prefectural Party Committee of Tongren Prefecture and the local economic, technological and cultural center. It has always been the border goods distribution center of the four provinces of Hunan, Hubei, Chongqing and Guizhou, known as "the gateway of eastern Guizhou".

Tongren City is located at 108° 50'~109° 28'E, 27° 26'~27° 53'N, with the territory of east–west length of 63km, north–south width of 48km, and area of 1 513.61km^2. It is adjacent to Fenghuang County of Hunan, Mayang Miao Autonomous County and Zhijiang Dong Autonomous County on the east, Wanshan SAR, Yuping Dong Autonomous County and Cengong County on the south, Jiangkou County on the west and the Songtao Miao Autonomous County on the north.

It is 63km from Tongren City to Phoenix by road trip, 97km to Mayang, 136km to Zhijiang, 41km to Wanshan, 81km to Yuping, 105km to Cengong, 49km to Jiangkou, 85km to Songtao, 420km to the provincial capital Guiyang, 582km to Changsha Hunan through Changde and 640km through Huaihua, 738km to Chongqing, and 311km to Zhangjiajie.

区位分析图

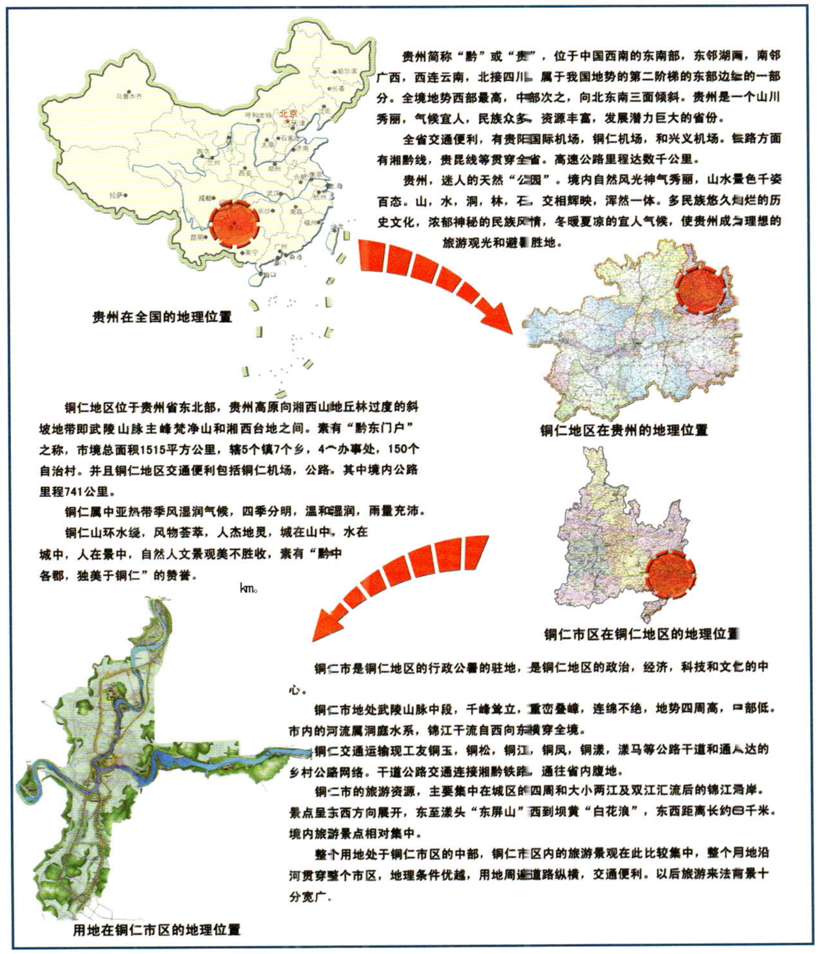

二、自然条件

1. 地质地貌

铜仁市地处云贵高原向湖南西部丘陵过渡的斜坡地带，属低中山丘陵河谷地貌，在武陵山脉主峰梵净山和湘西台地主峰雪峰山之间，河谷纵横，地势四周高中部低。境内河流较多，地表切割破碎，地面坡度大，具有地形崎岖的特点。中部川硐、茶店一带剥夷面保存较为完好，母岩风化物残积而成的风化壳，属红色黏土缓丘分布区域；边缘受河流切割而形成石灰岩峰林或峰丛山地，分布在六龙山、天生桥、鹭鸶岩、铜锣一带。

铜仁市境内山地面积占 55.9%，共 10.28hm^2，丘陵面积 32.4%，4.27hm^2，山间河谷、平坝面积 11.7%，0.6hm^2。最高点在东南部与湖南芷江县及万山特区交界处的米公山，海拔 1149.2m；最低点在东部漾头镇锦江出境处的下施滩，海拔 205m；全市平均海拔 600m 左右，500m 以下区域占土地总面积的 31.62%，城市中心区海拔 251.2m。

II Natural Conditions

1. Geological Features

Tongren City is located in the slope area of Yunnan-Guizhou Plateau to the western hills of Hunan Province, with a low and middle hilly and river valley landform between the main peak Fanjing Mountain of Wuling Mountains and the main peak Xuefeng Mountain of the platform in western Hunan, a crisscross network of river valleys horizon, and a terrain of basin. It has the rugged terrain features with many rivers, broken ground and steep slopes. The strip surface is well kept in the central area of Chuantong and Chadian, with weathering crusts formed by the weathering residual rock. It belongs to a region of slow red clay hills. The limestone peaks or mountainous region with cutting edges formed by the river scatter in the areas from Liulong Mountain, Tianshengqiao, Lusi Cliff to Tongluo.

The mountainous area accounts for 55.9% in Tongren City, a total of 102 800 hectares, hilly area of 32.4%, 42 700 hectares, and river valley and flatland area of 11.7%, 60 000 hectares. The highest point is 1 149.2 m above sea level, located in Migong Mountain at the junction of Zhijiang Hunan and Wanshan SAR on the southeast; the lowest point is 205m above sea level, located in the Xiashitan of Yangtou Town where Jin River goes outbound on the east. The city's average altitude is around 600m. The area below 500m accounts for 31.62% of the total area in the regional. The elevation of the central city is 251.2m.

现状分析图

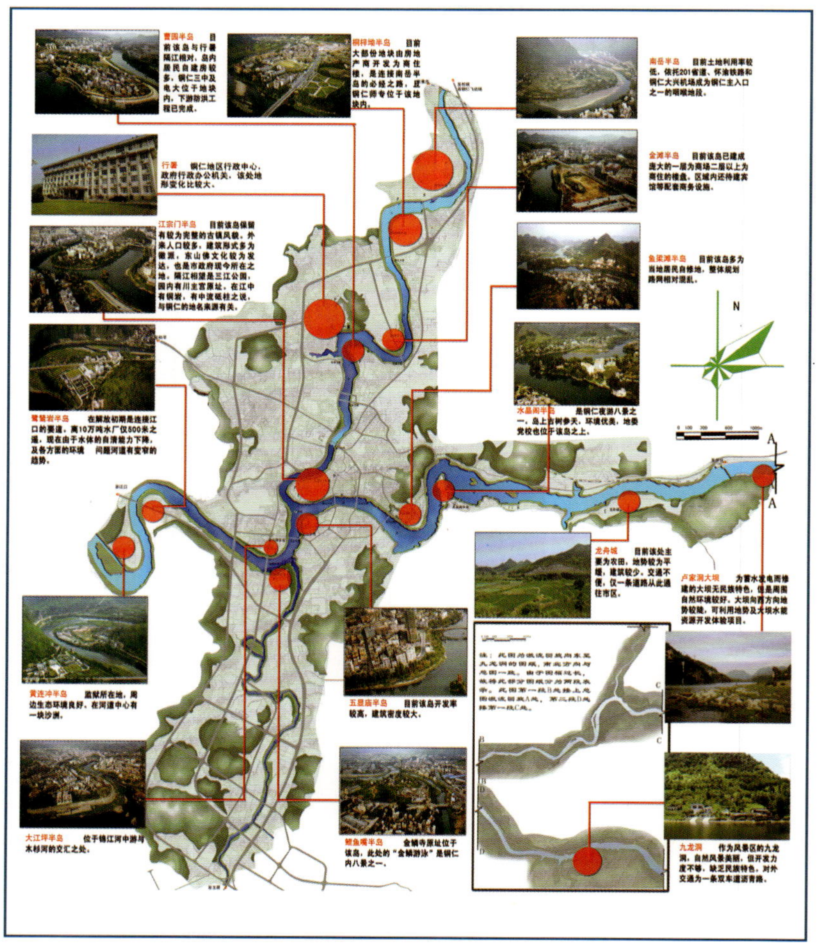

道路现状分析图

2. 气候水文

铜仁市气候属中亚热带温暖湿润季风气候，春温多变，夏季炎热，秋温速降，冬少严寒，四季分明，温和湿润，气候宜人，年平均气温 16.9℃，1 月平均气温 5.2℃，7 月平均气温 27.9℃。年平均降水量 1302.7mm，年平均日照时间 1171.4h，全年无霜期 280d。光能资源较为丰富，雨热同季，有利于农作物生长。灾害性天气主要有伏旱、春秋冷害、洪灾。

境内主要河流属洞庭湖水系，锦江干流自西向东横穿全境。河流系山区雨源型。

3. 生物资源

铜仁市植被属中亚热带湿润性常绿阔叶林、樟栲林类型。还有部分落叶阔叶林、针阔叶混交林以及亚热带竹林、石灰岩藤刺灌丛等。

2. Climate and Hydrology

The climate of Tongren City is the mid-subtropical, warm and humid monsoon climate, with changing temperature in spring, hot in summer, downhill temperature in autumn and less cold in winter. The weather is mild, humid and very good with four distinctive seasons. The annual average temperature is 16.9℃, with 5.2℃ in January and 27.9℃ in July. The annual average rainfall is 1 302.7mm and annual average sunshine time is 1 171.4h, 280 frost-free day days a year. It has abundant solar energy resources, with the rain and heat at the same season, which is conducive to crop growth. Severe weather mainly includes hot summer, cold spring and autumn, and flooding.

The main rivers within the city belong to th Dongting Lake water system. The mainstream of Jin River goes across the territory from west to east. The rivers are mountainous rain type.

3. Biological Resources

The vegetation in Tongren City belongs to mid-subtropical humid evergreen broad-leaved forest, castanopsis type. There are some deciduous broad-leaved forest, coniferous and broad-leaved mixed forest, subtropical bamboo, and limestone rattan and thorn scrub.

桥梁现状分析图

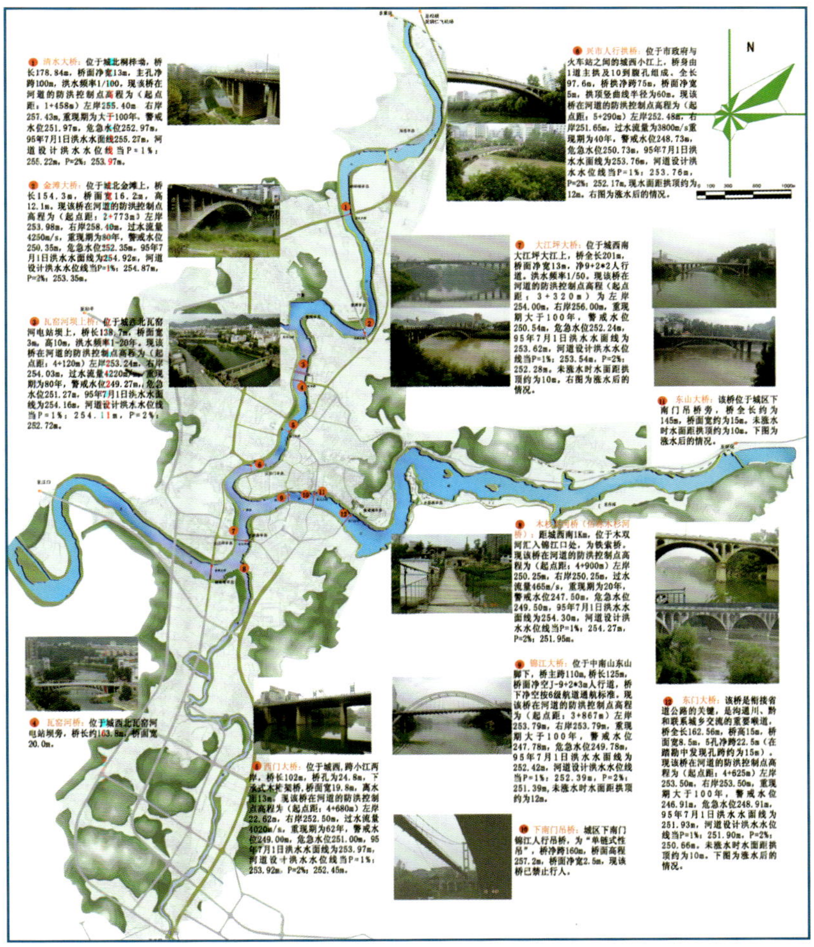

水文分析图

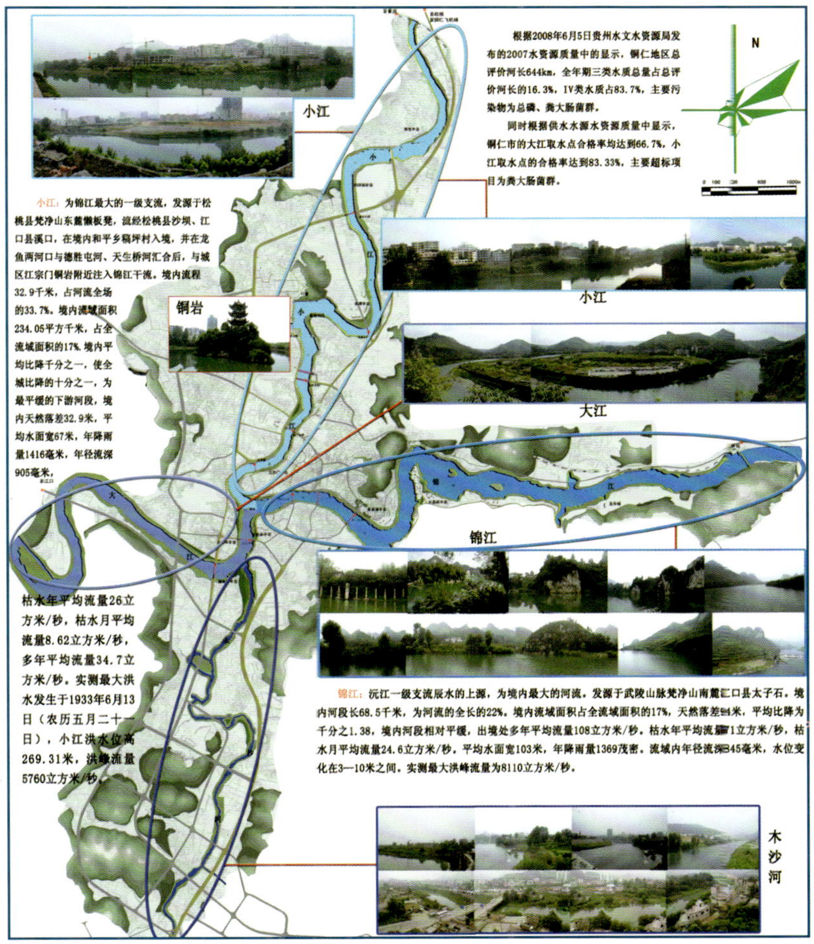

第二节 人文环境

一、历史沿革

铜仁原名铜人，相传元朝时有渔人在铜岩处潜入江底，得铜人三尊，挽而出之，符之名以此。考铜仁建置，自《禹贡》划分九江为荆州西南荒裔。周属楚。秦属黔中郡。汉为武陵郡。隋属沅陵郡。唐武后垂拱二年（686 年）始置万安县，隶锦州卢阳郡。天宝元年（724 年）改万安县为常丰县。宋初设于蛮，先属锦州砦，后并入沅州麻阳县。至元二十年（1283 年），置铜仁大小江等处蛮夷军民长官司，隶思州安抚使。明永乐十一年（1413 年）置铜仁府，隶属贵州布政使。万历二十六年（1598 年）裁铜仁司置铜仁县，为一府一县治。清光绪九年（1883 年），铜仁县治又从江口迁回铜仁，隶黔东道，道署设镇远。民国 9 年（1920 年）废道，铜仁县直属贵州省辖。民国 24 年（1935 年）设贵州省第九行政区督察专员公署，铜仁为驻地。民国 26 年（1937 年）裁，并入第一行政区。民国 32 年（1943 年）复设贵州省督察专员第六行政区，仍为驻地。

1949 年 11 月 11 日铜仁解放。12 月 1 日，成立铜仁县人民政府。同时，设贵州铜仁地区行政公署。1958 年 10 月，铜仁、江口、玉屏三县合并为铜仁县。1961 年 7 月又按原建置三县分开。

1987 年 8 月 21 日，经国务院批准，撤县建市。

二、社会经济情况

铜仁市现辖 5 个办事处，5 个镇，7 个乡、150 个村民委员会，21 个居民委员会，到 2006 年末，全市总人口达 36.12 万人，其中非农业人口 12.63 万人；少数民族人口 26.3 万人，以苗族、土家族、侗族为主。

2007 年，全市国民生产总值（GDP）达 34.09 亿元，比上年增长 15%，人均生产总值达到 9643 元。其中，第一产业完成 6.26 亿元，同比增长 6.4%；第二产业完成 12.74 亿元，同比增长 17.3%；第三产业完成 15.09 亿元，同比增长 17.2%。一、二、三产业分别拉动 GDP 增长 1.3、5.7、8 个百分点。全市财政收入完成 2.7 亿元，同比增长 28.2%；地方财政收入 2 亿元，同比增长 24.5%；全社会固定投资完成 22.57 亿元，同比增长 21.2%，其中，固定投资完成 18.21 亿元，同比增长 20.3%；社会消费品零售总额完成 12.27 亿元，同比增长 17.6%；全市城镇居民人均可支配收入 7672 元，比上年增长 8.4%；农民人均纯收入 2860 元，增长 6.1%。

Section II Human Environment

I History

Tongren formerly named Tongren (bronze man). It is said that a fishermen dived into the river from the Bronze Rock to get and pull up three bronze men in the Yuan Dynasty, so it got the name. As for the establishment of Tongren, Jiujiang is classified as the southwestern Jingzhou descendants from "Yu Gong". It belonged to Chu in Zhou Dynasty, Jinzhong County of Guizhou in Qin, Wuling County in Han, Yuanling County in Sui, and Luyang County of Jinzhou in the second year of Empress Wu Chuigong (AD 686) of Tang when Wan'an County was established. Wan'an was renamed as Changfeng in the first year of Tianbao (724). It was affiliated to Jinzhou and then integrated into Mayang County of Yuanzhou in Song dynasty when Man was set up. The military and civil head of Manyi administered Tongren from Yuan to the twentieth year of Yuan (1283), affiliated to the governor of Sizhou. The Tongren Prefecture was established in the tenth year of Yongle (1413) of Ming dynasty, governed by the Buzheng administrative office of Guizhou. Tongren County was established in the twentieth year of Wanli (1598) when there was a state and a county. The government of Tongren County moved from Jiangkou to Tongren in the ninth year of Guangxu (1883) of Qing dynasty, affiliated to Eastern Guizhou Road whose government was located at Zhenyuan. The Road was abolished and Tongren County was directly governed by Guizhou Province in the ninth year of the Republic of China (1920). The ninth administrative district of the Inspector government office of Guizhou Province was established in the twenty-fourth year (1935), and Tongren is the seat of the government. It integrated into the first administrative district in the twentieth year (1937). The sixth administrative district of the Inspector government office of Guizhou Province was re-established in the thirty-second year (1943), and Tongren is the seat of the government.

Tongren was liberated on November 11 1949. The people's government of Tongren County was establishment on December 1. Meanwhile, the government office of Tongren prefecture of Guizhou was established. The three counties of Tongren Jiangkou, Yuping incorporated to Tongren County in October 1958 and separated to the three original counties in July 1961.

It was approved by the State Council to become a city on August 21,1987.

II Social and Economic Situation

Tongren City has jurisdiction over 5 offices, 5 towns, 7 residential townships, 150 village committees, and 21 neighborhood committees. To the end of 2006, the city's total population reached 36.12 million, of which 12.63 million non-agricultural population and minority population of 26.3 million people, mainly including Miao, Tujia and Dong.

In 2007, the city's gross domestic product (GDP) reached 3.409 billion yuan, up 15% over the previous year, and per capita GDP reached 9 643 yuan. Among them, the first industry was 626 million yuan, up 6.4%; secondary industry 1 274 billion yuan, up 17.3%; tertiary industry was 1.509 billion yuan, up 17.2%. First, second and third industries are driving GDP growth by 1.3, 5.7 and 8 percentage points. The city's financial revenue was 270 million yuan, up 28.2%; local financial revenue 200 million yuan, up 24.5%; the fixed investment of entire society 2 257 billion yuan, up 21.2%, of which fixed investment completed 1.821 billion yuan, up 20.3%; total retail sales of social consumer goods 1.227 billion yuan, up 17.6%; the city's per capita disposable income of urban residents 7 672 yuan, an increase of 8.4%; and rural per capita net income 2 860 yuan, an increase of 6.1%.

三、基础设施状况

到 2006 年为止，铜仁市南出口、西出口、北出口公路已全部改造，全市公路总里程达到 1312km，已实现村村通公路。其中高等级公路铜（仁）贵（阳）高等级公路、铜（仁）麻（阳）公路均已竣工通车。

铜仁市境内有渝怀铁路经过，铁路通车里程约有 60km。

水路通航铜仁至芦家洞 7km，芦家洞至漾头 23km，瓦窑河至两河口 9km。

在民航方面，距铜仁市区 23km 处建有铜仁大兴支线机场，已开通铜仁往返广州、深圳、重庆、贵阳等多条航线。

Ⅲ Infrastructure Status

Up to 2006, the south, west and north exit roads of Tongren City have been modified, and the mileage of the county reached 1 312km. All villages are accessible by road. The Tong(ren)—Gui(yang) Highway and Tong(ren)—Ma(yang) Highway have been completed to traffic.

Yu-Huai railway goes through Tongren City with about 60km railway.

The navigable waterway includes 7km from Tongren to Lujiadong, 23km from Lujiacong to Yangtou and 9km from Wayao River to Lianghekou.

As for civil aviation, there is Daxing Extension airport which is 23km from the downtown of Tongren city. It has opened several air routes from Tongren to Guangzhou, Shenzhen, Chongqing and Guiyang and return routes.

四、文化旅游及服务业发展状况

通过开展“多彩贵州旅游形象大使选拔赛”和“游钓锦江——魅力铜仁”等宣传推介活动，提升铜仁的知名度。同时，加大对旅游景区（点）的基础设施建设。

Ⅳ Cultural Tourism and Service Development

Raise public awareness of Tongren through the publicity and promotional activities such as the "Colorful Guizhou Tourism Ambassador Competition", the "Swimming and Fishing in Jin River—Charm Tongren" and so on. At the same time, strengthen infrastructure construction in tourist areas (spots).

Chapter II Background Analysis

Taking the opportunity of the fourth round of urban development planning revision, Tongren City re-arranged and constructed the city style, from the appearance to the inner part, to form an unique image of the city as a whole, which is helpful to achieve the tourism development of Tongren City by leaps and bounds and comprehensively promote local social and economic development.

In addition, as far as the current situation of tourism development in the region where Tongren is located in is concerned, although there are tourist areas (spots) such as Fanjing Mountain, Phoenix ancient city in Western Hunan, Tongren Jiulong Cave, Zhenyuan ancient town around it, but Tongren city, with the heart of traffic, is not to give full play to its reception and distribution functions, connect and integrate local tourism-related industries, and improve the overall quality and image of tourism in the region. Now, the planning is considered at such a height, through the transformation of urban style to create a regional tourist central city and develop the regional tourism industry by leaps and bounds.

第二章 背景分析

铜仁市借着城市发展规划修编的机会，对铜仁城市风貌进行重新梳理和打造，欲从外形到内在上，都形成独具一格的城市整体形象。这对于铜仁市旅游业实现跨越式的发展，并带动当地社会经济的全面发展都是大有裨益的。

此外，就铜仁所在区域的旅游发展现状来看，虽然周边已形成了诸如梵净山、湘西凤凰古城、铜仁九龙洞、镇远古镇等旅游景区（点），但处于其交通核心位置的铜仁市并未能充分发挥其接待和集散的功能，起到连接、整合区内旅游相关产业，提升区内旅游整体品质和形象的作用。而如今进行的规划正是站在这样的高度上考虑的，通过城市风貌的改造，打造区域旅游中心城市，寻求区域旅游产业的跨越式发展。

第一节 区域背景分析

一、区位条件优越

铜仁市地处黔湘渝三地交界，是贵州省铜仁地区行政公署的所在地，区内已形成了集公路、铁路、航空线路于一体的立体式区域交通网，公路网可从铜仁到达贵阳、凯里、玉屏、万山、江口、松桃、凤凰、张家界等多地;铁路北通重庆，南抵湖南怀化、湘潭、株洲、长沙等地；航空线路已开通了从铜仁往返北京、贵阳、重庆、广州、深圳等城市的航班。可以说，铜仁市无论从区位优势、城市发展水平上，还是从交通通达性上，都有潜力成为黔湘渝三地交界中心城市。

就区域旅游资源的积累和发展的现状来看，地处黔湘渝三地交界的铜仁市周边拥有“人与自然保护圈成员之一”的梵净山自然保护区、“沈从文《边城》所描写的地方”凤凰古城、“世界自然遗产武陵源所在地”张家界等具有一流品位的资源，还拥有诸如铜仁九龙洞省级风景名胜区、江口太平河省级风景名胜区、铜仁傩文化、江口土家族民族文化、松桃苗族民族文化等一大批各式各样的自然及人文资源。可以说，就旅游开发价值而言，其古朴、自然、神秘的资源特性是极具市场吸引力的，而将铜仁市作为这些旅游开发项目的旅游中心接待和集散城市，不仅有利于铜仁城市的发展，还能充分带动区域内旅游业的整体发展。

二、区域引领潜力逐步显现

作为贵州省东部地区行政中心区域的铜仁，其黔湘渝三地交界的地理位置，公路、铁路、航空相结合的立体式交通网，都为其成为旅游接待、集散中心城市提供了充足的外部条件。

同时，铜仁市周边又拥有诸如梵净山、九龙洞、凤凰古城等高品位旅游资源，作为旅游吸引物而言，它们均具有巨大的市场吸引力和旅游开发价值，这为处于其中心的铜仁市打造成为该区域的旅游接待、集散中心城市提供了丰厚的资源。

“梵天净土，桃源铜仁”的提出，不仅代表着梵净山的佛文化在铜仁这片净土将得到弘扬，成为一个具有国家甚至国际级别的佛学研讨场所。同时也在彰显一个“大桃源”的概念，使“桃源”这一不只在铜仁拥有的形象，与同样拥有它的湖南、福建等地进行跨区域的联合营销，强化产品品牌形象，形成在全国范围内具有连锁性质的旅游产品。而铜仁市作为“梵天净土，桃源铜仁”这一形象的载体，运用适时得当的形象定位营销手段，在成就这一形象的同时，也塑造了其自身国内旅游文化产业名城的形象，成为引领区域发展的核心城市。

Section I Regional Background

I Advantageous Geographical Conditions

Tongren City is located at the junction of the three provinces of Guizhou, Hunan and Chongqing. It is the seat of the administrative office of Tongren Prefecture of Guizhou Province. Three-dimensional regional transport network of highways, railways and air lines has formed in the region. The road network reaches to Guiyang, Kaili, Yuping, Wanshan, Jiangkou, Songtao, Phoenix, Zhangjiajie and many other places; railway network links Chongqing on the north, and Huaihua Xiangtan, Zhuzhou, Changsha and other places on the south; Air Lines has opened a round-trip between Tongren and Beijing, Guiyang, Chongqing, Guangzhou, Shenzhen and other destinations. It can be said that Tongren City has the potential to become a central city at the junction of Guizhou, Hunan and Chongqin in terms of location advantages, the level of urban development, or from the transport accessibility.

As far as the abundance of regional tourism resources and the current development situation are concerned, Around the Tongren City, there are many tourist resources with first-class quality, such as the Fanjing Mountain Nature Reserve, one of the members of "Man and Nature Conservation Circle", the "Phoenix ancient city" which was described by Shen Congwen in his novel Border Town, and Zhangjiajie, the World Natural Heritage site of Wulingyuan. Besides, there is a large number of variety of natural and cultural resources, such as Jiulong Cave provincial scenic area of Tongren, Taiping River provincial scenic area of Jiangkou, Tongren Nuo Culture, Tujia ethnic culture of Jiangkou, Songtao Miao ethnic culture and so on. It can be said that, as for the value of the tourism development, the simple, natural and mysterious resources appeal to market. It not only benefit the urban development of Tongren but also to fully promote the overall development of tourism in the region if Tongren City is taken as the tourist central city of reception and distribution of the tourist development projects.

II Emergence of Regional Leading Potential

As the administrative central region of the eastern Guizhou, Tongren has the advantageous location as the junction of the Guizhou, Hunan and Chongqin and three-dimensional transportation network of roads, railways, and air lines which provides adequate exterior conditions to become a tourist central city of reception and distribution.

At the same time, they are many high-quality tourist resources, such as Fanjing Mountain, Jiulong Cave and Phoenix ancient city around Tongren. As tourist attractions, they all have huge market appeals and values of tourism development, which provide Tongren city with rich resource-based conditions to become a tourist central city of reception and distribution.

The image of "Brahma Pure Land, Peach Orchard Tongren" was proposed not only to represent that the Buddhist culture of Fanjing Mountain will be carried forward in this pure land of Tongren to become a national and even international Buddhist discussion site, but also to demonstrate the concept of "great Taoyuan (peach orchard)" so that "Taoyuan" is not only the image in Tongren but a national chained tourist product by cross-regional joint marketing with Hunan and Fujian provinces to strengthen brand image. As carrier of the image of "Brahma Pure Land, Peach Orchard Tongren", Tongren has positioned the image by proper and timely marketing. In the success of this image, its own image of the famous city of domestic tourist and cultural industries is shaped to be a core city leading the regional development.

第二节 市场条件分析

一、区域市场分析

1. 省内市场

鉴于贵州目前的经济发展水平，省内市民旅游活动范围主要还是集中在省内，其中省会贵阳更是贵州省的主体客源市场。据统计，贵州省总人口有3800多万人，其中，城镇居民超过600万人，贵阳市约占50%。加之，如今贵阳至铜仁已形成了集航空、铁路、高速公路、公路于一体的立体交通网，大大丰富了往返两地的方式、提升了客流量。可以说，古朴、自然而又充满神秘的铜仁，在基础设施完善、市场营销得当的前提下，定能吸引大量的省内旅游客源到来。

2. 周边市场

铜仁市地处黔湘渝三地交界的铜仁地区，是当地社会的经济、政治、文化、服务和交通中心，周边聚集了梵净山、九龙洞、凤凰古城等优质旅游资源，以及各种品级和类型的已开发和待开发的旅游资源，交通沿线上又串联了张家界市、镇远县等一批国家优秀旅游城市。这使铜仁市在具备成为区域旅游接待和集散中心的良好条件的同时，也具备了成为旅游线路中重要的旅游节点和区域旅游中心城市与城市旅游的目的地的潜力。因此，以铜仁市为中心构建的旅游圈，对铜仁市发展成为区域内乃至世界级的旅游休闲度假城市是大有裨益的，能实现在为周边市场带来大量客源的同时，还会对周边市场的客源产生强大的吸聚能力，最终从两个方面提升当地旅游市场规模。

二、铜仁市旅游发展分析

铜仁市历年旅游统计资料

年份	旅游接待量（万人次）	同比增长率（%）	旅游总收入（万元）	同比增长率（%）	入境旅游人次
2003	22.51	23.55	4,952.2	23.55	630
2004	30.36	34.87	7,590.0	53.27	1026
2005	41.30	36.03	12,390.0	63.32	1310
2006	51.63	25.00	16,107.0	30.00	371
2007	64.75	25.41	25,900.0	60.80	487

根据铜仁市旅游局提供的旅游统计资料显示，铜仁市作为梵净山旅游的主要依托城市，其旅游接待的能力还远远没达到同类城市的水平（2006年贵州西线旅游城市安顺的年接待量就已达739.05万人次，其王牌景区黄果树同年也达到了308.00万人次）。

究其原因，这与当地旅游业起步较晚、区域间交通条件不便、城市旅游接待设施不足等因素有着直接的关系。但同时也应认识到正是由于存在这些发展劣势，才为铜仁市旅游开发保存了大量优质的旅游资源和宽广的发展空间，为铜仁市实现以旅游业带动的跨越式发展提供了先决条件。

Section II Analysis of Market Conditions

I Regional Market Analysis

1. Inner-Provincial Market

Given the current level of economic development in Guizhou, the people of the province mainly focus on the scope of tourism inside, including the provincial capital Guiyang which is the main source market of Guizhou Province. According to statistics, the total population of Guizhou Province is more than 3 800 million, of which more than 600 million urban residents, about 50% in Guiyang. In addition, the three-dimensional transportation network of air lines, railways, highways and roads from Guiyang to Tongren has formed now, which greatly enriched the way between the two places and improved the traffic. It can be said that simple, natural and mysterious Tongren will be able to attract large numbers of tourists under the premise of necessary infrastructure and proper marketing.

2. Surrounding Market

Tongren City which is located in Tongren Prefecture at the junction of Guizhou, Hunan and Chongqing is the local economic, political, cultural, service and transportation center. There gathered high-quality tourism resources such as Fanjing Mountain, Jiulong Cave and Phoenix ancient city and various grades and types of developed and developing tourist resources around it. The traffic routes have linked a number of excellent tourist city such as Zhangjiajie city and Zhenyuan county. All of these make Tongren City not only have the good conditions to become a regional tourist center of reception and distribution but also have the potential to become an important joint and a regional tourist central city of the tourist routes and an urban tourist destination. Therefore, to build Tongren-centered tourist circle is quite helpful to develop Tongren City as the regional and the world-class tourism and leisure city, which can attract a large number of tourists to the neighboring markets and create powerful cohesive force for the tourist source of the neighboring markets to enhance the local tourism market.

II Development of Tourism in Tongren City

Tongren City tourism statistics over the years

Year	Tourist receipts (1000 person-trips)	Year growth rate (%)	total tourism income (Million yuan)	Year growth rate (%)	Inbound tourism (person-trips)
2003	225.1	23.55	49.522	23.55	630
2004	303.6	34.87	75.900	53.27	1 026
2005	413.0	36.03	123.900	63.32	1 310
2006	516.3	25.00	161.070	30.00	37
2007	647.5	25.41	259.000	60.80	487

According to the travel statistics provided by the Tourism Bureau of Tongren City, as a city mainly relying on Fanjing Mountain tourism, the tourist reception capacity of Tongren City is still far from the level of other similar cities (the annual reception volume of the western tourist city of Anshun in Guizhou has already reached 7 390 500 person-trips in 2006, and that of its most famous scenic area of Huangguoshu has reached 3.08 million person-trips in the same year).

The reasons include the local tourist industry with late start, inter-regional inconvenient transportation, insufficient urban tourist reception facilities, etc. However, it should also be aware of these development weaknesses which save a large number of high-quality tourist resources and a broad space for Tongren to develop tourism and provide Tongren with the prerequisites to achieve the leap-forward development led by tourism.

第三章 规划总则

第一节 规划依据

一、国家、省、市、县相关法律法规

二、项目委托书

Chapter Ⅲ General Planning

Section Ⅰ Planning Basis

Ⅰ National, Provincial, City and County Laws and Regulations

Ⅱ Letter of Authorization of the Project

总平面图

新增主要公共服务设施图

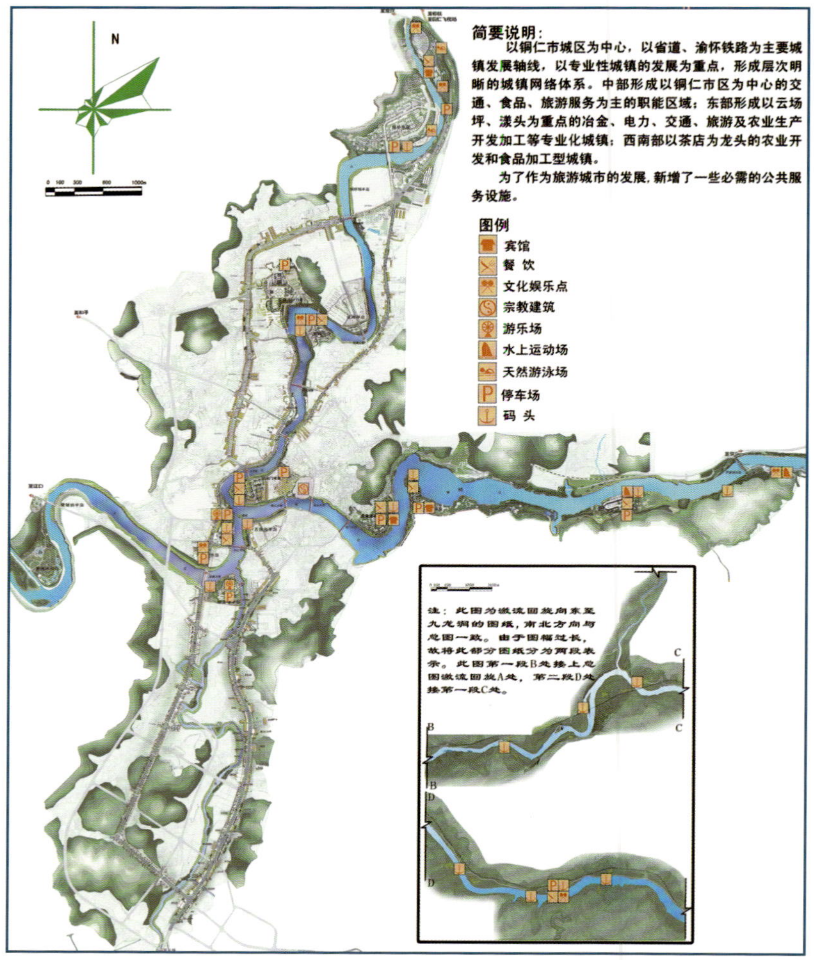

第二节 规划指导思想

一、生态优先的原则
二、全面协调的原则
三、以人为本的原则

Section II Planning Guidelines

I The Principle of Ecological Priority
II The Principle of Overall Coordination
III The Principle of People Oriented

交通游览线路图

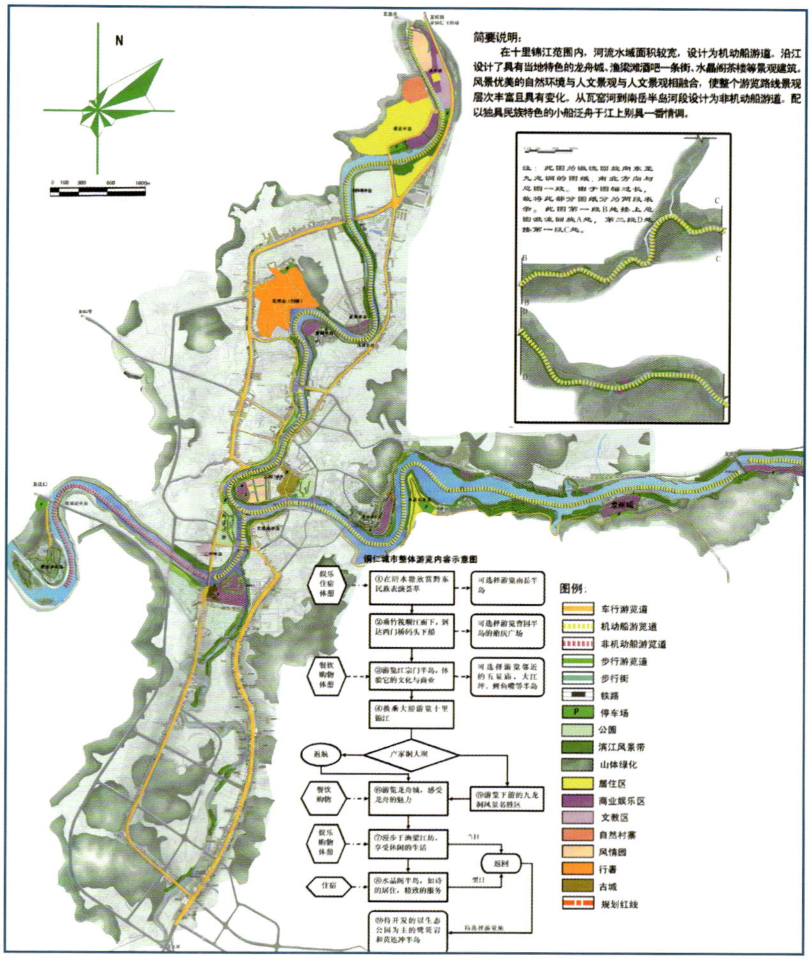

分期规划图

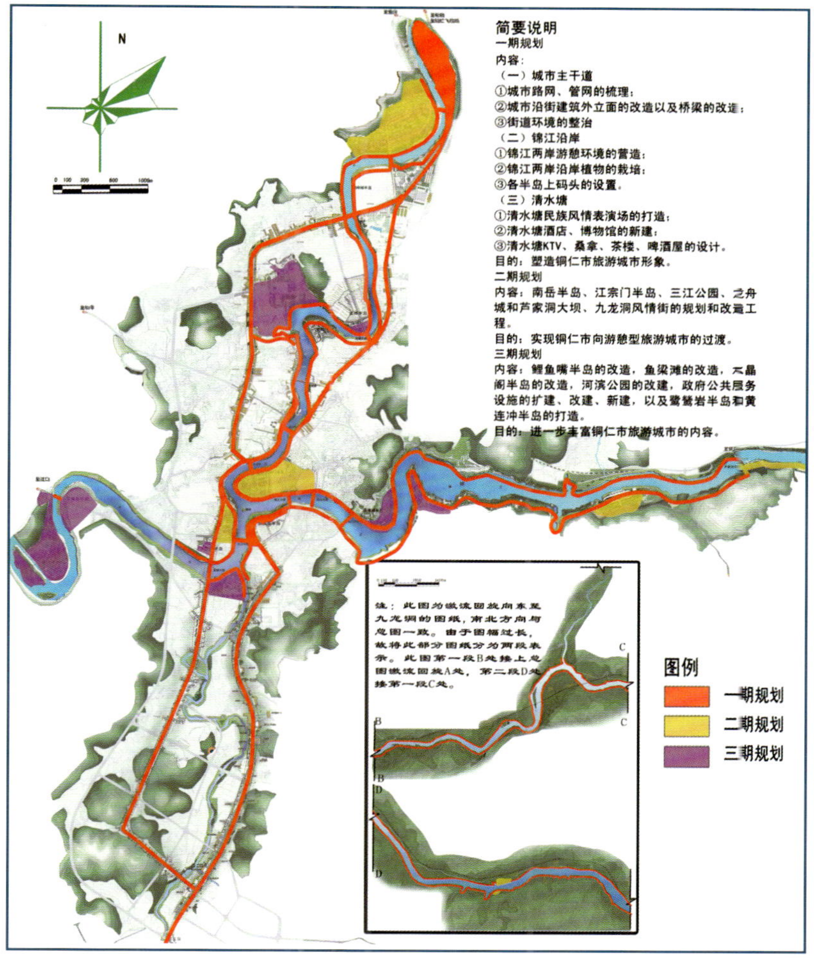

第四章 规划目标

一、近期目标——通过对城市主干道和锦江沿岸的改造和整治，塑造铜仁市旅游城市形象，形成黔湘渝三地交界的旅游接待中心城市。

二、中期目标——实现铜仁市接待型旅游城市向游憩型旅游城市的过渡。

三、最终目标——进一步丰富铜仁市旅游城市的内容，将铜仁市最终打造成为辐射黔湘渝三地交界的深度旅游目的地。

Chapter Ⅳ Planning Objectives

Ⅰ The Short-term Goal: go shape the image of tourist city of Tongren City to be a city of tourist reception center at the junction of Guizhou, Hunan and Chongqing by reconstructing and renovating the urban main road and Jin riversides.

Ⅱ The Medium-term Goal: to achieve the transition of Tongren City from host-based tourist city to a recreation tourist city.

Ⅲ The Ultimate Goal: to further enrich the contents of the tourist city of Tongren City, and eventually create a city of deep tourist destination at the junction of Guizhou, Hunan and Chongqing.

道路效果示意图

景点布置图

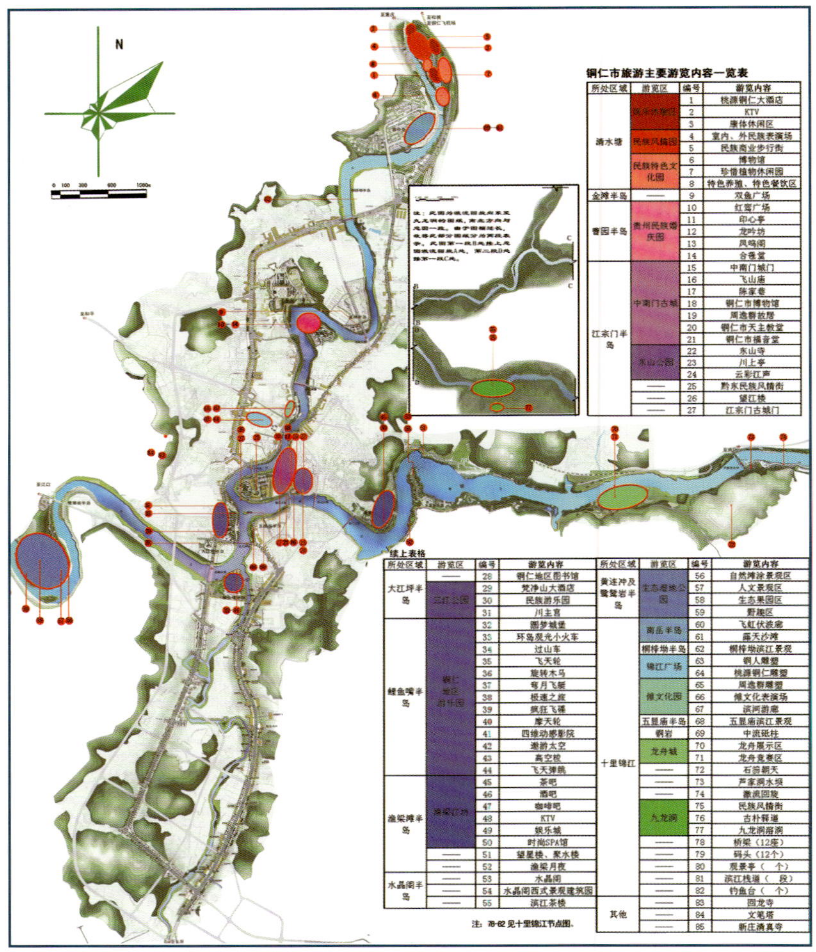

第五章 功能分区

一、主干道沿街建筑外立面改造区

二、十里锦江景观打造区

三、十二半岛规划区（本书中只对江宗门、渔梁滩以及南岳三个最具代表性的半岛进行介绍）

Chapter Ⅴ Functional Areas

Ⅰ Reconstruction Area of the Street Facades along the Main Roads

Ⅱ Construction Area of Ten-Mile Jinjiang Landscape

Ⅲ 12 Peninsula Planning Areas (only the three most representative peninsulas of Jiangzongmen, Yuliang Beach and Nanyue will be described in this book)

植物配置图

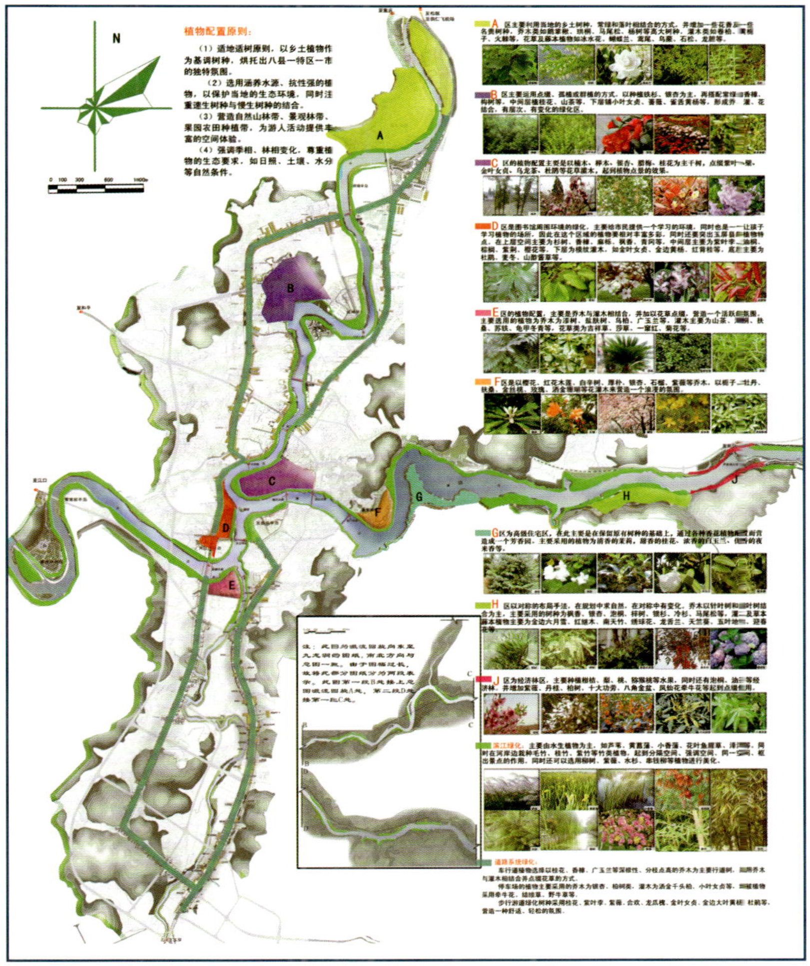

铜仁市全城鸟瞰图

Planning Idea of Streetscape Reconstruction of Tongren

With the overall shape of the city, the cultural connotation of the city will be promoted, the landscapes for sightseeing and rest increased, and cultural life of Tongren City enriched on the contents and forms, so as to form new tourist attractors, add another advantage for Tongren to create tourism city, and make the idea of creating tourism city more feasible.

沿街建筑改造
Rebuilding Street-Buildings

一、铜仁市沿街建筑改造规划思路

通过对主体城市风貌的塑造，提升城市的人文内涵，增添城市的游憩景观，从内容到形式上丰富铜仁市的文化生活，形成新的旅游吸引物，使铜仁市打造旅游城市又新添一个有利砝码，使打造旅游城市的想法更具可行性。

街道两侧矗立着的高低错落、鳞次栉比的一幢幢建筑，具有城市细胞中"最能代表城市格调，最能展示城市意蕴，最能彰显城市根源"的特性。它们犹如一个个音符，谱写着属于这个城市的主旋律，或轻快，或华丽，或热情，或内敛……一个属于这个时代的城市篇章，一曲凝固的华彩乐章。如将城市比做人，街道就是它的脸面；如将城市比作书，街道就是它的序言。

On both sides of the streets, buildings, as the urban elements with the characters of "the best representative of urban style, the most demonstration of urban charm, and the deepest root of urban origin", are standing densely and in order. They are like musical notes composing urban themes lively and passionately, which is the urban passage of this era and a stable magnificent musical passage. If city is a man, street is his face; and if city is a book, street is its preface.

铜仁作为一个多种文化共生的地区，各具特色的文化在这片土地上相互渗透、相互融合，形成了铜仁的文化。但也正是由于这种多文化的共生，反而导致了铜仁的文化没有了特色，缺少了主心骨，没有了引领，从而也难显示其独特的城市性格。因此，在规划设计中对铜仁城市的历史文化元素进行提炼和整理，以沿街建筑为载体，构建当地所独有风格，意在城市内涵的表达，以营造一个"来了就不想走"的城市环境氛围，与铜仁的城市定位互相呼应。

As a multicultural prefecture, Tongren possesses its culture of various cultural mingling. Nevertheless, this phenomenon of multicultural symbiosis causes Tongren culture lack of characteristics, and it is hard for Tongren to show its unique urban characteristics. Therefore, we aim to refine and select historical and cultural elements of Tongren City in the planning and design to construct local special style with the street architectures, so as to express urban connotation, create a environment atmosphere of "coming without any intention of leaving", and reflect urban orientation of Tongren.

二、铜仁市沿街建筑改造设计理念

现今的铜仁市主干道两侧的建筑，给人的感觉犹如一个不梳洗、不修眉，却浓妆艳抹的女子，虽然五官尚算标致，但总让人感觉显得不伦不类，缺乏应有协调，难以让人产生美的感受，留下较为深刻的印象。因此，在铜仁市沿街建筑的新建和外立面改造上，运用徽派、土家族、苗族、侗族等建筑的元素，加以提炼、糅合，形成铜仁自己的建筑风格，同时结合现代建筑的功能效用，打造“民族现代化，现代民族化”的铜仁市。在着色上，采用以暖灰调为主的色彩体系，使铜仁市的沿街建筑在“民族现代化，现代民族化”的基础上，又表达了一份内敛的热情。此外，在街道中段设置沿街休憩小广场、街心花园，以显现其浓郁的地方风格和独特的城市风采，使得人们来了就难以忘怀，成为中国城市设计的一个典范。

II Design Concept of Street Bulidings of Tongren City

The present architecture along the main avenue of Tongren City can not leave people a good impression. It is like a lady dressed up inadequately without the coordination it should have. Therefore, we should unify premise firstly, and refine and mingle the architectural elements of Hui, Tujia, Miao and Dong Ethnic groups to form Tongren culture, and, at the same time, combine the functions and effects of modern architecture to create Tongren City with the concept of "Modernization of Nationality, Nationalization of Modernity". The warm gray will be adopted as the main color to endow the street architecture with a connotational passion. In addition, at the sections of the streets, the small squares, street gardens which show thick local style and unique urban charm will leave visitors deep impression and make Tongren become one of the examples of urban transformation of China.

区位分析图

三、改造内容

1. 东太大道

(1) 东太大道现状

东太大道是连接火车站的主干道，道路两侧建筑密度较大，多为办公楼、公共服务建筑以及商住楼，建筑质量较好，但是多数建筑特色不明显，指向性不强。

Ⅲ Reconstruction Consent

1. Dongtai Avenue

(1) Current situation of Dongtai Avenue

Dongtai Avenue is the major road linking the rail station, with beautiful landscapes and dense construction. Most buildings along both sides of the street are used as offices of enterprises. With good quality but different styles of buildings, the street is lack of unity, wholeness and identification.

南段区位分析图

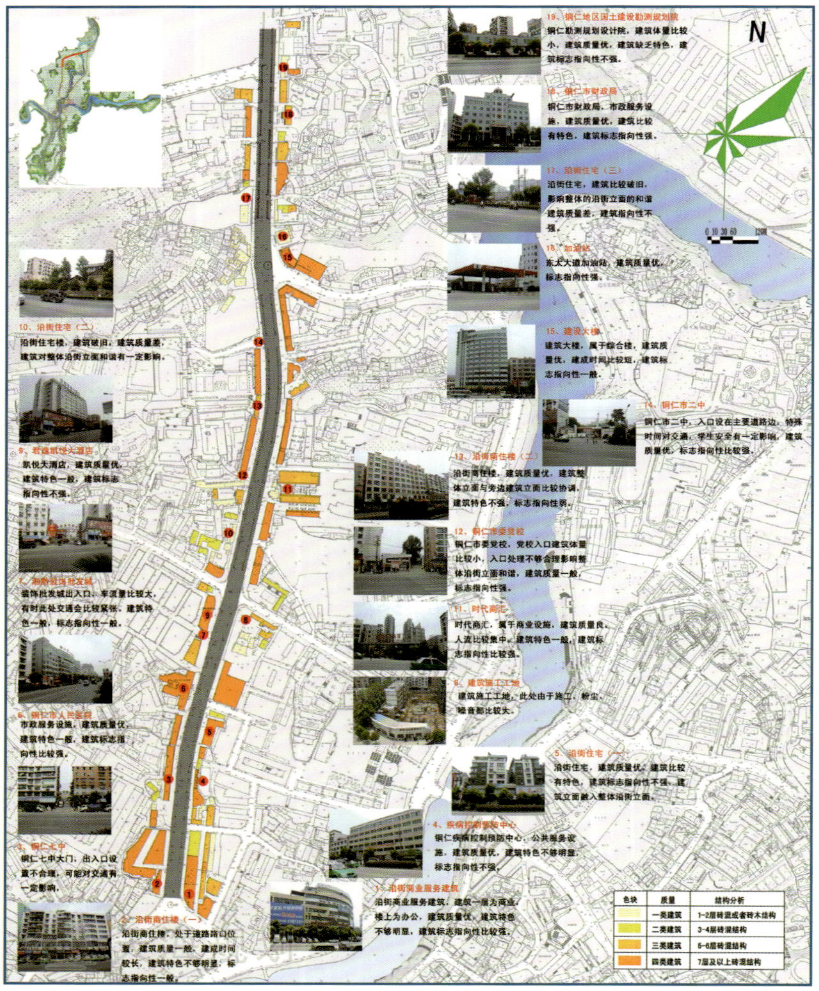

北段区位分析图

(2) 改造内容

① 主干道

东太大道改造路段长约3517.37m，规划改造建筑98栋，绿地率28.16%，道路两侧规划公共汽车站台10座，公共厕所5座，废物箱约61个。

② 沿街建筑

沿街建筑改造与铜仁市沿街建筑改造设计理念保持一致。

根据城市发展需要，如需利用道路两侧土地开发新建建筑，建筑风格建议同沿街建筑设计理念相符。

(2) Reconstruction Content

① Main Road

The reconstruction along Jinjiang Avenue is about 3 517.37 meters long. What needs to be reformed in the plan includes 98 buildings, 10 bus stations on both sides of roads, 5 public toilets, 61litter bins and greening rate 28.16%.

② Buildings along Street

Reconstructing streetscape follows the same design concept of reconstruction of streetscape of Tongren.

Any need of developing and constructing new buildings on both sides of the roads due to city development is kindly advised to be coordinated with our design concept.

总平面图

道路节点放大图

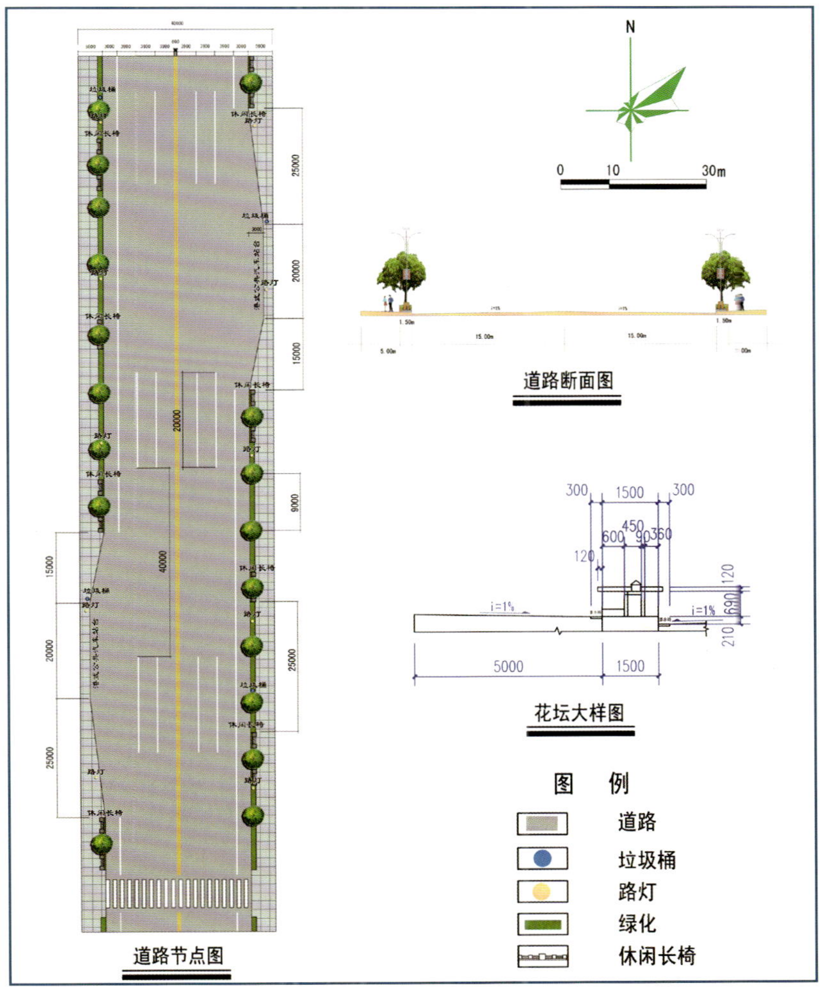

2. 锦江大道

(1) 锦江大道现状

锦江大道东侧临江，西侧建筑物较多，多为商住楼及公共服务建筑，建筑质量较高，但缺乏特色。

2. Jinjiang Avenue

(1) Current Situation of Jinjiang Avenue

It is near the river to east and with many constructions to west of which most are business buildings and public service buildings, high quality but without features.

区位分析图

总平面图

(2) 改造内容

① 主干道

锦江大道改造路段长约 1531.58m，规划改造建筑 38 栋，绿地率 25.76%，道路两侧规划公共汽车站台 6 座，公共厕所 4 座，路灯 61 座，休闲长椅 45 个，废物箱 32 个。

② 沿街建筑

沿街建筑改造与铜仁市沿街建筑改造设计理念保持一致。

根据城市发展需要，如需利用道路两侧土地开发新建建筑，建筑风格建议同沿街建筑设计理念相符。

(2) Reconstruction Content

① Major Road

The reconstruction along Jinjiang Avenue is about 1 531.58 meters long. What needs to be reformed in the plan includes 38 buildings, 6 bus stations on both sides of roads, 4 public toilets, 61 street lamps, 45 leisure benches, 32 litter bins and greening rate 25.76%.

② Buildings along Street

Any need of developing and constructing new buildings on both sides of the roads due to city development is kindly advised to be coordinated with our design concept.

道路节点放大图

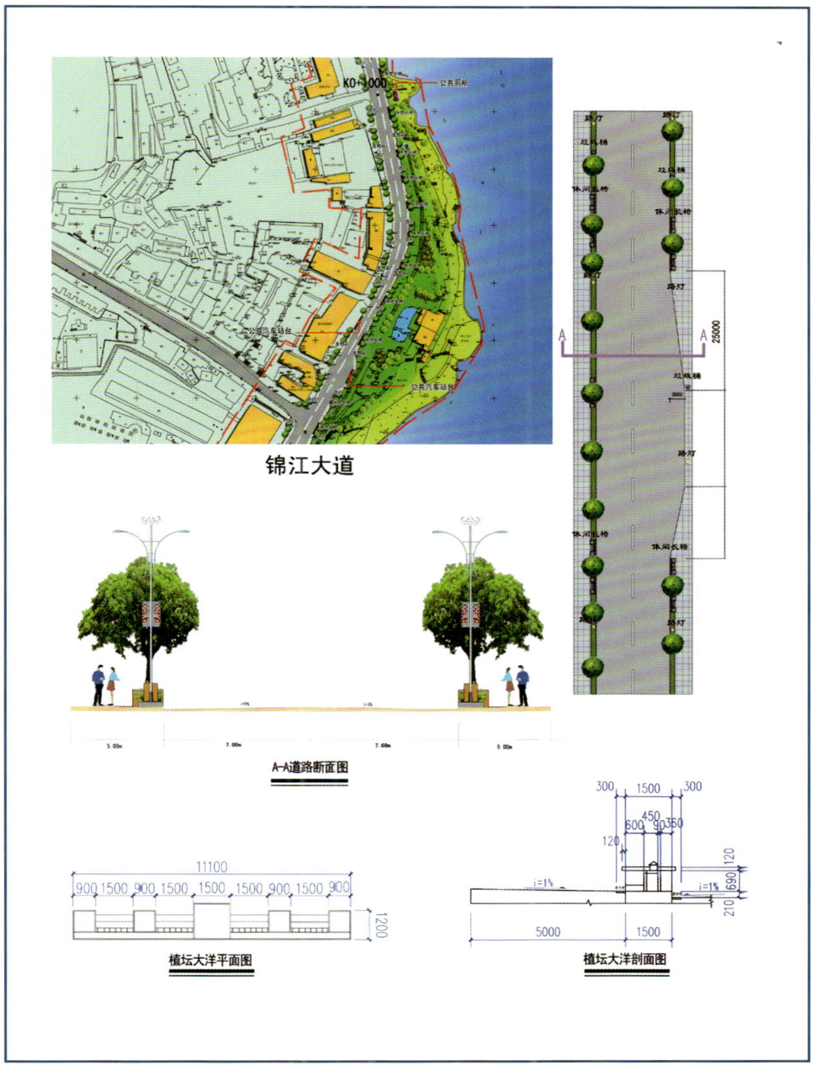

③ 锦江广场

锦江广场位于锦江大道西侧，原为一市民休憩广场，在此基础上将铜仁地方特色融入广场中，增加广场文化性、装饰性，将其改造为具有铜仁特色的城市休憩广场，与铜仁沿街建筑改造风格相适应的同时增加广场的景观效果。其中对锦江广场几大亮点加以说明。

1）整体风格

铜仁是一个融合了土家、苗、侗以及梵文化的城市，其文化博大精深，包罗万象，锦江广场作为一个市民休憩广场，以小品雕塑的形式将这些文化再现，装饰美化广场的同时时时刻刻提醒市民其文化内涵，深化市民的民族自豪感。另外还能为外地游客带来视觉冲击力及铜仁多元文化的具象体验。

2）小品雕塑

三圣雕塑——位于广场中部醒目位置，三圣雕塑与铜仁地名息息相关，此雕塑体现了铜仁的起源、文脉。

四面鼓牌坊——位于三圣雕塑后方，四面鼓是铜仁苗族特有的乐器，演奏时鼓声震天，气势磅礴，古时曾进京演出。以此为主题，代表独特的铜仁苗族文化。

长号牌坊——位于广场西面阶梯上，以乐手吹奏长号为造型，代表铜仁人民吹响迎宾的号角迎接四方宾客。

白虎图腾——位于广场中心，三圣雕塑下方。白虎是土家族图腾，以此雕塑代表土家文化。

梵文化主题雕塑——位于广场上方，老干部活动中心前方。雕塑中心为一“梵”字，外围为一“桃”形，寓意铜仁市主题形象“梵天净土，桃园铜仁”。

④ 傩文化园

傩文化园位于锦江大道西侧，濒临锦江。

铜仁地区是傩文化的故乡之一，千百年来傩文化以其古朴神秘的色彩受到世人的关注和青睐。在锦江大道旁原锦江公园的基础上增设傩文化园，将铜仁文化融入城市生活，丰富景观效果的同时为当地居民及游客带来新的、特色鲜明的休憩景点。

③ Jinjiang Square

Jinjiang Square, originally a square where citizens get rest, is located to the west of Jinjiang Avenue. The area includes the whole North and South Riverside Road. It is a riverside landscape zone along both sides of the Jinjiang River. There are commercial pedestrian street, high living area, green spots, and entertainment and fitness places in this area. In the future, it will be an important urban living area in Jinjiang Airport.

1) Overall Style

Tongren is a city integrating with Tujia, Miao, Dong and Fan culture. The whole peninsula faces water with hills in its back and enjoys good natural ecological system as well as the potential for landscape development. With the completion and penning to traffic of the Nanyue Number one bridge, Nanyue peninsula has turned itself from an isolated status into a closely related extension with transportation in Tongren city.

2) Accessory Sculpture

Sansheng Sculpture—it is standing on an obvious location of the square. Sansheng Sculpture, showing the de has many relations with the name of Tongren.

Simiangu Arch—it is behind Sansheng Sculpture. Simian Drum is a kind of instrument of Miao with thundering sound and power when played. Thus, it is used as the theme to represent Miao culture of Tongren.

Changhao Arch—it on the west step of the square, is in the shape of a player playing trombone which means that Tongren people welcome guests with playing trombone.

White tiger totem—it is located on the center of the square and beneath Sansheng Sculpture. White tiger, as the totem of Tujia ethnic group, represent Tujia culture.

Fan culture theme sculpture—it is located up the square and in front of Old Leaders Activity Center. The sculpture, with the shape of a "peach", has a character "梵" in the center, which displays the theme image of Tongren.

④ Nuo Culture Park

It is to the west of Jinjiang Avenue and near Jinjiang River.

Tongren Prefecture is one of the homes of Nuo culture which has drawn people's attention by its mystery and primitiveness. Thus, adding a Nuo Culture Park in the Jinjiang Park to integrate Tongren culture into urban life will enrich landscape effect and bring new and special scenic spot to visitors and local people.

①入城景观大门效果图

①

改造前

②

①街面建筑改造效果图 ②街面建筑改造效果图 ③街道透视图

①街面建筑改造效果图 ②街道透视图 ③街面建筑改造效果图 ④街面建筑改造效果图

②

③

④

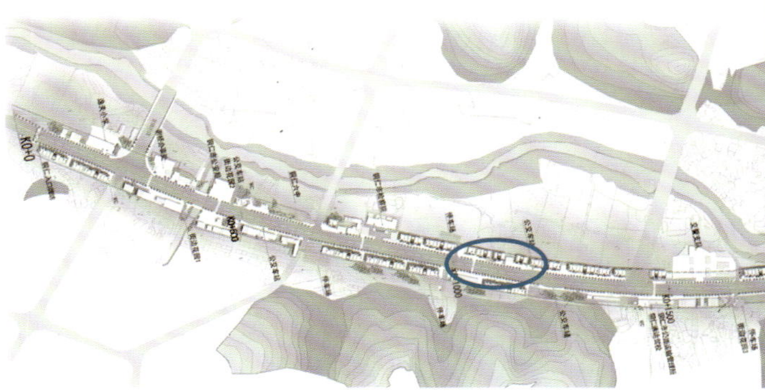

①街道透视图 ②街面建筑改造效果图 ③街面建筑改造效果图

①

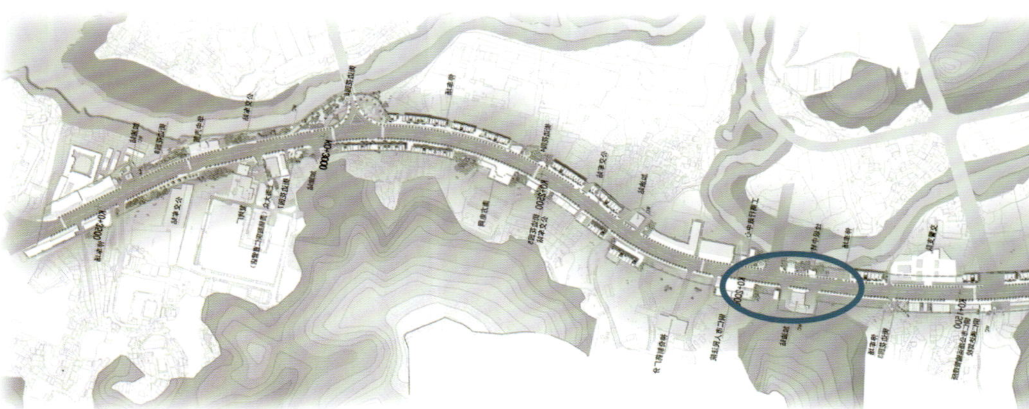

改造前

②

改造前

③

①街面建筑改造效果图 ②街面建筑改造效果图 ③街面建筑改造效果图

①

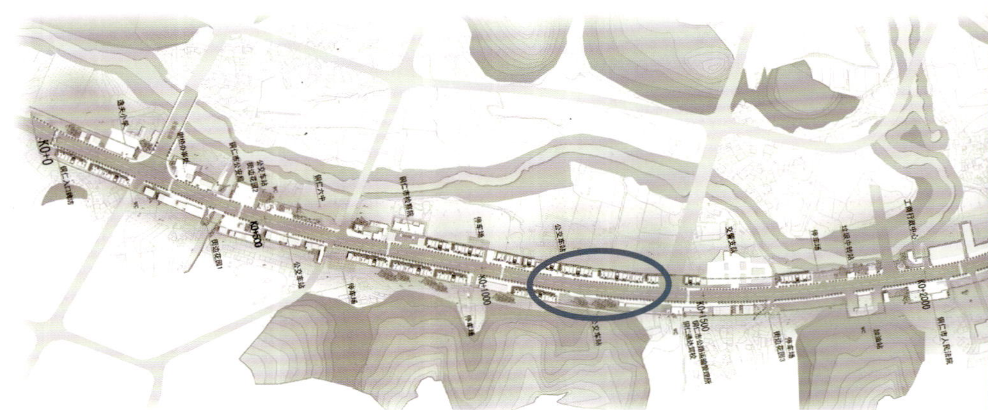

改造前

②

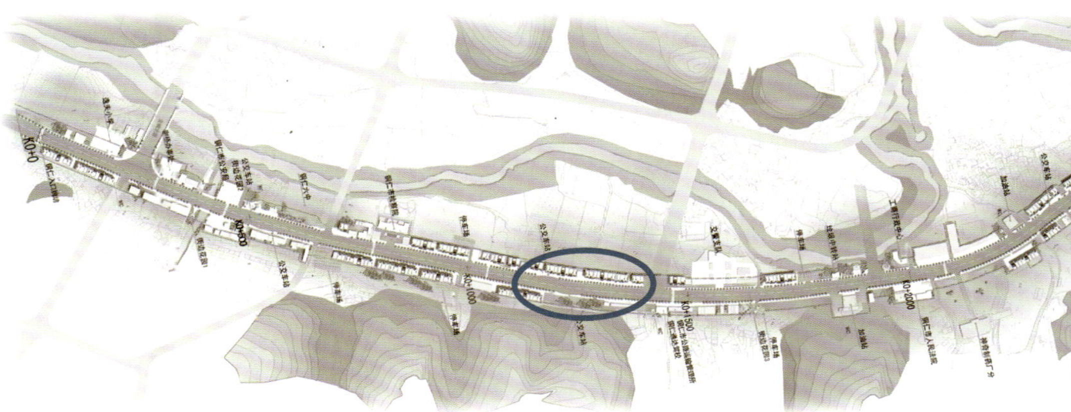

改造前

①街面建筑改造效果图 ②街道改造效果图

沿街建筑立面对照图

改造前

改造后

平面图

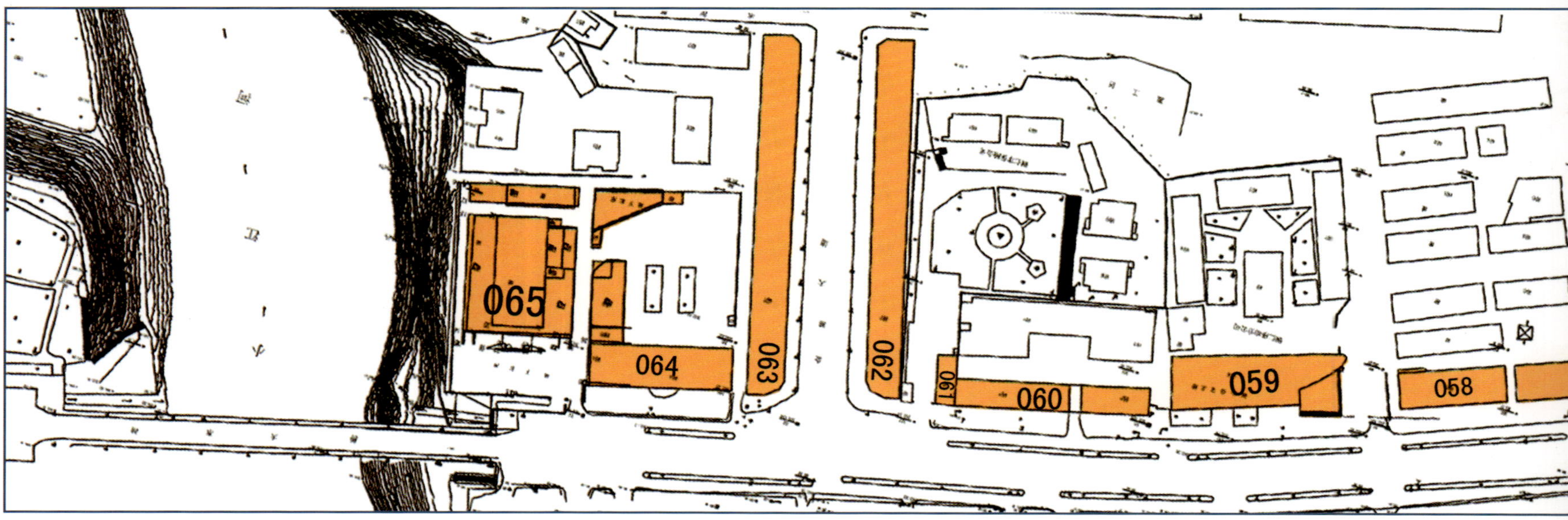

DTR-057
DTR-056
路口
DTR-055
DTR-054
DTR-053
DTR-052
DTR-051
DTR-050
DTR-049
DTR-048
路口
DTR-047
DTR-046

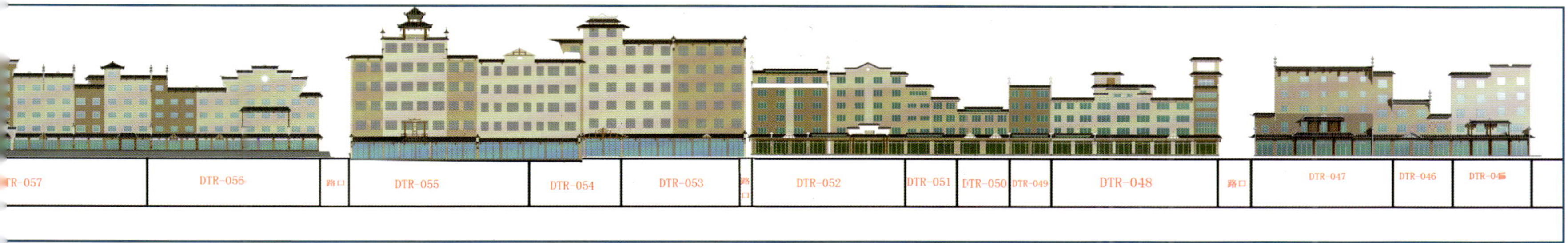
DTR-055
路口
DTR-055
DTR-054
DTR-053
DTR-052
DTR-051
DTR-050
DTR-049
DTR-048
路口
DTR-047
DTR-046

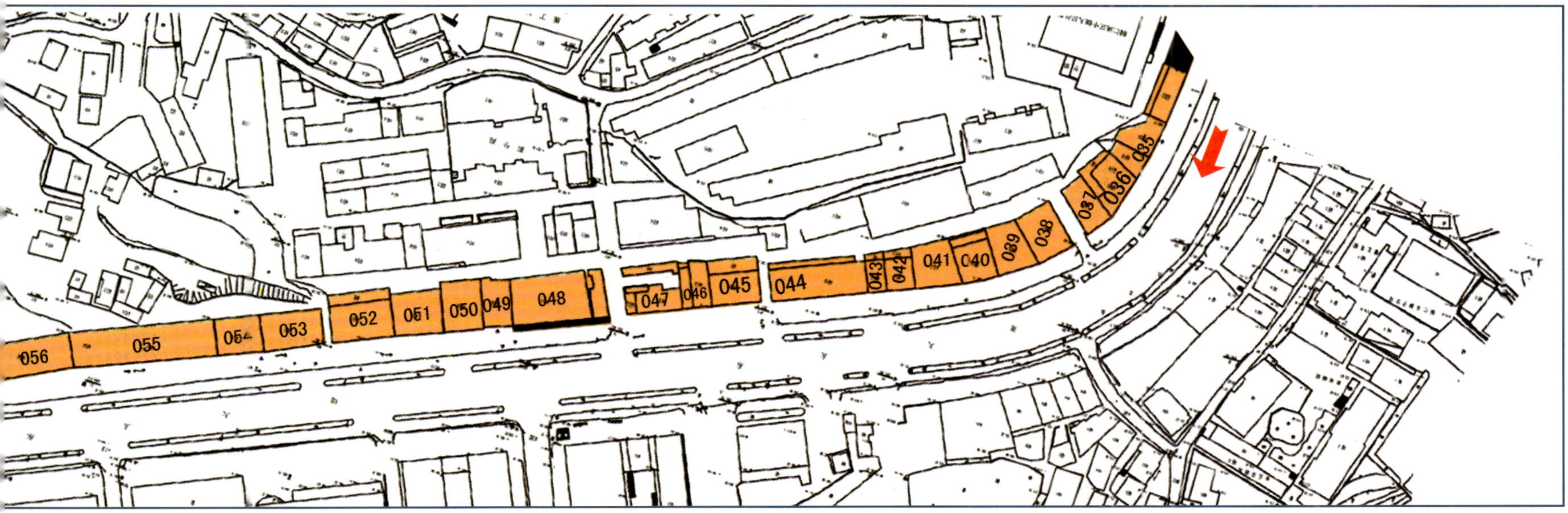
056
055
053
052
051
050
049
048
047
046
045
044
043
042
041
040
039
038
037
036
035

江宗门半岛规划设计

Planning and Design of Jiangzongmen Peninsula

一、景区概况

江宗门半岛位于铜仁市区中部，处于城区小江河下游与锦江交汇处，占地面积约为 15.32hm^2,半岛依山傍水,三面环江。铜仁市商业圈的中心地带，铜仁市政府位于半岛上，半岛古镇气息浓郁，但是随着城市的发展，大量现代建筑涌现，使得整体风格欠缺协调性。

在规划中考虑到江宗门是整个铜仁市、铜仁地区的起源，铜仁地区发展到今天离不开江宗门，因此需要铜仁地区“八县一市一特区”来共同见证铜仁的发展，同时将“八县一市一特区”的文化浓缩于江宗门半岛，让半岛感受从此源流出的铜仁文脉气息。

因此在半岛的顶弧端规划黔东历史风情街区,分别以“八县一市一特区”命名并体现当地文化。

I Landscape Overview

Jiang Zongmen peninsula is located in central Tongren city, and lies on the intersection between the lower Xiaojiang River in the city and Jinjiang River, with hills in its back and facing river in the other three sides. Tongren municipal government, the commercial center of the city, is located in the peninsula with a strong flavour of old towns. However, with the development of the city, a large number of modern architectures appeared leading to the overall style lack of coordination.

Considering Jiang Zongmenas the origin of the whole Tongren city and Tongren regions, and the developments of Tongren regions are unable to do without Jiang Zongmen, thus Tongren regions including "Eight countries, one city and one special district" need to be together witness the developments of Tongren, while the cultures of "eight countries, one city and one special district" are concentrated in Jiang Zongmen peninsula which will let the peninsula feel the smell of context sourcing from the Jiang Zongmen peninsula.

Therefore, the landscape blocks with the history of East Guizhou are programmed in the utmost tip of the peninsula, separately named as "eight countries, one city and one special district", and reflect the local cultures.

二、规划思路 II Planning Ideas

江宗门半岛对于铜仁市来说，不仅是铜仁市商业圈的中心地带，还是铜仁市城市的起源和根本，可以这样说，没有江宗门的昨天，就没有铜仁市的今天。江宗门对于铜仁市而言，不仅是一个半岛，她是铜仁的根，铜仁的源，铜仁的文化从这里萌芽，铜仁的发展从这里起步，她承载着铜仁市发展的所有、见证了铜仁市发展的今昔，是铜仁最具历史底蕴和文化内涵的城市中心，对其的规划设计旨在对其历史文化的恢复、重现和提升。

Jiang Zongmen peninsula is not only a commercial center, but also the birth place and root of Tongren City. We can say that if there were no fundament of Jiang Zongmen, there would be no development of Tongren City. In a word, Jiang Zongmen is the city center with the deepest historic and cultural connotation of Tongren as it has carried and witnessed the whole process of development of Tongren. The planning and design of this region is to recover, reappear and promote its history and culture.

区位分析图

现状分析图

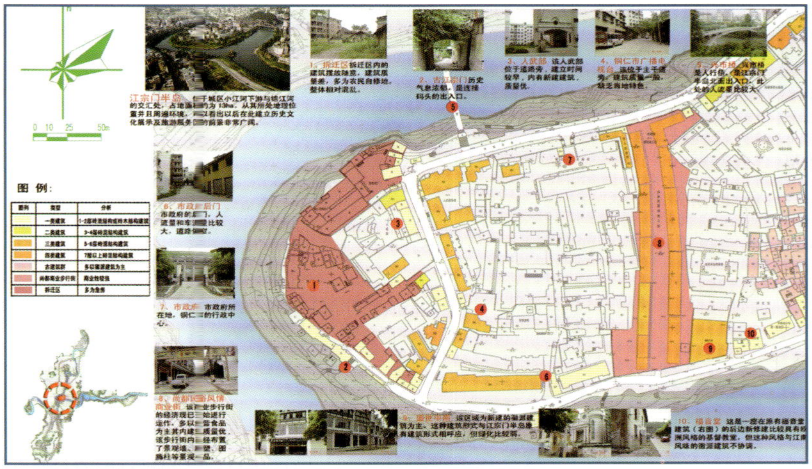

高程分析图

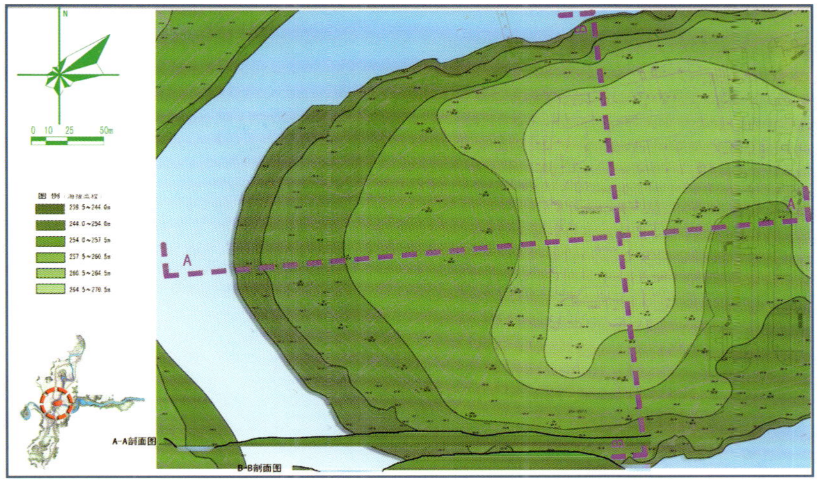

三、功能布局 Ⅲ Functional Layouts

针对江宗门半岛在铜仁市的特殊地位，以及铜仁文化的多元特点，在对其进行功能布局时有了多种方案的考虑。

In view of the special status of Jiang Zongmen peninsula in Tognren city and multiple features of Tongren cultures, we consider many programs when carry out functional layout of the peninsula.

总体上根据江宗门半岛的现状和特点，设置东山文化园区、中南门古城维护区、黔东民族风情街区、望江楼、滨江休憩景观区等几个主要功能区域。

According to the status and characteristics of the peninsula, we set Dongshan Cultural Park District, Mid–South Gate of Old Town maintenance district, Qiandong Ethnic Block, Wangjiang House, Riverside Landscape Park and several other major functional areas.

总平面图

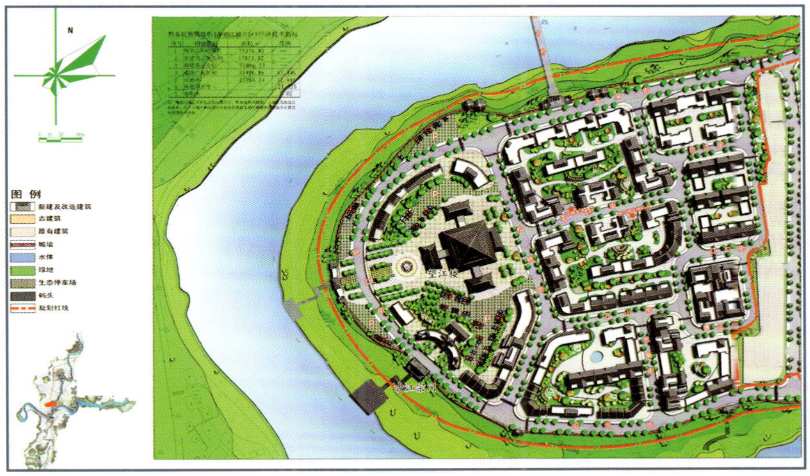

功能分区图

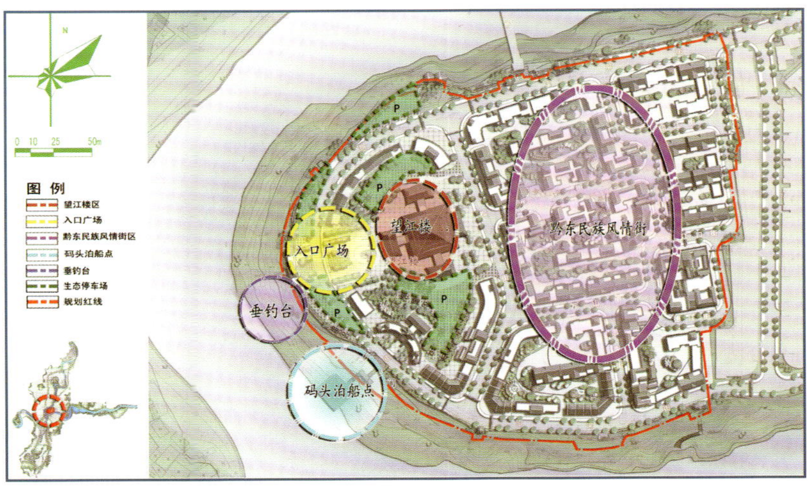

公共服务设施

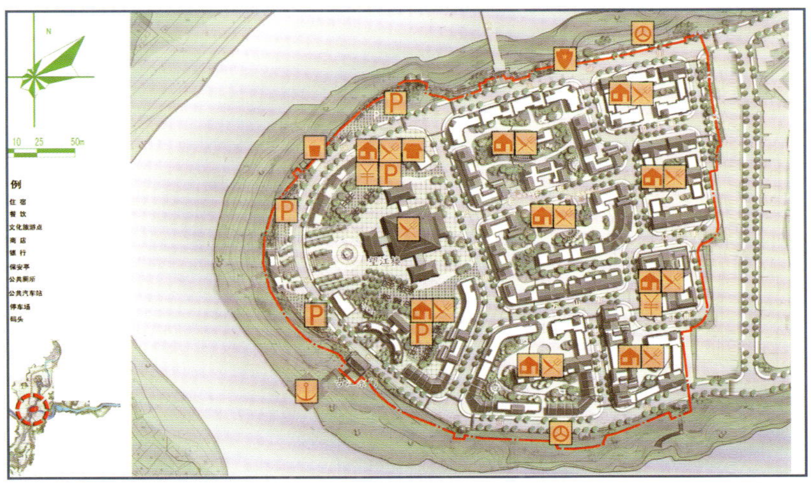

四、重点项目

1. 东山文化园区

以东山公园为主体，位于江宗门半岛的东山之上，自古就是铜仁市香火鼎盛之地，现有东山寺立于其上，但规模已大不如前。因此，本次规划以历史文献资料和城市发展趋势为导向，在东山寺现有规模的基础上，进行修缮，重修大雄宝殿和真武殿，修复东山寺主殿，增设静心斋和素食斋，务求恢复其昔日鼎盛之状。

2. 中南门古城维护区

包括了古城区和立面改造区，其中还含有铜仁市博物馆，其主体铜仁中南门古城与东山寺同为国家级文物保护单位，具有极高的历史文化价值，对这一片区的规划，以维修和保护为主，维护其古城风貌。

3. 黔东民族风情街区

位于江宗门半岛现铜仁市政府所在地，与高程控制区相邻，其以铜仁地区的“八县一市一特区”为基准，对街区进行划分，无论从建筑外形上，还是从文化内涵上，都对铜仁民族风情进行多层面、立体化的展现，务求将其打造成为汇聚铜仁民族风情浓缩精华之地。

黔东民族风情街区中，部分建筑为新建建筑，部分建筑在原有建筑基础上改造而成，无论是新建还是改造，每一组团街区分别代表了铜仁地区的一个县（市），蕴含了这个地区的文化特性，展现了这个地区的人文风貌。建筑底层的商铺与街名相照应，展示销售其对应地区的土特产品和民族工艺品；楼上为住宿用商品房，相信其所浓缩的文化对消费者，乃至整个铜仁而言，价值已远远超过其本身。

黔东文化街区的一大亮点在于街口牌坊的设置上，在每个街区的入口处设置一醒目的牌坊，牌坊分别以各个县市的文化为主题，建筑风格为内容，特色鲜明，区分各个县市的同时让人耳目一新。

4. 望江楼

江宗门半岛顶弧，是临江而望三江汇流和跨鳌亭的绝佳观憩点，具有良好的景观视线，同时留有江宗门古城门遗迹。因此，在规划设计过程中，设置一集休憩、餐饮、观景等多种功能于一体的主体建筑——望江楼，作为铜仁市的高端餐饮消费场所；楼前设一广场，作集散之用；广场左侧为江宗门半岛的游客服务中心，其正前方为江宗门古城门（需进行修复），其下为江宗门码头。

IV Key Projects

1. Dongshan Cultural Park District

Being the major part of Dongshan Cultural Park District, the Dongshan Park is located on the Dongshan Mountain of Jinag Zongmen peninsula, with flourishing incense from ancient times, but the scale is not as large as they were. Thus, this plan, guided by the historical documents and the development trend of the city and on the basis of current scale of Dongshan Temple, will repair and reconstruct Daxiong shrine and Zhenwu shrine, renew the main hall of Dongshan Temple and add Jingxin House and Vegetarian House in order to restore its former prosperous status.

2. Mid-South Gate of Old Town Maintenance District

This district includes old town district and reconstructive district, in which there is the Museum of Tongren City. Being the main part of this district, the Mid-South Gate of Old Town and Dongshan Temple are state-level key units of cultural relics protection with high historical and cultural values. Therefore, the planning for this area is major in the maintenance and protection, and to maintain the styles and features of the ancient city.

3. Qiandong Ethnic Block

This district is located in the seat of present Tongren City government in Jiang Zongmen peninsula, near the vertical control area. The blocks are divided according to the standard of "Eight countries, one city and one special district". The landscapes of Tongren nationality are multidimensional and vividly displayed from not only the building appearances but also the cultural connotations, in order to make it as the concentrated land collecting the landscapes of Tongren flavor.

In Qiandong Ethnic Block, some of the buildings are new constructions; and the others are reconstructed on the basis of the existing buildings. Whether new constructions or reconstructed buildings, every group of the district presents a country (city) in Tongren district, includes its cultural characteristics, and displays the human culture of this district. The shops in the building floor are consistent with the name of the blocks, displaying and selling the native produces and ethnic crafts with correspondence to the regions. Upstairs with the commercial housing, it is believed that it's the value of enrichment culture for the consumers and even the Tongren city is more than its own.

One bright spot of East Guizhou national landscape block is setting the memorial arches in blocks. A smart memorial arch is set in the entrance of each block. The theme of the memorial arches is according to cultures in each county and city, and the architectural style with distinctive characteristics which are easy to distinguish each country and city while refreshing.

4. Wang Jiang House

The top arc of Jiang Zongmen peninsula is the excellent view point with Sanjiang sinking flow and Kuabie Pavilion. It has good line of sight while maintains the remains of the ancient city gate in Jiang Zongmen. Therefore, in the planning and design process, we set the main building, Wang Jiang House, as the High-end food and beverage establishments with leisure, dining, viewing and other functions; A square used for distribution is set in front of the building; The service center of Jiang Zongmen peninsula is in the left side of the square, the front of the square is ancient city gate of Jiang Zongmen (need to be repaired), and the Jiang Zongmen wharf is under the square.

道路分析图

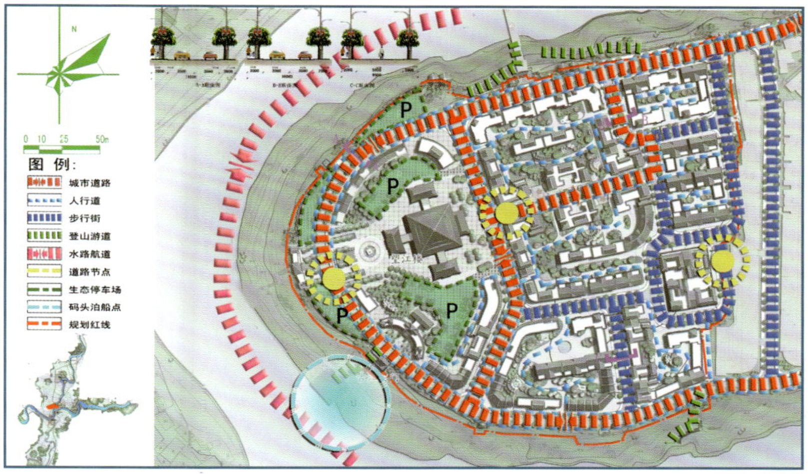

绿地分析图

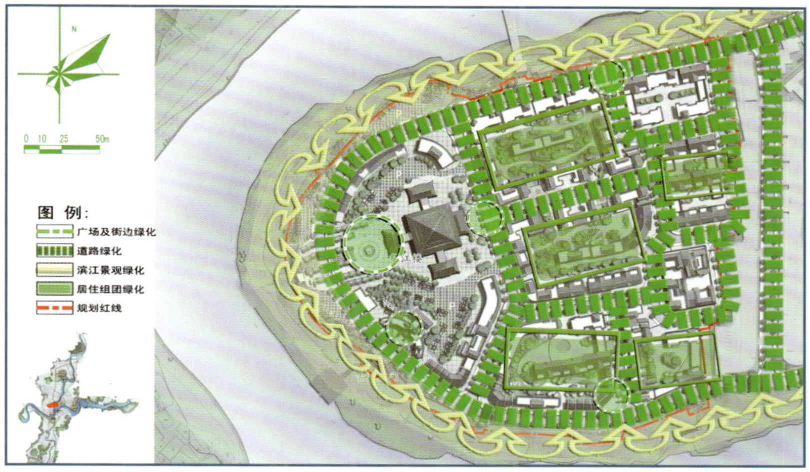

景观分析图

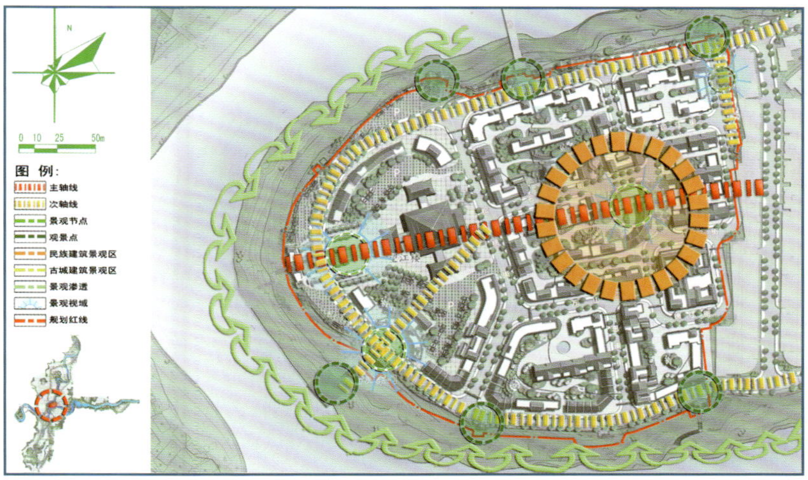

植物配置图

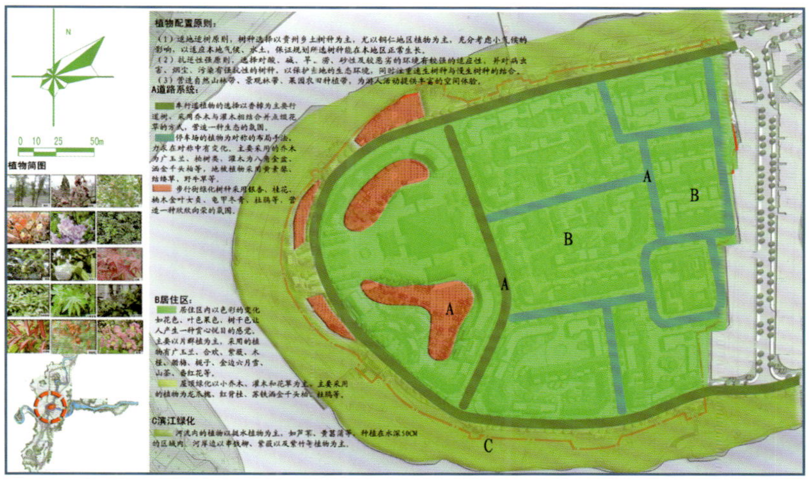

江宗门半岛鸟瞰图

①

②

③

①望江楼效果图 ②中南门入口 ③铜岩塔楼 ④沿街建筑效果图 ⑤沿街透视图 ⑥入口牌坊 ⑦入口牌坊

十里锦江景观设计
Planning and Design of Ten-mile Jinjiang River

Ⅰ General Situation

The main stream of Jinjiang crosses the urban area of Tongren from west to east. Jinjiang, the big river, and the small river divide Tongren city into 12 peninsulas. Jinjiang to Tongren is more than the river of life to Tongren. What is of greater significance is the three rivers flow across the city and afflux toward a point to form the fabulous landscape. The winding river course divides Tongren city into 12 peninsulas, as a basis for taking shape of the ecology river, or the humanity river, or the landscape river, which is a perfect union of the shape, spirit, and artistic that can evoke people's infinite imagination and arouse the infinite creativity.

一、概况

锦江干流自西向东横穿铜仁市区全境。锦江与大江、小江将铜仁市划分为 12 个半岛。锦江之于铜仁，不仅是铜仁的发源之河、生命之河， 更大的意义是在于其三江汇流穿城而过的景观，蜿蜒的河道将铜仁市划分出 12 个半岛，并以此为基础，形成“让人有无限遐想，能发挥无限创意空间，‘形、神、意’完美结合”的生态之江、人文之江、景观之江”。

二、规划思路

滨江风光带建立在铜仁市境内锦江河段的基础上，其由西向东贯穿铜仁市全境，上游的大小两江汇集于铜仁城中，并呈“S”状环城逶迤东去。沿江两岸奇峰林立，峭壁嵌绿，碧波粼粼，山色青青。丰富的自然景观、独特的人文景观及珍贵的文物古迹、雕梁的亭台楼阁等，如一颗颗珍宝般镶嵌在锦江两旁，构成了一幅“清水出芙蓉，天然去雕饰”的美丽画卷。诗人廖经天对铜仁锦江情有独钟，并留下了“四面青山楼外楼，新妆巧扮最风流，多情最是锦江水，依依一步一回头”这脍炙人口的诗句。

对锦江滨江风光带的打造，是“以水为魂，营造浪漫铜仁”的点睛之笔，这一线的贯穿和点亮，不仅能成为铜仁市城市旅游发展的一大基石，同时也将是铜仁市提升城市品位、形成城市形象的一大法宝。此外，还是将铜仁城市进行包装、营销，形成具有核心竞争力的城市旅游产品。因而，对锦江滨江风光带的设计可谓是匠心独运，以水为媒，最大限度地突出铜仁的文化，彰显铜仁的风情。

Ⅱ View of Development Plan

The riverside scenic belt is located in Jinjiang course in Tongren. It runs through the whole city of Tongren from west to east. The upstream of the big and small rivers converge in the city and assume the "S" shape running around the city, meandering toward the east. Both sides paint with its clear water and cluster of pinnacles, making a unique traditional Chinese picture. The rich natural landscape, the unique humanity landscape and the precious cultural relics and historic sports dot around the Jinjiang banks, forming a beautiful picture that "clear lake can grow into the beautiful hibiscus, and modified removed from the same nature". The poet Liao Jingtian showed special preference to Jinjiang, leaving behind the verse which enjoys great popularity– "With row upon row of mountains and houses; With more and more charms of new looks. Jinjiang water fondly, Is moving looking back one step one glimpse".

The construction of Jinjiang scenic belt is the finishing touch of constructing romantic atmosphere by lighting ventilation fluid. The scenic belt is not only a cornerstone for the tourist development of Tongren, but also a magic weapon of promoting the city taste and forming the city image. In addition, it's a product of packing and marketing Tongren, making the city become a tourist city with core competitiveness. Therefore, the design of Jinjiang scenic belt is unique and super, highlighting Tongren's uttermost culture and displaying Tongren's style by the medium of water.

三、功能布局

在规划设计当中，主要在以下几个方面下功夫：游船、码头、芦家洞大坝、跨鳌亭、桥梁、休憩景观、滨江景观，对铜仁市的锦江河段从形式到内容上进行丰盈。

Ⅲ Functional Layout

In the planning and design, the following areas worth concentrating efforts: boat, dock, reed house hole dam, cross–ao kiosks, bridges, open landscape, and riverside landscape. Enrich the charm of the Jinjiang course from the format to the content.

四、重点项目

1. 码头

有游船自然就会有码头。因此，在对锦江滨江风光带进行规划设计时，着重地考虑了码头，并以码头为媒介，将铜仁地区“八县一市一特区”的人文元素融入其中，使江上除了渔舟外，还充满了浓郁的地域文化。

每个码头精炼地将各县（市）的文化运用门头、小品、浮雕、扣件等形式，通过艺术化的加工，表现在世人的面前，既饱含风情又易于理解，集景观、文化、功能于一体。

各个码头在融入“八县一市一特区”的人文元素的同时还兼顾了所处半岛的整体定位，人文元素的运用与半岛主题有机结合。

从清水塘顺锦江而下，分别是：

以松桃文化为代表的清水塘码头；

以印江文化为代表的曹园半岛码头；

以思南文化为代表的江宗门半岛码头；

以玉屏文化为代表的大江坪半岛码头；

以万山文化为代表的鲤鱼嘴半岛码头；

以江口文化为代表的五显庙半岛码头；

以沿河文化为代表的渔梁滩半岛码头；

以德江文化为代表的水晶阁半岛码头；

以铜仁本地文化为代表的龙舟城码头；

以石阡文化为代表的芦家洞大坝码头。

2. 滨江景观

沿江两岸，除码头和休憩景观外，还有许多的滨江景观，如摩崖石刻、雕塑小品、景观塔等，景观虽小，但小而精、小而巧，无一不在述说着铜仁的文化、铜仁的历史，展示着铜仁的过往、铜仁的风采。

同时还拥有运用激光灯和 LED 灯等现代科技手段设计的灯光夜景效果，向人们展现了一幅幅富含铜仁文化意蕴且流光溢彩的画卷，让锦江的夜游又平添几分情趣，增添几许兴致。

(1) 五显庙半岛

由于五显庙半岛建筑密度较大，因此选择对沿江建筑进行外立面的改造，改造风格与铜仁市沿街外立面改造总体风格保持一致。

五显庙半岛规划面积 46389m^2，规划改造道路长度为 599m，规划改造建筑 14 栋，绿地率为 36.59%；道路两侧规划有公共汽车站台 2 座，公共厕所 1 座，垃圾站 10 座。

(2) 鲤鱼嘴半岛

对鲤鱼嘴半岛主要进行沿江景观的改造，将之打造为城市亲水平台。主要包括沿街景观绿化的配置，休闲步道及沿江观景亭的修建。

IV Key Projects

1. Docks

Where there is tourism, there are docks. Therefore, in the planning and design of Jinjiang scenic belt, serious consideration is taken into docks. By the medium of docks, we mingle with the humanity element of "8 counties, 1 city, 1 special zone". Thus, apart from fishing boats, the docks are rich in local culture.

Each dock transfers the culture of each county(city)in the form of door display, comedies, relief, fasteners, etc. via the artistic process, which is full of flavor and easily to be understood. It's a unity of landscape, culture, and function.

Each dock mingles with the humanity element of "8 counties, 1 city, 1 special zone", and in the mean time it give attention to the overall positioning of the peninsula. The use of humanity elements is in combination with the theme of the peninsula.

Down from Qingshui Pond with the flow:

Qingshui Pond dock represented by Songtao culture;

Caoyuan Peninsula dock represented by Yinjiang culture;

Jiangzhongmen Peninsula dock represented by Sinan culture;

Dajiangping Peninsula dock represented by Yuping culture;

Carp mouth Peninsula dock represented by Wanshan culture;

Wuxian Temple Peninsula dock represented by Jiangkou culture;

Yuliangtan Peninsula dock represented by Yanhe culture;

Crystal Court Peninsula dock represented by Dejiang culture;

Dragon-boat City dock represented by Tongren the local culture;

Lujiadong Dam dock represented by Shiqian culture.

2. Riverside Landscape

Apart from some docks and open-space for recreation, there are many riverfront landscapes along the riversides, such as cliff, sculpture pieces, landscape tower, and so on. They are small but refined, telling the culture and history of Tongren and displaying the unique charm and style of Tongren all without exception.

At the same time, Tongren has the marvelous night lighting view, which is designed by modern technological means, such as the use of laser lights and LED lights. It shows people the dazzling picture of Tongren which is rich in cultural deposits, adding a hint of taste to the night scene of Tongren.

(1) Wuxian Temple Peninsula

Due to the dense scattered buildings on the peninsula, the transformation of the facades of buildings along the river is to be made, in order to consistent with the style of the facades of buildings along the streets of Tongren.

The projected size of the Wuxian Temple is 46 389 square meters. The length of projected paths to be transformed is 599m. There are 14 buildings to be transformed. The projected green coverage rate is 36.59%. There should be two bus stations, one public toilet, and 10 refuse collection points along the sides of the streets, as it is planned.

(2) Carp Mouth Peninsula

The transformation of Carp mouth Peninsula is mainly the landscape along the river. It will be transformed to a waterfront platform of the city. It consists of the setting of afforestation along the street sides, and the construction of leisure trails and viewing pavilions along the street sides.

(3) 休憩景观

锦江滨江风光带除了为游船上的旅游者创造景观外，也应为沿江游览的旅游者和当地居民提供一个休闲观光的平台。因此在沿江合适的区域设置垂钓台、休闲栈道、观景楼等休憩景观，在满足旅游者和当地居民休闲观光需求的同时，也是在锦江滨江风光带上增添了一道与众不同的风景线。

3. 游船

锦江承载了铜仁市的发展，而这条铜仁的生命之河的人文内涵是靠江上往来的船只支撑起来的。换句话说，锦江上的船只能在一定程度上从一个侧面折射出当地的人文内涵。因此，对锦江滨江风光带而言，船只在其中的作用是毋庸置疑的，它在向世人传递相关的信息的同时，也让人切实从中体验到了这种文化，这对于以锦江为纽带的铜仁市城市旅游发展，是十分必要的。

在设计中，规划设计各具功能、大小不一、造型各异的船只共计 50 艘（包含人力船和机动船）。其中在功能上，以观光和餐饮两大功能为主；在类型上，分为大型（能容纳 20 ~ 30 人）、中型（能容纳 10 人左右）和小型（能容纳 2 ~ 5 人）三种；在造型上，有仿古的、现代的、民族风格的、卡通形象的。

4. 桥梁

铜仁市是一座多水的山城，整个城区与山水相依相融，水将城划分，桥将城相连，桥成为铜仁市必不可少的一道景观。凭高远眺，十桥跨江的景致尽收眼底，实是不可多得。而城市中的桥，作为现代文化的产物，多少显得过于冰冷，缺乏铜仁的本土特色，无法彰显其最大价值，因而对桥加上了文化特点浓郁的扣件，使其功能性得到保证的同时，景观性得到加强、文化性得到彰显。

分别对各桥细部加以土家、苗、侗元素，使其与铜仁城市改造风格一致，更具铜仁本土特征。

5. 跨鳌亭（铜岩亭）

作为铜仁城市名称来源之地铜岩上的景观建筑，可谓是屡毁屡建，其意义并不在于建成时间的长短，而在于所蕴涵的历史人文底蕴。基于这样的认识，在设计中对现有跨鳌亭进行改建，重塑儒、释、道三尊铜人立于亭中，同时对跨鳌亭进行与城市风格一致的外观改造，旨在弘扬其深厚意蕴，同时起到联结、协调周边环境的作用。

改建后的铜岩亭采用徽派建筑的马头墙及少数民族建筑风格，该建筑的精髓在于亭子中部，以马头墙将亭子划分为三个部分，分别竖立儒、释、道三尊铜人，以此诠释“铜仁”地名的由来，带出铜岩于铜仁之地位。

(3) Open-space for Recreation

The riverside landscape of Jinjiang not only provides tourists on boat with the beautiful view, but also offers a sightseeing platform for the local residents. Therefore, the construction of open-space for recreation like fishing stations, leisure trails, sightseeing pavilions, etc. meets the recreation and sightseeing needs of the tourists and local residents. Also it's a distinctive scenery line in the scenic belt of Jinjiang.

3. Cruise

The development of Tongren relies on Jinjiang, and the humanistic connotation is represented by the boats on Jinjiang. Therefore, as for the scenic belt of Jinjiang, there's no doubt for the importance of boats. They transmit the related information to the world, at the same time make people feel the culture, which play a vital role for Tongren's tourism development.

We plan to design 50 boats of various functions, sizes, and shapes (including human-powered boats and motor boats). As for the wide range of functions, mainly sightseeing and dining. In terms of sizes, there are 3 types-large (can accommodate 20 to 30 people), medium (can accommodate 10 people) and small (can accommodate 2 to 5). Speak of the shapes, there are antique, modern, ethnic style, cartoon image style.

4. Bridges

Tongren is a mountain city rich in water. Water divides the city into parts, and bridges link the parts. Thus, bridges become an indispensable landscape of Tongren. Standing on the high bridge, it offers you a panoramic view of "10 Bridges Crossing the River". As the products of modern culture, bridges in the city seem a little rigid, lacking of Tongren's local characteristics, which cannot present its maximum value. Thus, in order to guarantee the function of bridges and to strengthen their sightseeing and cultural characteristics, the unique fasteners with deep culture deposits are added to the bridges.

Add Miao and Dong elements to each part of the bridge, so that the style will consistent with the style of the transformed Tongren, with more local features.

5. Cross-ao Pavilion (Copper Rock Pavilion)

The name of the city Tongren is derived from the landscape on the land of copper. It was repeatedly destroyed repeatedly built. The significance of the city does not lie in the length of its completion time, but lies in the historical humanity heritages that the city contains. Based on this understanding, in the design of the reconstruction of the existing cross-ao Pavilion, we need to remodel three bronze figures of Confucianism, Buddhism and Taoism standing in the pavilion. And reconstruct the Cross-ao Pavilion simultaneously, so that it will take on a new outward appearance which consistent with the transformed city style. Its purpose is to promote its profound significance, and plays the role of linking and coordinating the surrounding environment.

The reconstructed Copper Rock Pavilion adopts Huizhou wharf wall and ethnic architectural style. The essence of the building lies in the central part of the pavilion. The wharf wall divides the pavilion into three parts; each part is erected by bronze figures of Confucianism, Buddhism and Taoism respectively to interpret the origin of the name Tongren.

6. 锦江夜景

夜景是十里锦江的重头戏，穿城而过的锦江在灯光的映衬下更能体现出其浪漫旖旎的色彩，通过夜景氛围的营造，将十里锦江打造为中国梦游第一江——一个可以晚上旅游的城市。

夜的锦江中百舸争流，绚丽多姿；江面流光溢彩，美轮美奂；两岸火树银花，五彩斑斓；桥面万紫千红，宛若游龙；遍地华灯若天河群星，满天繁星似人间灯火。上下浑然一体，五彩交相辉映，俯仰顾盼，情景各异，如梦如幻，如诗如歌，撩人耳目，动人心脾。

乘坐一叶扁舟，泛舟江上，在月影、渔火、灯光的映衬下，品茶、赏月、观景，仿若超脱世俗的纷争，涤荡心灵的尘埃，给人前所未有的平静和松弛。

傩文化广场前，巫傩面具造型的灯光幻影从水中缓缓升起，在增加夜游灯光效果的同时营造梦幻诡异的效果。

江宗门半岛前的锦江水域中，一朵朵睡莲幽静地漂浮着，睡莲上的梵文与坐佛在灯光幻像下若隐若现；江宗门半岛上梵文化博物馆在灯光的照耀下越显庄严肃穆，二者相互映衬，让人感觉已然置身于一个梵光、梵语的极乐世界，经受着梵文化的洗礼。

夜幕下的铜岩亭灯火通明，为铜仁古老的地标带来了新的神韵。

夜色中的东山寺古建筑群在灯光的照耀下古朴而静宜，让人感叹几百年来的沧桑巨变。

一轮明月悄然升空，渔梁滩苏醒过来，伴随着一簇簇不时划破夜空的绚烂礼花，为您呈现出“渔梁月夜”的精美景致：无论是岸上、江边帆船上还是江面游船中，灯火阑珊，彻夜通明，或露天，或包间，或三五亲朋把酒谈心，或一众玩伴激情狂欢，在夜色的映衬下，在灯光的浸浴中，总能找到属于游客的性情流露，午夜梦回，不禁感叹，好一个十里锦江不夜城！

与渔梁滩的激情张扬形成强烈对比的是夜幕中的水晶阁，它内敛而安逸，静静地为游客带走满身疲倦，洗去满身尘埃。

灯光照耀下的龙舟城恢宏依旧却多了几分魔幻色彩。

夜色中的儿童游乐园被特效灯光装点得犹如一座童话中的城堡，让人不禁怀疑当游人渐渐散去的时候，儿时童话故事中的人物是否会在这里登场，儿时童话故事中的情节是否会在这里上演。

6. Night Scene of Jinjiang River

The night scene of Jinjiang is the highlight of Jinjiang. Tinged with the lights, Jinjiang manifests its romantic atmosphere and charm. The night scene of Jinjiang makes Jinjiang the first Chinese sleepwalking city—a city that can travel on night.

The night scene of Jinjiang is gorgeous. Dazzling fireworks explodes in the sky. The sky is studded with twinkling stars. Like a dream and a poem, the beauty is no doubt breath-taking, arousing people's imaginary and moving us deeply.

Taking a skiff on the river against the shadow of the moon shade, fishing-boat lights, and the background lights, people drink tea, appreciate the moon and the sights around. The disputes seems to be transcendental, the dust of the soul is cleaned up. What you can do is to enjoy the unprecedented tranquil and the relaxation.

In front of Nuo culture plaza, the witch Nuo mask shaped light illusory raises from the water phantom slowly, which increase the effects of night lighting effects to create a strange fairyland.

Before the Jinjiang waters in Jiangzhongmen Peninsula, water lilies are floating quietly. The Sanskrit on water lilies and the seated Buddha seems between visible and invisible against the ghost-like light. The Buddhist culture museum on Jinagzhongmen Peninsula seems more dignified and solemn. Stay here, you will feel you throw yourself into a Elysian Fields full of Buddhist light and Sanskrit, undergoing the Buddhist culture baptism.

Under curtain of night, the copper rock pavilion is brilliantly illuminated throughout the night, bringing new verve to the old Tongren.

The Dongshan Ancient Temple looks magnificent at night when it is illuminated, making people lament the profound changes over the centuries.

The round bright moon raises quietly, and Yuliangwan Beach wakes up, along with the clusters of the splendid fireworks pierced the night sky from time to time. A beautiful scenery of "fishing Liang moonlit night" is presented: Whether on the shore, in the sailing ship or in the river cruise, the lights wane, but brightly lit all night. Several friends have some delicacies and talk about trivial things either in the open-air, or in a room. Some can put on a passionate carnival. In the moonlight, one can find the real self back and couldn't help sighing: What a wonderful 10-mile Jinjiang that never sleeps!

The Crystal Pavilion in the night forms a striking contrast to the passionate Yuliangtan. It is restrained and comfortable, disposing of tourists' weary and fatigue and washing the dust away.

The Dragon Boat City which shines in the lights looks grand and has more magic flavor.
In the dim light of night, children's playground decorated by the special effect light looks like a castle in a fairy tale. People would wonder whether the characters in the childhood fairy tales will stage here, or whether the childhood fairy tales' plot will perform in here when the tourists disperse gradually.

一个可以晚上旅游的城市（铜仁市全城鸟瞰图）

⑤

⑥

①锦江透视图 ②景观码头 ③沿江景观透视图 ④沿江景观透视图 ⑤沿江景观透视图 ⑥沿江景观塔楼

①沿江景观局部图 ②沿江栈道局部效果图 ③沿江栈道局部效果图 ④沿江景观塔楼

④

三都篇 Sandu Profile

城市设计篇（第二版）

变城记

City Changing

前期策划
Prophase Planning

Sandu County is located in the "Mountains of the Moon, Thunder Mountain" hinterland, which is in the southeast of Qiannan Autonomous Prefecture of the Buyi and Miao Nationalities. It's the only Shui autonomous county in the country and more than 60% population of the Shui nationality live in sandu.

Sandu is located in the golden district of the integrated tourist area in south Gui Zhou. It's the central scenic region of international travel route from Li (ping)—Cong (jiang)—Rong (jiang) to Guilin, Guangxi; in the six integrated tourist area of Guizhou Province, "Libo integrated tourism" and "Carey—Zhenyuan integrated tourism area"; It will form Guiyang—Kunming and Kunming—Guangzhou Express Railway and Kunming—Bangkok international highway in the railway in the future, which will string Guangzhou, Guilin, Guiyang, Kunming, Bangkok and other important tourist cities, and even spread to the Indian Ocean and the Pacific, the "world-class tourist routes" will pass through Sandu. Sandu forms its unique advantage for the well-kept human and cultural features and its rich multinational resources.

However, how to grasp this opportunity to turn this advantage into a winning situation, and help to achieve a leapfrog development in the local economy and society, making more benefits for local people remains a very important problem that the local government has to face and consider urgently currently.

三都县位于贵州省黔南布依族苗族自治州东南部，地处“月亮山、雷公山”腹地。是全国唯一的水族自治县，全国60%以上的水族人口居住在三都。

三都处于贵州南部综合旅游区的黄金地段，为黎（平）——从（江）——榕（江）至广西桂林国际精品旅游线的中枢景区，位于贵州省六大综合旅游区“荔波综合旅游区”和“凯里——镇远综合旅游区”中，未来还将形成贵昆、贵广快速铁路及昆曼国际公路将串起广州、桂林、贵阳、昆明、曼谷等重磅旅游城市，甚至辐射到印度洋与太平洋，“世界级旅游线路”将穿越三都。

境内水族人文风貌保持良好，加上多民族资源丰富，形成了三都县独特的优势。

然而，如何把握这一机遇，将优势转换成胜势，使当地经济、社会得以借此机遇实现跨越式发展，为当地民众谋得更多的实惠，则是当地政府当前必须面对和急需思考的一个重大问题。

SWOT 分析

一、优势

1. 优越的旅游区位

三都位于贵州南部综合旅游区的黄金地段，为黎（平）——从（江）——榕（江）至广西桂林国际精品旅游线的中枢景区，位于贵州省六大综合旅游区“荔波综合旅游区”中。

2. 独特的水族风情

三都水族自治县是全国唯一的水族自治县，这里聚居着全国 60% 以上的水族人民，保存有浓郁的水族民风民俗，资源具有独特性。

3. 资源保持完好

由于三都是全国唯一的水族自治县，水族人民长期聚居，保留了较为完整的水族民风民俗，加之三都县交通不便利，旅游开发处于初始阶段，受外来文化冲击较小，因此资源本源性强。

二、劣势

1. 基础设施较为落后

目前三都县对外交通主要依赖于 321 国道和 206 省道，交通较为拥挤，且路况较差，进入三都耗时过长，在一定程度上制约了三都的可进入性。

2. 配套设施不完善

三都县现有能够接待游客的大小酒店 20 多家，其中准三星级宾馆 3 家，其他多为民俗接待村寨。三都旅游住宿设施总量不足，档次较低，条件较差，功能不完善且缺乏当地特色。

目前三都县只有旅行社 1 家，持证导游 2 个，明显与三都县旅游业的发展不相适应。

餐饮业缺乏三都特色，水族特色。

文化娱乐设施不完善，总量缺乏，种类单调。

旅游商品生产销售未形成规模，营销不到位，旅游商品种类不足。

3. 旅游产品单一

由于旅游开发起步较晚，旅游投入不足，基础设施不完善等原因，致使三都县旅游资源利用开发严重不足，旅游产品单一，旅游业发展相对沛后。

4. 产品特色没有得到凸显

三都拥有世界级的旅游资源并且资源的特色鲜明，垄断性强，却未对资源进行深入挖掘和利用，资源利用仅仅停留在初级的观光展示阶段，致使资源被浪费，难以区别于周边资源，无法形成自身特色。

SWOT Analysis

Ⅰ Advantage

1. Favorable Tourism Location.

Sandu is located in the golden district of the integrated tourist area in south Gui Zhou. It's the central scenic region of international travel route from Li (flat)—Cong (river)—Rong (river) to Guilin, Guangxi; in the six integrated tourist area of Guizhou Province, "Libo integrated tourism".

2. The Unique Charm of the Shui Nationality

Sandu Shui autonomous County is the only Shui autonomous County in the country, where more than 60% of the Shui people live, here preserves rich folk customs of the Shui nationality, and the resources are unique.

3. Good Condition of Resources

Sandu County is the only Shui autonomous County in our country, the Shui people inhabit here for a long period of time, and keeps a quite complete Shui folk customs. What's more, transportation in Sandu County is not convenient, tourism development is in the initial stage, and foreign culture shock is relatively small, so the original resource is strong.

Ⅱ Weakness

1. Backward Infrastructure

Now the main traffic of Sandu County relies on the 321 national highway and 206 provincial highway, the traffic is congested and and the road condition is poor, it takes too much time to enter Sandu. To a certain extent, it restricts the access of Sandu.

2. Incomplete Facilities

There are more than 20 hotels that can welcome tourists in Sandu County now, in which has three quasi–three–star ones, the others are mostly receiving folk village. The tourism accommodation facilities in Sandu is not enough in number, poor in condition, incomplete in function and lacking of local characteristic.

At present Sandu county has only one travel agency, two qualified tour guide, and that is obviously unsuited to the development of Sandu tourism.

The catering industry lacks Sandu characteristics and the Shui features.

The cultural recreation facility is not complete enough, and has poor variety.

Tourism commodity marketing is not in scale, marketing is not good enough, and the variety of tourist commodities is poor.

3. The Simplex of Product

As the tourism development started late, lack of tourism investment, inadequate infrastructure and other factors, led to the serious shortage of the use of resources in Sandu County and Tourism Development with single tourism product, so tourism development has relatively lagged behind.

4. The Product Characteristics have not been Highlighted

Sandu County has world–class tourism resources with distinct features, monopolistic in nature, but the resource haven't been made full use of, the resource utilization is merely in the primary stage of tourism display, resulting in a waste of resources, difficult to distinguish it from the surrounding resources, and can not create its own characteristics.

三、机遇

1. 金融危机下我国旅游业受到前所未有的关注

(1) 党中央国务院下拨 4 万亿人民币紧急救市，提振内需。

(2) 国家将旅游作为拉动消费，扩大内需的战略产业和重要突破口。

(3) 城乡居民收入继续增长，居民旅游消费需求潜力依然巨大。

(4) 政府对旅游业发展采取了积极关注的态度。

(5) 国家旅游局提出了“大力发展国内旅游”的战略思想。

2. 旅游业成为贵州省国民经济新的增长点，并被确立为后续支柱产业

目前，贵州省已开展经营或已进行规划的有 477 个旅游区，全省旅游星级宾馆饭店数量达 240 家、旅行社达 194 家、旅游定点接待汽车 810 辆，旅游直接从业人员 7 万余人。全省重点景区基础设施条件明显改善，旅游产业规模不断扩张。2007 年，贵州全省共接待旅游者 7963.23 万人次，同比增长 28.77%，全年实现旅游总收入 512 亿元人民币，净增 124 亿人民币，比 2006 年增长 47%；相当于全省 GDP 的 18%，相当于第三产业增加值的 45%，成为贵州省国民经济新的增长点。并且各项指标均居西部 12 省区之首，因此，旅游业正在成为贵州的后续支柱产业。

3. 多民族特色文化是贵州旅游业发展的重点

无论是 2002 年贵州省委省政府出台的《关于加快旅游业发展的意见》、《贵州省国民经济和社会发展十一五规划纲要》，还是今年年初贵州省旅游局召开的旅游业应对金融危机发布会上，都提到了要大力发展民族旅游，以原生态民族文化为平台，力争把贵州省建成多民族特色文化和喀斯特高原生态旅游的重要目的地，成为中国西部旅游热点地区，旅游业成为贵州省新的支柱产业。由此可见民族文化是贵州省发展旅游业当仁不让的重点，这为三都这一全国唯一的水族自治县深入挖掘水族文化，发展民族旅游打下了坚实的基础。

4. 三都县“旅游兴县”战略的提出为旅游业发展带来契机

为响应省委、省政府提出大力开发贵州南部综合旅游区的战略部署，近年来，三都县县委提出了“科教兴县、农业稳县、工业强县、城镇旅游活县、民族文化塑县”五大战略，围绕这一战略，三都县相关部门紧抓“城镇旅游活县、民族文化塑县”这一主线，大力推进旅游业发展，将“必须把旅游产业作为加快我县经济建设的战略重点”、“做强做大旅游业”写进 2009 年政府工作报告的“五个必须”和“八项任务”中。 三都将举全县之力，浓墨重彩勾画民族生态旅游开发的美好蓝图，循序渐进分步实施，确定用 5 年时间全力以赴建设贵州南部民族生态旅游强县，倾情打造“民族”和“生态”这两张王牌。充分说明三都县委县政府在发展旅游业上的决心，为三都县旅游业的发展带来契机。

Ⅲ Opportunity

1. Tourism Industry Attracts Unprecedented Attention Under the Financial Crisis

(1) The central State allocates 4 trillion yuan for emergency rescue, to boost domestic demand.

(2)The country will take travel as the strategic industry and the important breakthrough to boost consumption and enlarge domestic demand.

(3) The urban–rural residents' income continues to grow, the potential demand of resident tourism consumption remains huge.

(4) The Government has adopted an attitude positive concern on the development of tourism.

(5) The National Tourism Administration proposed the strategic thinking of "strive to develop domestic tourism".

2. Tourism Industry Become New Growth Point for the National Economy in Guizhou Province, and was Established as Subsequent Pillar Industries

At present, the Guizhou province has launched operation or planning of 477 tourist areas, the number of province's tourism star hotels is up to 240, travel agencies up to 194, tourism front car 810, direct travel employees more than 70 thousand. Infrastructure conditions of the province's key attractions are obviously improved, tourist industry keeps expanding. In 2007, Guizhou province had received 79.6323 million tourists, an increase of 28.77%, the tourism revenues was 512 billion yuan, increasing about 124 million RMB, 47% more that in 2006; it occupies 18% of the provincial GDP, equivalent to 45%, of the increased value of the tertiary industry, and becomes a new national economy in Guizhou province.What's more, Guizhou ranks the first among the 12 provinces in west on all index, therefore, tourism is becoming a subsequent pillar industries in Guizhou province.

3. Multi–national Culture is the key to the Development of Tourism in Guizhou

Both the "views on accelerating the development of tourism", "economic and social development in Guizhou Province Eleventh Five–Year Plan" introduced by provincial government in 2002, and the Tourism Bureau in Guizhou Province early this year held in the tourism industry to address the solving financial crisis news conference mentioned the need to vigorously develop the national tourism, take original national culture as platform, and strive to build our province into a multi–ethnic culture of and the Karst plateau important eco–tourism destination, and become a tourist hot spots in Western China, the tourism industry becomes a new pillar industry in our province. This shows that ethnic culture is the important part in the development of tourism in our province; this has laid a solid foundation for Sandu County, the only Shui autonomous county in this country, to take advantage of the Shui culture and develop the national tourism.

4. The Raise of "Tourism Makes the County Prosperous" Strategy in Sandu County has Brought Business Opportunities for the Tourism Development

In response to the provincial government's strategic plan to made great efforts to develop an integrated tourism area in southern Guizhou, in recent years, the Sandu County Committee have put forward the five strategies "technology and education makes the county rich, agriculture makes the County steady, Industry makes County strong, urban tourism makes the county active, ethnic culture shapes the county" Around this strategy, relevant departments of the county stick to the main line "urban tourism makes the county active, ethnic culture shapes the county", and vigorously promote the development of tourism, and wrote "have to take tourism industry as the strategic key point as speeding up county economic development", "strengthen the tourism" into the 2009 government work report "five must" and "eight tasks". Sandu will make the whole county's effort to write the blueprint of developing the ethnic eco–tourism, to put it into practice step by step, make sure to use 5 years to establish southern Guizhou national eco–tourism county, take greatest effort to establish the two trump cards "nation" and "ecology". These fully shows that Sandu county government's determination on developing the tourism, and brings opportunities on the development of tourism in Sandu County.

5.“两高”开通将为三都带来前所未有的区位机遇

厦蓉高速公路贵阳至水口段和贵广快速铁路（简称“两高”）将贯穿黔南州三都水族自治县70多公里，随着“两高”的开工建设，素有“像凤凰羽毛一样美丽”的三都水族自治县，如同一颗璀璨的五彩明珠镶嵌在贵州南部与东南沿海相连的大通道上，其重要地位和区位优势将逐步凸显出来。

四、挑战

1. 来自周边竞争者的挑战

首先，来自同属黔南州荔波自然风光的挑战。

其次，面临紧邻三都的黔东南苗侗风情的挑战。

实际上，周边竞争者对于三都来说是挑战更是机遇，三都县旅游资源具有较强的垄断性，与周边地区旅游业的发展不存在同质竞争的问题，相反可以作为荔波自然风光的有力补充，同时也可以丰盈黔东南少数民族风情，构成多民族特色旅游圈。

2. 核心资源流失的挑战

首先，作为三都县核心资源的水书面临失传的危险。

其次，外来文化对水族文化的冲击。

再次，旅游业发展对生态环境的不利影响。

5. Unprecedented Regional Opportunities of Sandu by the "Two High"

The section from Guiyang to Shuikou of Xiarong Highway and Guiguang railway are called "two high" which will go through Sandu Shui Autonomous County of Qiannan in more than 70 kilometers. With the construction of the "two high", Sandu, enjoyed the fame of "as beautiful as the feather" will demonstrate its regional and locational strength gradually like a precious pearl embellished in the road linking the south of Guizhou and the Southeast seashore cities.

IV Challenge

1. The Challenge of Natural Scenery of Libo, Qiannan

First, the challenge from natural scenery in Libo, which also belongs to QianDongnan.

Secondly, the challenge from the Tong and Miao nationality's charm which is next to Sandu.

In fact, the surrounding competitors are opportunities more than challenges for Sandu, Sandu County's tourism resources has strong monopoly, and there is no competition of the same quality in the development of surrounding tourism, on the other hand, they can be used as a powerful supply for Libo's natural scenery, at th while at the same time making the charm of national minority in east and south Guizhou more abundant, thus forming a tourism area with multi-national characteristics.

2. The Challenge of Losing Core Resources

Firstly, the Shui character, as the core resource of Sandu county, is facing the danger of failing to be handed down from past generations.

Secondly, foreign culture has a strong strike on the Shui culture.

What's more, the development of tourism has negative impact on the ecological environment.

定位 Orientation

Take the only Shui Autonomous County in China as the foundation, Shui culture as the core, original eco-tourism resources as the center of Chinese Shui culture, at the same time take this as trump card and form a tourism center which affects the south of Guizhou Province.

以全国唯一的水族自治县为依托，以水族文化为核心，以原生态旅游资源为内涵的中国水族文化中心，同时以此为拳头，形成辐射黔南的特色旅游集散中心。

目标定位

通过对三都县县城沿街建筑改造，从视觉上塑造水族文化旅游品牌，使三都水族文化旅游的形象初具雏形，形成贵州多民族特色文化旅游的有机构成部分。

通过对三都水族文化的深入挖掘，将三都打造为“世界水族文化中心”。

进一步丰富三都旅游内容，并利用水族文化中心的影响力，形成水族文化旅游品牌，最终将三都县打造成为国内知名，涉足境外，具有水族特色，辐射周边县市的黔南旅游集散中心。

Aim Orientation

Through the reform of street architecture in Sandu County, create the Shui culture tourism brand visually, so that the images of Sandu Shui cultural tourism will have an embryonic figure, and form the organic components of multi-ethnic cultural tourism characteristics in Guizhou.

Build Sandu as "center of the world's Shui culture" by developing the Sandu Shui culture.

Further enrich the contents of Sandu tourism and use the influence of Shui Cultural Centre to the form the tourism brand of Shui culture, and eventually establish Sandu as a travel distribution center well-known country, involving foreign tourism,with Shui nationality characteristic and affecting surrounding counties and cities.

开发战略

利用文化内涵，挖掘核心价值。水族文化历史悠久，神秘古朴，水族是我国 17 个拥有自己文字的民族之一，水书以其内容博大精深，记载了水族古代天文、地理、宗教、民俗等文化信息，被誉为象形文字的“活化石”。“水族马尾绣”是水族最古老又最具民族特色的一种刺绣工艺；水族端节是水族众多传统节日中规模最大、覆盖范围最广的节日，历时 49 天，是世界上最长的节日。水书、水族马尾绣以及水族端节被列入我国首批非物质文化遗产名录。三都作为全国唯一的水族自治县保存了完好的水族文化，这一文化贴合贵州以少数民族文化为重点的旅游发展战略的同时又有别于当前贵州唱主角的苗侗文化，是三都的特色，也是贵州的又一特色，因此将水族文化运用到三都的旅游城市打造中定能形成三都的核心价值。

紧抓核心价值，变“无形”为“有形”。受现代思维的影响，当今水族人民、水族文化受到了一定程度的影响，因此在三都旅游城市打造的过程中要深入挖掘散落在民间的非物质文化遗产和具有民族特征的遗落元素，将这些组件变成实体，形成当地特色的同时也对水族文化的保护起到一定的促进作用。

城市改造凸显三都城市张力。三都作为全国唯一的水族自治县，理应充分展现这一独具特色的风貌。而现在三都县城与中国成百上千的普通县城没有区别，这样的县城怎么能让到此旅游的游客直观感受到其丰富多彩的人文内涵呢？为此，应运用符号化的表现手段，对三都县城主要街景进行改造，营造出浓郁的民族氛围，使之成为既具景观价值又有视觉冲击力的景致，为打造黔南旅游集散地埋下伏笔。

Development Strategy

Using cultural content to excavate core values. The Shui culture has a long history with mysterious simplicity; the shui nationality is one of the 17 nations that have their own characters. The Shui character is known as "living fossil" of hieroglyphic, with its extensive and profound content, and records cultural information of the Shui nationality's ancient astronomy, geography, religion, folk custom and so on; the "Shui aquarium Horsetail Embroidery" is the oldest embroidery technique with ethnic characteristic in the Shui nationality; the Duan Festival is the largest of the many traditional festivals in the Shui nationality, covering the widest range and lasts 49 days, is the world's longest holiday. The Shui characteristic, Shui aquarium Horsetail Embroidery and the the Duan festival are included in our first batch of intangible cultural heritage. Sandu, as the only Shui Autonomous County in the country, has well-preserved the Shui culture, this culture fits Guizhou's tourism development strategy of taking minority nationality culture as key point and at the same time different from the Miao and Tong culture which currently play a leading role in Guizhou, this is the feature of Sandu, and also another feature of Guizhou, therefore applying Shui culture to the developing of Sandu tourism city will definitely form Sandu's core values.

To seize the core values, change the "invisible" as "visible" Under the influence of modern thought, nowadays the Shui residents and culture are affected to certain extent, so in the process of developing Sandu tourism city, we should find the non-material cultural heritage and elements with ethnic characteristic, turn these components into a solid form and also play a promoting role in the protection of Shui culture when forming local characteristic.

City reforming reflects city tensile force of Sandu. Sandu, as the only Shui autonomous County in the country, should fully display this unique style. Now Sandu is the same with thousands of ordinary town in China, how can this town makes tourists have intuitive sense of its rich humanistic connotation? So symbolic means of expression should be used to reform the main street scenery in Sandu County, creating a strong ethnic atmosphere, making it a scenery with both have landscape value and visual lash, to make foundation for creating a tourist and leisure center.

设计元素（水族建筑）
Desigr Element

设计元素（非物质文化）
Design Element

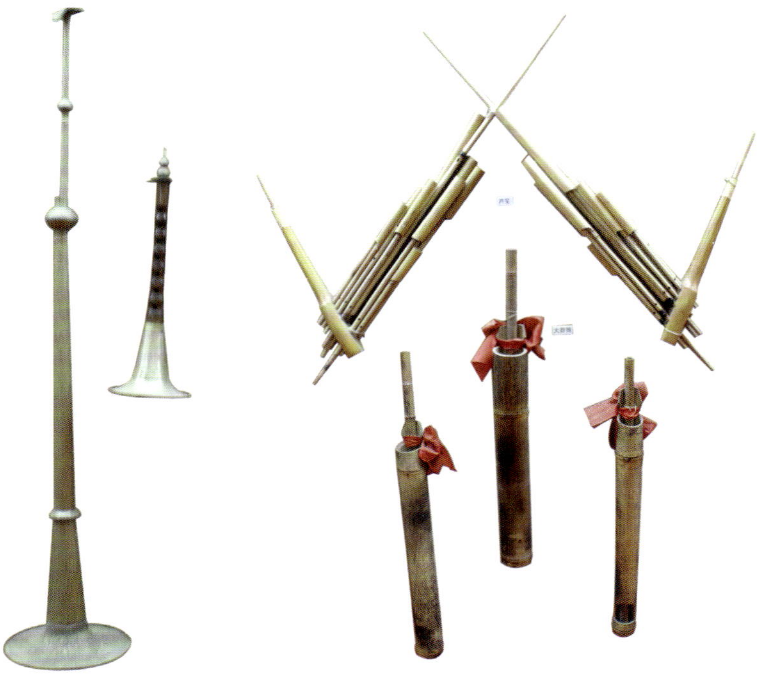

城市总说
Outline

第一章 总则

第一节 规划依据

国家、省、市、县相关法律法规

《中华人民共和国城乡规划法》（2008）
《中华人民共和国环境保护法》（1989）
《中华人民共和国土地管理法》（2004 修订）
《中华人民共和国建筑法》（1997）
《中华人民共和国工程建设标准强制性条文（城乡规划部分）》（建标 [2000]179 号）
《城市用地分类与规划建设用地标准》GBJ 137—90
《城市规划编制办法》（2005，建设部令第 146 号）
《关于加强城市总体规划修编和审批工作的通知》（建规 [2005]2 号）
《城市道路设计规范》CJJ 37—90
《城市用地分类代码》CII 46—91
《风景名胜区规划规范》GB 50298—1999
《旅游规划通则》GB/T 18972—2003
《旅游发展规划管理暂行办法》（1999）
《贵州省城镇体系规划》（2001 ～ 2020）
《中共贵州省委 贵州省人民政府——关于加快旅游业发展的意见》（黔党发（2002）20 号）
《贵州省旅游发展总体规划》
《贵州旅游业发展“十一五”规划纲要》
《贵州省旅游管理条例》
《黔南州城镇体系规划》（2003 ～ 2020）
《黔南州旅游产业发展振兴纲要》

《黔南州城镇体系规划》（2003 ～ 2020）
《黔南州旅游产业发展振兴纲要》
《三都县县城总体规划（修编）》（1997 ～ 2010）
《三都水族自治县城市发展总体规划》（2007 ～ 2025）
《三都水族自治县县城老城片区控制性详细规划》（2008.9）
《三都水族自治县国民经济和社会发展第十一个五年规划纲要》
《三都水族自治县旅游发展规划》（2008 ～ 2020）
《三都水族自治县县志》
《三都水族自治县综合农业区划》
《三都统计年鉴》
三都水族自治县部分乡镇规划和相关行业规划
其他相关法规、规范

Chapter I General Rules

Section I Planning Basis

Relevant Laws and Regulations at National, Provincial, City and County Level

Urban–Rural Planning Law of the People's Republic of China (2008)
Environmental Protection Law of the People's Republic of China (1989)
The Land Administration Law of the People's Republic of China (2004 Revised Edition)
Construction Law of the People's Republic of China (1997)
Engineering Construction Standards of the People's Republic of China Compulsory Provisions for Urban and Rural Planning (Construction Standards [2000]No.179
Standard for Classification of Urban Land and for Planning of Constructional Land (GBJ 137—90)
Regulations on Town Planning methodology (Ministry of Construction Order No. 146, 2005)
Notice on Enhancing the Overall Urban Planning and Approval Work (Jiangui [2005] No. 2)
Rules of planning design of transport on urban road (CJJ 37—90)
Urban Land Classification Code (CII 46—91)
Rules of Scenic Area Planning (GB 50298—1999)
Tourism Planning Rules (GB/T 18972—2003)
Provisional Regulations for Tourism Development Planning and Management (1999)
Urban System Planning of Guizhou Province (2001~2020)
Advice on Quickening the Development of Tourism Industry issued by Guizhou Provincial Committee of the CPC and the People's Government of Guizhou Province (Qian Committee (2002) No. 20)
General Plan of Tourism Development of Guizhou Province
Outline for Eleventh Five Year Tourism Plan in Guizhou Province
Tourism management regulations in Guizhou Province
Urban System Planning of Qiannan prefecture (2003~2020)
Outline for the Development and Rejuvenation of Tourism Industry in Qiannan prefecture

Urban System Planning of Qiannan prefecture (2003~2020)
Outline for the Development and Rejuvenation of Tourism Industry in Qiannan prefecture
The Overall Township Planning in Sandu County (revised edition) (1997~2010)
The Overall Planning for Urban Development in Sandu Shui Autonomous County (2007~2025)
Old Urban District Regulatory Detailed Planning in Sandu Shui Autonomous County (September, 2008)
National Economy and Social Progress Outline for Eleventh Five year Program in Sandu Shui Autonomous County
Urban Development Planning in Sandu Shui Autonomous County (2008~2020)
County Annals for Sandu Shui Autonomous County
Comprehensive agricultural divisions in Sandu Shui Autonomous County
Sandu Statistical Yearbook
Planning of Part of Towns and Relevant Industries in Sandu Shui Autonomous County
Other Relevant Rules and Regulations

第二节 规划范围及年限

一、规划范围

1. 规划设计深度达到“修建性详细规划”的片区

(1) 北出口片区
(2) 河滨北路西段萝卜寨、中营片区
(3) 河滨北路步行街片区
(4) 河滨南路片区
(5) 南出口片区
(6) 河滨南路片区
(7) 环城北路片区
(8) 麻光三郎片区

Section Ⅱ Scope and Period of Planning

Ⅰ Scope of Planning

1. The Depth of Planning and Design Extends to "the Constructive Detailed Planning" Area

(1) North Exit Area
(2) Zhongying Area, Luobu Village, North Riverside Road
(3) Pedestrian Street Area, North Riverside Road
(4) South Riverside Road Area
(5) South Exit Area
(6) South Riverside Road Area
(7) North Huancheng Road Area
(8) Maguang Sanlang Area

区位分析图

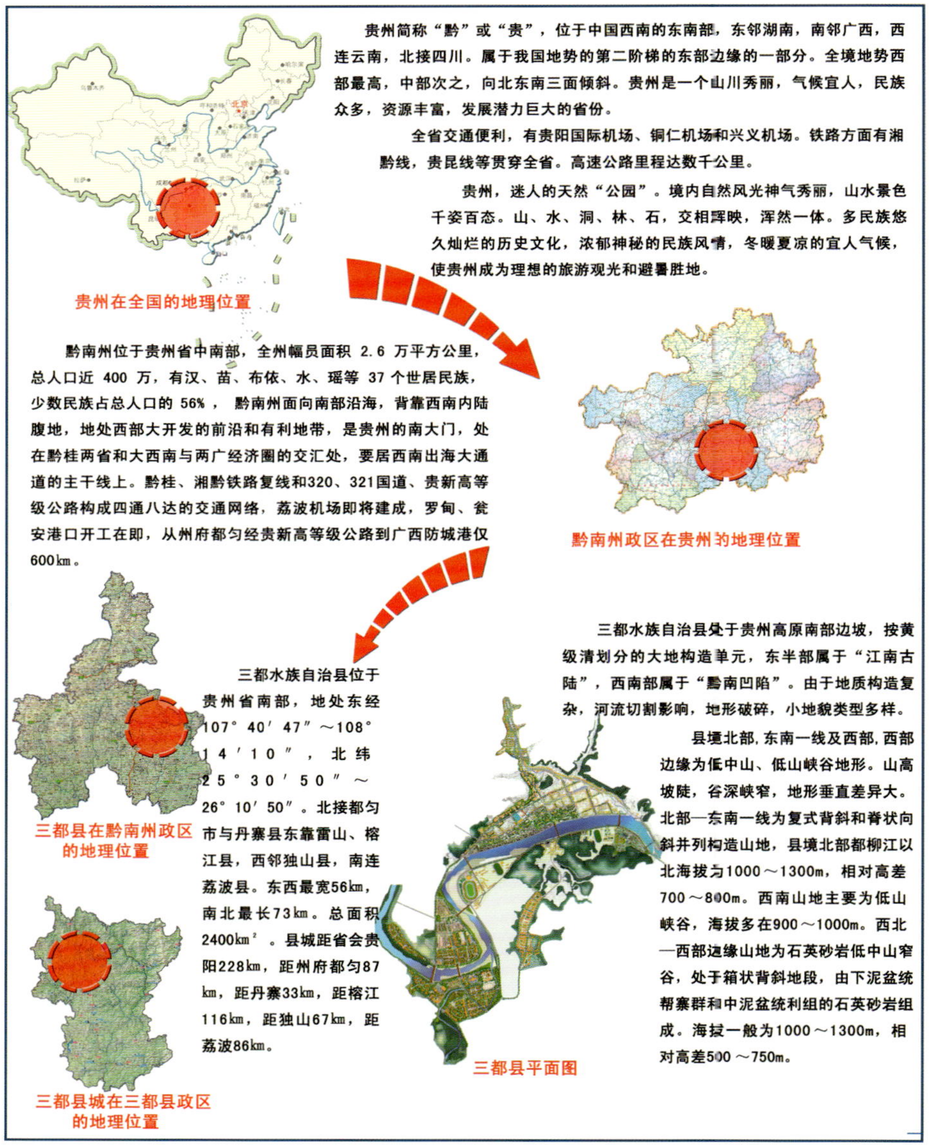

2. 旧城区风貌改造，深度达到“穿衣戴帽”的片区

建设东路、建设西路、中山路、都江路、深圳路、团结东路、团结西路、振兴路、民族路、商贸路、县府路、河滨北路东段、环城东路、凤凰大道、原大操场处（行政中心对面）、上坝小区（搬迁安置地）、文化路两侧建筑物及明确设计深度达修建性详细规划片区内的个别无序的邻街建筑物。

3. 创意设计部分

厦蓉高速公路巫腰匝道口、可视坡头和规划范围内主要节点的相关配套设施。

二、规划年限

根据《三都水族自治县城市发展总体规划》，规划期限调整为：

1. 规划期：2009 ~ 2025 年，17 年。

2. 近期：2009 ~ 2014 年，6 年。

3. 远期：2015 ~ 2025 年，11 年。

Ⅱ Period of Planning

In accordance with Overall Urban Development and Planning in Sandu Shui Autonomous County, the adjustment of period of planning is as follows:

1. planning period: 2009~2025, 17 years.

2. short-term period: 2009 ~ 2014, 6 years.

3. long-term period 2015 ~ 2025, 11 years.

2. Changes in Old City With Its Depth Reaching "Dressing Up" Area

East Jianshe Road, West Jianshe Road, Zhongshan Road, Dujiang Road, Shenzhen Road, East Tuanjie Road, West Tuanjian Road, Zhenxing Road, Minzu Road, Shangmao Road, Xianfu Road, East quarter in North Riverside Road, East Huancheng Road, Fenghuang Highway, the original Palestra Grande (opposite administration center), Shangba Community (relocated place), buildings on both sides of Wenhua Road and individual unordered adjacent buildings whose clear design depth reaching the constructive detailed planning Area.

3. Creative Design

Relevant supporting facilities of primary node at Wuyao road junction on Xiarong express way, visible head of slope and within the scope of planning.

用地现状分析图

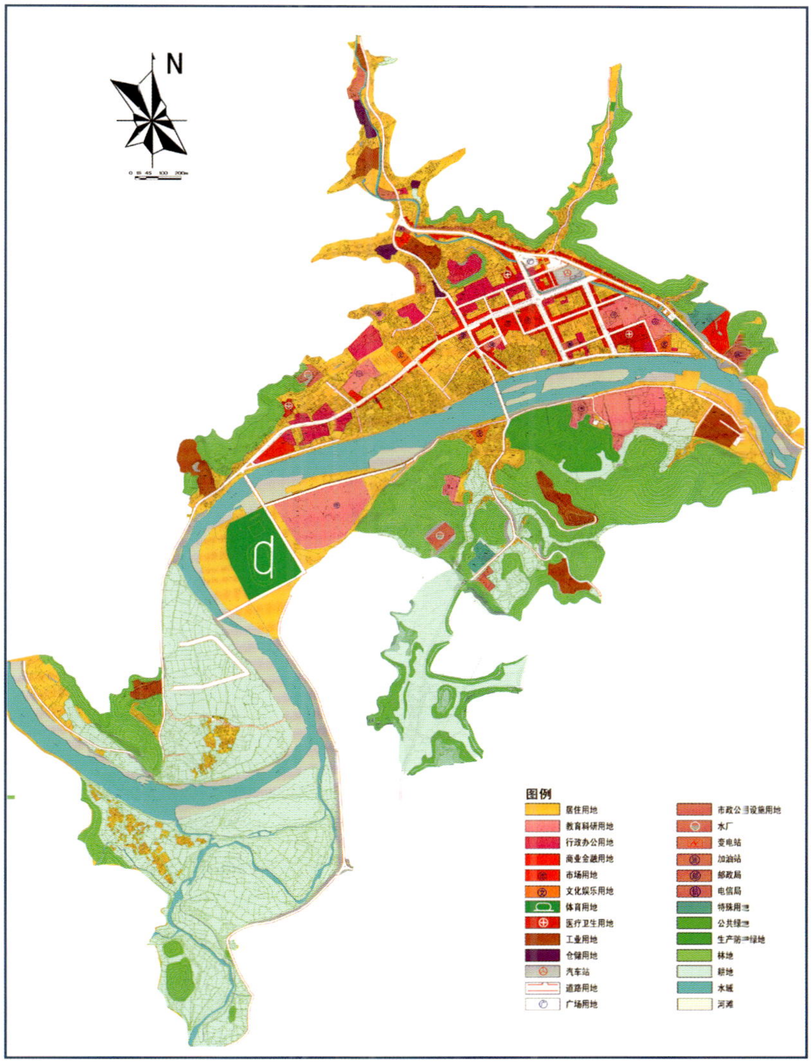

第三节 城市规划规模

三都水族自治县共辖 10 镇 11 乡，县域城镇分布密度为 0.88 座 /100km^2。2006 年年底县域总人口为 32.20 万人，其中非农业人口 20001 人，城镇人口 7.77 万人，城镇化水平为 24.13%，根据《三都水族自治县城市总体规划》预测，2010 年全县总人口达 33.60 万人，城镇人口 9.71 万人，城镇化水平达 28.90%，县城人口 5 万人；2025 年全县总人口达 38.70 万人，城镇人口 18.35 万人，城镇化水平达 47.42%，县城人口 9 万人。

Section Ⅲ Scale of Urban Planning

Sandu Shui Autonomous County has 10 towns and 11 countries with distribution at 0.88 towns per 100km^2. The total population at county level is 322 000 at the end of 2006, among which 20 001 people are nonagricultural population and 77 700 people urban population with urbanization level at 24.13%. According to predictions made by Urban Master planning in Sandu Shui Autonomous County, by the year of 2010, the total population in the whole county will reach 336 000, among which urban population will reach 97 100 and county town population 50 000 with urbanization level at 28.90%. In 2005, the total population in the whole county was 387 000, urban population 183 500 and county town population 90 000 with urbanization level at 47.42%.

第四节 基本概况 Section Four Basic Information

一、自然地理 Ⅰ Nature and Geography

1. 地理位置 Geographical position

三都是全国唯一的水族自治县，位于贵州省黔南布依族苗族自治州东南部，地处“月亮山、雷公山”腹地，地跨东经 107° 40' ~ 108° 14'，北纬 25° 30' ~ 26° 10' 之间。东邻榕江、雷山，南接荔波，西接独山、都匀，北连丹寨。东西宽 56km，南北长 78km，距省城贵阳 230km，距州府都匀 85km，全县总面积 2400km^2。

Sandu, the only Shui Autonomous County in the country, is situated at Southeastern part of Qiannan Bouyei and Miao Autonomous Prefecture in Guizhou province. It is located at hinterland of "the Moon Mountain and the Leigong Mountain" extending across 25° 30'~26° 10'E and 107° 40'~108° 14'N. The county is next to Rongjiang and Lei Mountain to its east, Libo to its South, Dushan, Duyun to its West and Danzhai to its North. It is 56km wide in its East and West, 78km in its South and North, 230km to Guiyang province, and 85km to its Capital City Duyun. The total area of the county is 2 400km^2.

2. 地质地貌 Geological features

三都县处于云贵高原的东南斜坡，地势自西北向东南倾斜，平均海拔在 500 ~ 1000m 之间，最高为西北面的更顶山，海拔 1665.5m；最低处是坝街附近的都柳江出境处，海拔 303m。境内山岭连绵，溪流交错，其间夹着若干起伏的丘陵和平坝，在总面积中耕地占 9.4%，林地占 55.6%，草山占 29.7%，水面占 1.3%，有“九山半水半分田”之称。

Sandu County is located at Southeastern slopes of the Yungui Plateau. Its topography leans from Northwest to Southeast with an average altitude at 500 to 1 000 meters. The highest position of the county is the Gengding Mountain in its Northwest with an altitude at 1 665.5 meters while the lowest position is outbound places of the Duliu River near the Ba Street with an altitude at 303 meters. The county enjoys the continuous mountain ranges and the braided river streams accompanied by the rolling hills and basins within its borders. Within its total area, the arable land accounts for 9.4%, woodland 55.6%, grassy hill 29.7% and water surface 1.3%, which renders it the name of "nine hill half water half minutes of cropland".

道路现状分析图

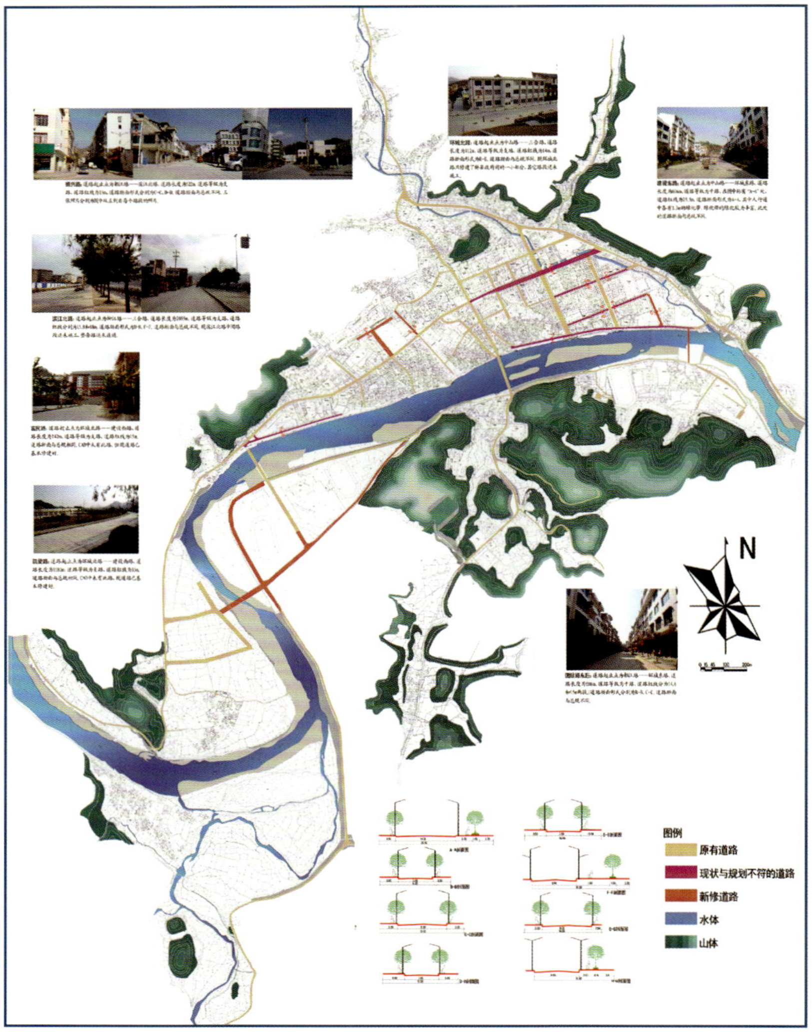

3. 气候条件

三都县属中亚热带温润季风气候类型，特征为夏长冬短，春秋分明，夏无酷暑，冬无严寒，年平均气温18℃，无霜期平均为328d，年平均降雨量1349.5mm。

3. Climatic Conditions

Sandu County belongs to subtropical humid monsoon climate, featuring long winter and short summer, distinct spring and autumn seasons, no heat in summer and no cold in winter. The average temperature is 18℃ with average frost free period of 328 days and average annual rainfall at 1 349.5mm.

4. 水文状况

三都县境内大小河流42条，水力资源丰富，理论储量25万千瓦，目前仅开发2524千瓦。穿境而过的都柳江是珠江的重要支流，是县内最大的河流，境内长83.5km，落差197 m，流域控制面积为1680km^2，占全县总面积70.6%，有集水面积20km^2以上的一、二级支流31条汇入，极具开发电力、水运、旅游的潜力。

4. Hydrological Conditions

As there are 42 rivers within its border, Sandu county bounds in water power resources with 250 C00kW theoretical reserves, of which only 2 524kW has been exploited at present. The Duliu River across border is an important tributary of the Pearl River and the largest river within the county. It is 83.5km within its border with 197m disparity. The watershed control area is 1 680km^2, taking up 70.6% of the total area. There are 31 first-level and second-level tributaries with contributing area over 20km^2 merging into the river, which gives the river the potential to exploit Electric power, water transportation and tourism.

5. 矿产资源 Mineral resources

三都县境内物产丰富，地下矿藏有汞、锑、金、铅、锌、铜、铁、硫黄、煤、金刚石等。其中，锑矿、硫铁矿、金矿、煤、铅锌等矿产资源具有较高的开发价值。

Sandu is rich in products with mineral resources such as Mercury, antimony, gold, lead, zinc, copper, iron, sulfur, coal and diamond buried underground, among which, mineral resources like antimony ore, pyritic, gold mine, coal lead, zinc, etc enjoys relatively higher exploitation value.

6. 林产资源 Forest resources

三都县境内拥有丰富的森林资源，森林覆盖率50.08%，主要树种有杉树、马尾松、油桐、油茶、麻栎、柏树、枫树、楠木、樟木、桫椤、楠竹、漆树、杜仲等，是全省十大林业县之一，全县有11片林区。

Sandu is also rich in forest resources with the forest coverage standing at 50.08%, The mian trees are firs, Pinus massonian, Aleurites fordii, german oak, cypress, Maple, Phoebe nanmu, camphor wood, Polystrochyous aphanaminis, bamboo, sumac Eucommia Bark, etc. The county has 11 forest areas which makes it one of the top ten forestry counties in the whole province.

用地范围

二、人文环境

1. 政区人口

全县辖10镇11个乡270个村委会4个居委会2413个村民小组，总人口31.47万人，其中少数民族人口30.47万人，占总人口的96.4%；水族人口20.24万人，占总人口的64.8%，全国60%以上的水族人口居住在三都，县境内共居住着汉、水、布依、苗、瑶等14个民族。

2. 历史沿革

三都在夏朝属梁州东南裔；商属于荆州西南裔；周属越领牂牁国地；秦属象郡且兰县地；汉属牂牁郡；晋属牂牁郡的东南角和武陵郡的西南角等处交界地；南朝初属牂牁郡东南角和齐熙郡西北角等处交界地；南朝梁、陈属南牂牁郡；隋属牂牁郡。唐初开始在三都境内设治。

1949年12月6日三都解放，1950年1月14日三都水族自治县人民政府成立。1956年9月11日，国务院作出决定，撤销三都水族自治县，设置三都水家族自治县，同年12月21日，将三都水家族自治县更名为三都水族自治县。1957年1月2日，三都水族自治县正式成立，县治设三合镇，隶属黔南布依族苗族自治州。此后，县名、县治一直沿袭至今。

3. 社会经济

近年来，三都县坚持以邓小平理论和“三个代表”重要思想为指导，认真落实科学发展观，在科学分析、深入研究的基础上，坚持又好又快发展的主题，更新发展观念，加大招商引资力度，盘整资源，大力实施“科教兴县、农业稳县、工业强县、城镇旅游活县、民族文化塑县”五大战略，全县经济社会实现了持续、协调、快速发展，呈现出速度比较快、效益比较好的发展态势。

4. 基础设施

(1) 道路交通：抢抓西部大开发、新阶段扶贫开发的机遇，加大投入，加快建设，基础设施条件进一步改善。县境内公路总里程达1080km，村通路率达85%。形成了纵横交错，干支相连，四通八达的公路交通网络。

(2) 电视电信：“村村通”广播电视工程全面实现通村目标，工程覆盖21个乡镇，受益群众1050户，每户可收看37套卫星电视节目。全县拥有固定电话用户38800户，乡镇程控电话率达100%。全县21个乡镇的274个村（居）委会实现了移动电话“村村通”，通信覆盖率达100%。联合通信也实现了“乡乡通”，移动网络信号覆盖了全县21个乡镇的152个村（居）委会。

II The Human Environment

1. Population

The entire county has 10 towns, 11 villages, 270 village committee, 4 residents' committee and 2 413 village groups. The total population is 314 700, among which, ethnic minorities population is 304 700, accounting for 96.4% of total population, and Shui population 202 400, accounting for 64.8% of total population. More than 60% Shui people in the country live in Sandu and another 14 ethnic groups such as Han, Shui, Bouyei, Miao, Yao live within county.

2. History

Sandu belongs to Southeast Liangzhou in Xia Dynasty; Shang belongs to Southwest Jingzhou; Zhou belongs to Yueling Zangke Kingdom; Han belongs to Zangke prefecture; From Jin Dynasty to the Southern Dynasties, they belongs to the boundary of Southeast corner and Southwest corner of Zangke Prefecture; the Southern Dynasties belongs to the boundary of Southeast corner of Zangke Prefecture and Northwest corner of Qixiang Prefecture; Liang and Chen belongs to South of Zangke Prefeture; Sui belongs to Zangke Prefecture. Start from the beginning of Tang dynasty, it established rules within the border of Sandu.

Sandu was emancipated on December 6, 1949 and the people's government of Sandu Shui Autonomous County was founded on January 14, 1950. On September 11, 1956, the State Council decided to repeal Sandu Shui Autonomous County and set Sandu Shui jiazu Autonomous County instead. On December 21 of the same year, the State Council renamed the county back to Sandu Shui Autonomous County. On January 21, 1957, Sandu Shui Autonomous County was officially established. The county set Sanhe town subjected to Qiannan Bouyei and Miao Autonomous Prefecture. Since then, the county has carried on its tradition to this day.

3. Social Economy

In recent years, Sandu county has pursued Deng Xiaoping Theory and "Three Representatives" as its guidance, conscientiously implementing scientific outlook on development, adhering to the theme of sound and fast development on the basis of scientific analysis and profound research, renewing development ideas in order to attract overseas investment, make good use of resources, and vigorously fulfill five strategies of "developing the county with education and science, stabilizing the county with agriculture, enhancing the county with industry, reviving the county with urban construction and tourism, as well as shape the county with national culture". The county has realized stable, coordinated and rapid development in its social economy, presenting a fast and effective growth trend.

4. Infrastructure

(1) Road Traffic: We should grasp the opportunity to develop west regions of China and alleviate poverty in the new stage, increase investment, accelerate construction and further improve infrastructure conditions. The total mileage of the road within the county reaches 1 080km with accessible pathways accounting at 85%, which has formed a crisscrossing and intersecting transport network, extending to all directions.

(2) Television and telecommunications: "Every Village" radio and television project has realized its goal and has covered 21 villages and towns, which has benefited 1 050 households with each household receiving 37 satellite TV programs. The county has 38 800 households of fixed-line users with program-controlled telephones in villages and towns standing at 100%. 274 village/resident's committees in 21 villages and towns have accomplished "every village" mobile phones with a communication coverage rate at 100%. United Telecommunications have also realized "every township" mobile phones with mobile network coverage at 21 townships and 152 village/resident's committee of the whole county.

第二章 旅游资源分析与评价

第一节 资源现状

一、俊秀神奇的自然资源

1. 都柳江风光带

都柳江省级风景名胜区位于贵州省黔南布依族苗族自治州三都水族自治县东北部。321 国道沿途经过。都柳江景区总面积约 127km²。其间溪流瀑布、峡谷沙滩兼容，峰峦绵延、梯田层层、林海苍茫、云雾环绕、鸟语花香。森林覆盖率为 70% 以上。景区以典型的岩溶地貌及其形成的山、水、洞、瀑为主要景观，融民俗文化、历史文化为一体，地方特色浓郁，水族文化得天独厚，是人们休闲度假的理想场所。

2. 尧人山国家森林公园

尧人山国家森林公园位于县城东南 12km，方圆 350km²。公园内峰岭连绵、幽谷叠翠、山高谷深、瀑布众多，原始森林浩瀚无垠，野生动物种类繁多。"产蛋崖"、"风流草"、"月亮树"、斗鱼等奇观更增添了尧人山的神奇色彩。

3. 风流草

风流草是三都独有的珍树灵草，每当听到情歌，其两羽嫩叶会随着歌声摆动，歌声激昂时，叶片竟然扭成一团，似相互拥抱的亲密情侣在翩翩起舞。歌声停止，两羽嫩叶慢慢舒展开来，直至恢复原状，又仿佛是依依不舍的热恋情人在诉说着离愁别绪。让人奇怪的是，如果不是唱情歌，无论你如何叫唤，这些草却纹丝不动。

4. 产蛋岩

产蛋岩位于三都县县城东约 1 里处秀丽的都柳江畔。这里的山崖崖壁内，镶嵌着形状各异、大小不一的"石蛋"。据专家考察，每隔 30 年，每当狂风暴雨之夜，雷声大作之时，成熟的"石蛋"就自然"产"出落地，令人称奇。

5. 晴雨石

位于距三合镇 95km 的九阡镇姑右村附近，因为它能预报天气晴雨，故通称为"晴雨石"。如天气明朗，烈日当空，万里无云。"屏幕"呈黄白色；如果"屏幕"是暗色，则预示风雨即将来临。雨中则显黑色，久雨出现由黑转白时，必定阴雨转晴。

Chapter II Analysis and Assessment of Tourist Resources

Section I Present Situation of Resources

I Majestic Nature Sceneries

1. Duliu River Scenic Zone

Duliu River scenic spots and historical interests at provincial level are located at Northeastern part of Sandu Shui Autonomous County of Qiannan Bouyei and Miao Autonomous Prefecture in Guizhou province with the national highway 321 coming along. The total area of Duliu River scenic spots is about 127km². Visitors can enjoy in scenic spots the beauty of streams and waterfalls, mixed beaches and canyons, rolling peaks and ridges, terraced layers, surrounded clouds and mists, birds' twitter and fragrance of flowers, etc. The forest coverage rate is above 70%. The scenic area takes the typical karst landform as well as mountains, rivers, caves and waterfalls as its principle landscape, integrating folk and historical culture with strong local characteristics and unique culture of Shui people. It is an ideal place for people to relax on holidays.

2. National Forest Park at Yaoren Mountain

National Forest Park at Yaoren Mountain is situated at 12km Southeast of the county, that is, 350km² at its surrounding area. Within the park, there are rolling peaks, range upon the range of secluded valleys, high mountains with numerous waterfalls, vast boundless of virgin forests and a great variety of wild animals. The Greatest wonders such as the "Egg cliffs", "merry grass", "moon tree" and rumble fish add magical colors to the Yaoren Mountain.

3. Merry Grass

Merry grass is a kind of rare and precious grass in Sandu. Each time when it hears a love song, its two pieces of tender leaves sway with the song. It the song is full of emotion, the leaves will be intertwined with each other as if intimate lovers were embracing each other and dancing together. When the song ceases, the two pieces of leaves will unfold slowly till they resume to their original status as if they were passionate lovers telling each other the sorrow of parting unwillingly. What is amazing is that no matter what you are calling out, the grass will stay absolutely still.

4. Egg Cliffs

Egg cliffs are located at beautiful banks of Duliu River 1 li away from east of Sandu county. Here, a great variety of "stone eggs" with different sizes are embedded in these cliffs. According to specialists, it is incredible that every 30 years when at the stormy night with great thunder, those ripe "stone eggs" will be born naturally.

5. Barometer Stone

Barometer Stone is located near Guyou village, Jiuqian town, 95km away from Sanhe town. It is called "Barometer Stone" because it can forecast the weather. When the weather is clear with a blazing sun and cloudless, the "screen" will present yellowish-white color. If the "screen" is dark color, it predicts that storm is around the corner. When the "screen" changes its color from dark to white, the weather is bound to become clear.

二、古朴神秘的水族文化

1. 水书

"水书"是我国水族的独特文字，水语称"泐（le）睢"，是类似甲骨文和金文的一种古老文字符号。其内容博大精深，记载着水族古代天文、地理、宗教、民俗等文化信息，被誉为象形文字的"活化石"。水族的文字水书不亚于纳西族的东巴文，具有极高的文物价值和文献价值，现已引起了国内外专家学者的极大关注。

水书被列入我国首批国家级非物质文化遗产名录。

2. 水族马尾绣

马尾绣是水族最古老又最具民族特色的一种刺绣工艺。马尾绣顾名思义是以马尾为原料的刺绣工艺，其制作过程繁琐复杂，成品古色古香，华美精致，结实耐用。马尾绣的刺绣图案古朴、典雅、抽象并具有固定的框架和模式。

马尾绣绣品在水族人民的生产生活中具有深远影响。马尾绣背带、马尾绣尖角鞋、马尾绣童帽等，在水族生活礼仪中具有特殊的意义。

水族马尾绣已被列入我国首批国家级非物质文化遗产名录。

3. 水族端节

水族端节是水族众多传统节日中规模最大、覆盖范围最广的节日，相当于汉族的春节，以水族自己的历法来推算，水族历法以农历九月初一为岁首，八月为岁尾，水族端节从农历八月的第一个"亥"日开始到农历十月的"亥"日结束，从头至尾长达 49 天，被誉为世界上最长的节日。水族端节被列入我国首批国家级非物质文化遗产名录。

4. 水族卯节

水族卯节被誉为"东方情人节"，流行于三都县部分村寨。每年水历九、十月（公历五、六月）择一卯日举行。当天，青年男女打着伞，手拿花帕到卯坡唱歌、跳舞和游玩。晚上，人们汇集村寨广场，击铜鼓、敲皮鼓、吹唢呐、演出传统的花灯剧等。还邀请客人饮宴，是水族青年男女择偶的重要活动。

水族传统节日有端节和卯节之分，由于不同区域的水族习惯略有不同，水族有"过端不过卯"、"过卯不过端"的习俗。

II Simple but Mysterious Culture of Shui Nationality

1. Shui Characters

"Shui character" is a kind of unique character, naming "le sui" in Shui language which is a kind of ancient letter symbols similar to inscriptions on bones or tortoise shells of the Shang Dynasty and inscriptions on ancient bronze objects. Its content is extensive and profound, recording cultural information such as ancient astronomy, geography, religion, and customs of Shui nationality, which is hailed as "living fossils" of Hieroglyphics. The letter symbols of Shui nationality are no second to Dongba script of the Naxi nationality, boasting high value of cultural relics which has aroused great concern from specialists both at home and abroad.

Shui character was listed into the first list of national intangible cultural heritages.

2. Mawei Embroidered Material of Shui People

Mawei embroidered material is a kind of most ancient embroidery crafts of Shui people with national characteristics. As its name shows, its original material is Mawei (horse tail). Though its working process is very complicated, the finished products are brimmed with antique flavor, delicate and luxurious, strong and durable. Mawei embroidered pattern feathers simple, elegant, abstract with fixed frames and models.

Its embroidered pattern has brought about significant influences on lives and productions of Shui people. Mawei embroidered braces, shoes and beanies enjoy a unique and special significance in living etiquette of Shui people.

Mawei embroidered material of Shui people has been listed into the first national intangible cultural heritage.

3. Duan Festival of Shui People

Duan Festival of Shui people is the largest festival with the widest coverage among many traditional festivals, which is amount to the spring festival of Han people. Calculated from the lunar calendar of Shui people, every September 1st in the lunar calendar is the beginning of the year while August in the lunar calendar is the end of the year. Duan Festival of Shui people starts from the first day of "Hai" day in August and ends with the day of "Hai" day in October, lasting 49 days from beginning to the end. Duan Festival of Shui people was listed into the first national intangible cultural heritage.

4. Mao Festival of Shui People

Mao Festival of Shui People has been hailed as "Eastern Valentine's Day", which is popular in some villages of Sandu County. The festival will be held in some day of "Mao" in September or October in the Shui calendar (May or June in lunar calendar). On that day, young men and women hold umbrellas and handkerchiefs in their hands, singing, dancing and playing on Mao Slope. In the evening, people gather together at village plaza, beating the drum, blowing the suona and performing the lantern show. They will also invite guests to dinner. All these are important activities for young people to choose their spouses.

The traditional festivals of Shui people divided into Duan Festival and Mao Festival. Owing to slightly different customs in different regions of Shui people, they have the customs of "celebrating the only either one festivals" during the time.

5. 水族民居

水族传统民居为“干栏”式建筑，这种建筑一般是两层或三层，最高的是四层，多用杉松木建造，屋顶用杉树皮或瓦片覆盖。楼下喂养牲畜家禽，楼上住人。房屋间数为单数，忌取双数，多为三间、五间或七间。

6. 水族服饰

水族青壮年妇女平时上衣穿右衽大襟半长衫，衣长过膝，下穿青布长裤，腰系青布围腰。袖口、裤脚绣有花边。

7. 水族音乐

水族的民间音乐有其独特的民族风格，尤其是民歌，形式多样，可分为双歌、单歌、蔸歌、调歌和诘歌五种。

水族的乐器主要有铜鼓、大皮鼓、芦笙、唢呐、胡琴等。其中铜鼓和大皮鼓最为流行。

8. 水族舞蹈

水族是一个能歌善舞的民族，水族的传统舞蹈主要有铜鼓舞、斗角舞、芦笙舞。

第二节 资源评价

一、旅游资源类型多样

据《三都水族自治县旅游发展规划》调查，三都水族自治县的主要资源涵盖中华人民共和国国家标准《旅游资源分类、调查与评价》规定的八大主类的全部；涉及二十五个亚类，占全国三十一个亚类的 80.65%；基本类型占全部 155 种基本类型的 38.70%，共拥有 471 个主要资源单体，除天象与气候景观类资源单体数量较少外，其他各类资源均较为丰富。

二、资源组合较为合理

三都县旅游资源种类较丰富，既有丰富的自然旅游资源也具有地方特色的人文旅游资源；以尧人山国家森林公园和都柳江风光带为代表的自然旅游资源以及以水族文化为代表的人文旅游资源，不仅品位较高，而且组合合理，互补性强，使得旅游资源相得益彰，适宜旅游产品的深度开发。

三、人文资源特色突出

《三都水族自治县旅游发展规划》调查显示三都县五级旅游资源三处，均为人文旅游资源，四级旅游资源十七处，其中十一处为人文旅游资源，从数据上说明三都人文旅游资源占据优势；另外从资源特色上看三都县的人文资源多数为水族文化资源，三都县作为全国唯一的水族自治县，使得原生态的水族文化资源具有垄断性、独占性，特色突出。

5. Shui Dwellings

Traditional dwellings of Shui people are houses with the Ganlan style, which have two or three floors with the fourth floor as the highest. The houses are built with China fir and roofed with fir skin or tiles. Shui people feed livestock and poultry downstairs and live upstairs. The number of the rooms is odd as they are forbidden to use the even numbers. There may be three rooms, five rooms or seven rooms.

6. Shui Costumes

The middle aged women of Shui people usually wear the half-long garment with big front over knees, black cloth trousers, and black cloth belly band. The cuff and the bottom of trousers are embroidered with laces.

7. Shui Music

Folk music of Shui people boasts a unique national style, especially folk songs with a great variety of songs like couple song, single song, dou song, etc.

The instruments of Shui people are mainly bronze drum, big leather drums, Lusheng, suona and huqin, among which, bronze drum and big leather drum are most popular.

8. Shui Dance

Shui is a nation skilled at singing and dancing. The traditional dances of Shui people mainly have timbale dance, wrestling dance and Lusheng dance.

Section Ⅱ Resource Assessment

Ⅰ Various Types of Tourist Resources

According to the survey conducted by Tourism Development Plans of Sandu Shui Autonomous County, the main resources in the county cover the eight major types specified in Classification, Survey and Assessment of Tourism Resources of P. R. China and National Standards; involve 25 subtypes, accounting for 80.65% in 31 subtypes in the country. The basic type accounts for 38.70% of total 155 basic types, together with 471 principle objects of tourism resources. Some other variety of resources are relatively abundant except a relatively few number of astronomer and climate landscape resources.

Ⅱ Relatively Reasonable Resource Profiles

Tourist resources in Sandu County are relatively rich with both abundant natural tourist resources and humane tourism resources with local characteristics. Natural tourist resources represented by Yaoren Mountain national forest park and Duliu River scenery as well as human tourist resources represented by culture of Shui people are not only endowed with high tastes but also enjoys reasonable profile and strong complementarities, which has brought out the best in one another and fitted profound exploitation in tourist products.

Ⅲ Outstanding Features of Human Resources

According to the survey conducted by Tourism Development Plans of Sandu Shui Autonomous County, three tourist resources at level-five are human tourist resources. There are seventeen tourist resources at level-four, among which, eleven are human tourist resources. This data shows that human tourist resources in Sandu enjoys great advantages, In addition, viewed from resources characteristics, most of Sandu County human resources belong to cultural resources of Shui people. Sandu County, as the only Shui Autonomous County in China, has the monopolistic, exclusive and unique original ecological cultural resources of Shui nationality.

第三章 形象定位及总体目标

一、战略定位

中国水族文化中心
黔南特色旅游集散中心

二、形象定位

凤羽三都，心随水动

“凤羽三都”表现的是三都的美丽。三都是像凤凰羽毛一样美丽的地方，它的美主要从两个方面体现：首先是自然美。旖旎的都柳江，浩瀚的尧人山国家森林公园，瑰丽多姿的中和瀑布，雄伟壮丽的雪花洞，神秘莫测的产蛋岩，闻歌起舞的风流草……美得秀丽，美得壮阔、美得奇特。其次是人文美。神秘莫测的水书，世界最长的节日——水族端节，东方情人节——水族卯节，华美精致的水族马尾绣，庄严神圣的水族习俗……美得深奥，美得浪漫，美得厚重。用“凤羽”形容三都，既包含了三都与凤凰之间的美丽传说，彰显了三都的文化内涵，又不着痕迹地表现了三都的美丽，使三都形象生动唯美，特色鲜明。

“心随水动”从意境上强调三都核心价值——水族文化的强大感染力。心动于水书的神秘莫测，心动于马尾绣的精湛工艺，心动于端节的悠长氛围，心动于卯坡的依依情愫，心动于水族同胞的热情好客，心动于水族历史的古朴厚重……每一次与水族文化的接触，都能触动到旅游者的心灵深处；每一次与水族文化的接触，都能使水族文化渗透到旅游者的血液骨髓，这便是三都核心价值的体现，是三都独一无二的主题形象。

除此之外，此“水”还包含另一层意思，即三都水族自治县的母亲河——都柳江。旖旎的都柳江同样会使旅游者频频心动：心动于都柳江醉人的山光水色，心动于都柳江两岸的历史人文，心动于都柳江蕴藏的神秘氛围（风流草、产蛋岩、晴雨石等）……

总之，来到像凤凰羽毛一样美丽的三都，便会怦然心动——为了神秘古朴的水族文化，为了秀丽多姿的都柳江。

三、目标定位

通过对三都县县城沿街建筑改造，从视觉上塑造水族文化旅游品牌，使三都水族文化旅游的形象初具雏形，形成贵州多民族特色文化旅游的有机构成部分。

通过对三都水族文化的深入挖掘，将三都打造为“中国水族文化中心”。

进一步丰富三都旅游内容，并利用水族文化中心的影响力，形成水族文化旅游品牌，最终将三都县打造成为国内知名，涉足境外，具有水族特色，辐射周边县市的黔南旅游集散中心。

Chapter III Image Orientation and Overall Objective

I Strategic Orientation

Chinese center of Shui culture
Qiannan Characteristic travel logistic center

II Image Orientation

Phoenix fluttering in Sandu and heart moving with the water

"Phoenix fluttering in Sandu" demonstrates the beauty of Sandu. Sandu is a place as beautiful as the feather of phoenix, which embodies from two aspects. The first aspect is its natural beauty: Charming Duliu River, vast national forest park on Yaoren Mountain, magnificent waterfalls, splendid Xuehua cave, mysterious egg cliff, dancing Merry grass...it is fascinating, grandiose and fabulous. The second aspect is human beauty: mysterious Shui shu, the longest festival in the world-Duan festival, Eastern Valentine's Day-Mao festival, exquisite Mawei Embroidery, solemn and sacred Shui customs...it is profound, merry and dignified. By using "Phoenix fluttering" to depict Sandu, it not only includes the beautiful legendary between Sandu and phoenix and presents cultural connotation of Sandu but also demonstrates the beauty of Sandu, which makes Sandu county vivid and lively with distinct characteristics.

"Heart moving with the water" emphasizes the core value of SANDU from artistic conception-the strong infectivity of Shui culture. The heart moves with mysterious Shui character, with long atmosphere of Duan Festival, with sincere feeling of Mao Festival, with enthusiasm and hospitality of Shui fellowmen, with simplicity and dignity of Shui history...each time, the contact with Shui culture can touch the bottom of tourists' heart; each time, the contact with Shui culture can make Shui culture penetrate into the blood and bones of tourists. All these are the core value as well as a unique theme image of Sandu.

In addition, this "water" also includes another meaning, that is, the mother river of the Sandu Autonomous County-Duliu River. Graceful Duliu River can also make tourists'hearts moved: heart moving with intoxicating landscapes of Duliu River, with historical figures on both sides of Duliu River, with its mysterious atmosphere (Merry grass, egg cliff, barometer stone, etc.)...

In a word, the visit to Sandu as beautiful as the feather of phoenix can make one's heart palpitating with excitement-for mysterious and simple Shui culture, for graceful and charming Duliu River.

Ⅲ Target Orientation

To shape Shui cultural tourist brand visually by reconstructing the buildings along street in Sandu County which enables it to take the shape and form an organic part of characterized cultural tourism of multi-ethnic group in Guizhou province.

To further enrich the tourist content of Sandu, make good use of the influence of Shui cultural centre, and form Shui cultural tourist brand-all these will make Sandu a domestic well-known, overseas involved, Shui characterized, as well as surrounding radiated travel logistic center in Qiannan.

总平面图

设计说明：

本规划以甲方提供的CAD地形图为基础，将规划区域分为老城区北出口安置区、老城区沿街立面改造区、老城区规划区、麻光三郎新区、南出口片区和环城北路片区六大区域。由于水文资料、地勘资料的不全，本规划方案的防洪堤、拦河坝等防洪设施的设计仅作参考，不作后期施工图的依据。

第四章 功能布局

Chapter IV Functional Layout

一、老城区北出口安置区

北出口片区是"水族迎宾街"，是体现水族风韵，展示水族文化的主要窗口，该区主要为拆迁安置片区。

Ⅰ Old Wown North Exit Settlement Area

North exit area is the "Shui nationality Yingbin Street", a main window embodying the charm of Shui nationality and showing the Shui culture of Sandu County. It is mainly an area of settlement.

二、老城区沿街立面改造区

老城区沿街立面改造是用科学的规划规范原有建筑，使其突出城市街道特色，呼应城市定位，提升城市人文历史内涵，增加城市的都市品质。将原有的"散、乱、差、杂"通过改造以适应城市发展的需要。

Ⅱ Old Town Street Facade Reconstruction Area

Old town facade reconstruction employs scientific planning to standardize the original buildings to highlight the characteristics of city streets, respond to the developmental orientation, enhance cultural and historical connotations, and increase the city's urban quality. The original "scattered, chaotic, poor and mixed" phenomena were changed to meet the needs of urban development.

三、一江两岸片区 Ⅲ Riverbanks Area

该片区包括河滨北路、河滨南路全段，是沿都柳江重点打造的两岸滨江景观带，该片区集商业步行街、高尚生活小区、绿化景点、休闲健身于一体，是未来三都县城生活的一个重要城市节点。

The area includes the whole North and South Riverside Road. It is a riverside landscape zone along both sides of the Duliu River. There are commercial pedestrian street, high living area, green spots, and entertainment and fitness places in this area. In the future, it will be an important urban living area in Sandu County.

四、麻光三郎新区 Ⅳ Maguang Sanlang New Area

麻光三郎新区是三都的城市拓展区，是三都城市发展的又一亮点。它集城市功能（歌剧院、综合大楼）、五星级休闲度假酒店、民族风情街及都市高端小区为一体。

Maguang Sanlang new area is an expansion area of Sandu and it is another bright spot of urban development. It combines the urban functions (opera houses, complex buildings), five-star leisure hotels, national customs streets and high-end urban residential districts.

功能结构分析图

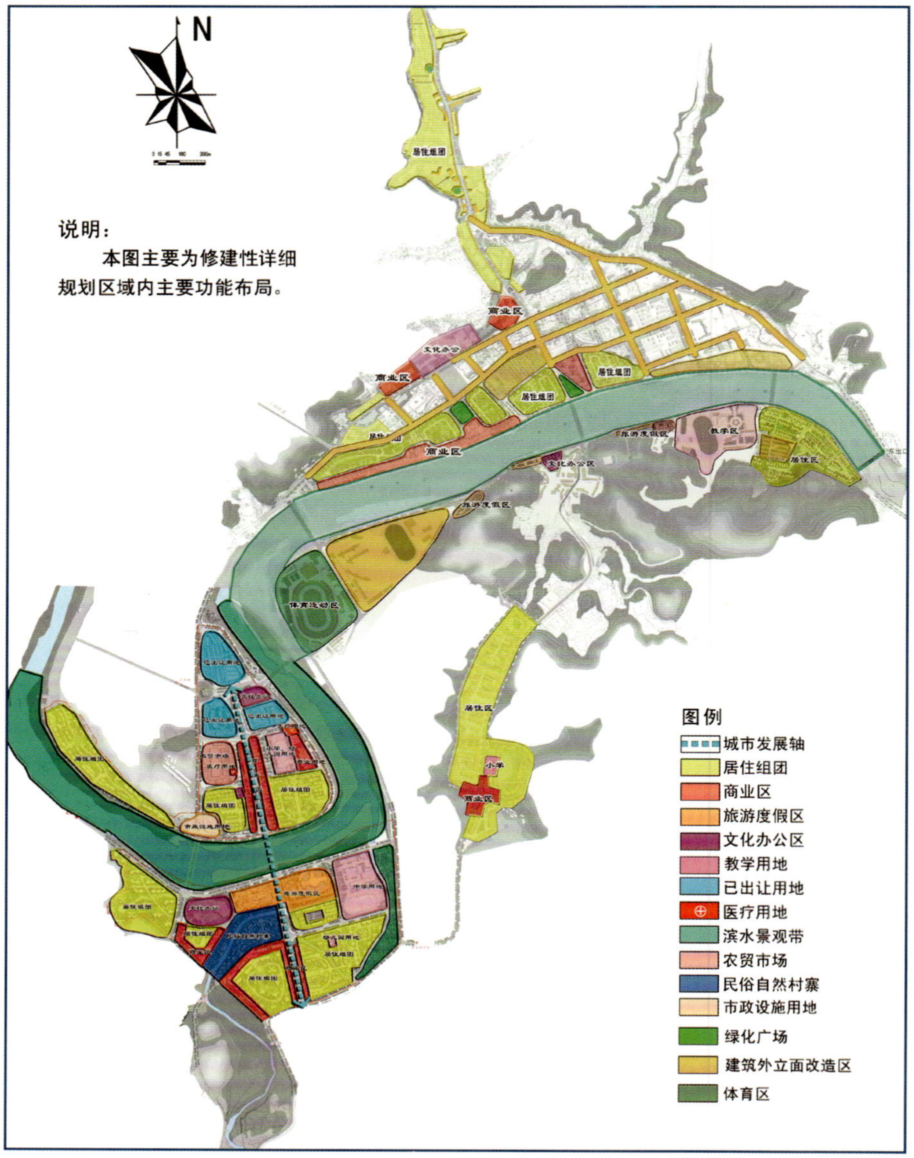

五、南出口片区

根据南出口片区的区位条件、土地开发利用程度，将南出口片区定位为三都县城住宅新区。

六、环城北路片区

在现有布局的基础上，规划配置相关附属设施，完善片区功能，包括商业中心、医院以及商住等设施，并配合周围的山体景观形成山、城一体的立体景观格局。

V South Exit Area

According to the regional conditions and the land utilization degree, the south exit area is orientated the new residential district of Sandu county.

VI North Ring Road Area

Based on the existing layout, we will plan and configure the related auxiliary facilities and improve its functions, including business centers, hospitals and residential facilities, and have the area match the surrounding mountain landscape to form a mountain-city integrated solid landscape.

公共服务设施规划

第五章 分期规划

Chapter V Planning Stages

根据 2009 年 4 月 14 日《三都水族自治县建设局关于对接县城规划设计有关事宜的函》和《三都水族自治县县城城市规划设计合同》的内容，按委托方的要求而做的分期规划，具体内容为：

According to the Letter of the Planning Matters on Connecting the County Town issued by the Construction Bureau of Sandu Shui Autonomous County on April 14, 2009 and the Urban Planning and Design Contract signed by Sandu Shui Autonomous County, and the requirements of the entrusting party, the planning stages can be divided into the following:

一期规划

Planning Stage I

1. 内容：北出口片区、河滨南路、河滨北路西段（河堤往北 38m）、南京东路两侧建筑外观、博物馆设计。

1. Contents: Designing the building exteriors of the North Exit Area, the South Riverside Road, the western part of the North Riverside Road (38 meters going north to the riverbank) and the East Nanjing Road and the museum.

2. 目的：实现都柳江两岸及北出口的打造。

2. Aims: Constructing the sides of the Duliu River and the North Exit.

二期规划 Planning Stage II

1. 内容 Contents:

(1) 修建性详细规划片区：河滨北路西段连萝卜寨、一江两岸、东出口片区、南出口片区、河滨南路片区、麻光至水文站片区、三郎片区、中营片区、抚水路、能建公园可视坡头。

The detailed constructive planning areas: the western part of the North Riverside Road with radish Village, the two Riverbanks, the East Exit Area, the South Exit Area, South Riverside Road Area, the area from Maguang to the Hydrological Station, the Sanlang Area , the Zhongying Area, the Fushui Road, and the visual slope of the Nengjian Park.

(2) 旧城区风貌改造：南京西路、中山路中段（公安局至种子公司段）、都江路、深圳路、团结路、振兴路、民族路、市西路（商贸路）、都尚路（含规划拆迁路段）、三合路（解放路至南京西路）、凤凰路南段、原大操场处、上坝小区（搬迁安置地）、文化路、环城东路、河滨北路东段、北出口页岩砖厂至交梨磅房路段。

The old town reconstruction: the West Nanjing Road, the middle Zhongshan Road (from the Public Security Bureau to the Seeds company), the Dujiang Road, the Shenzhen Road, the Tuanjie Road, the Zhenxing Road, the Minzu Road, the Shixi Road (the Shangmao Road), the Dushang Road (including the section planned to remove), the Sanhe Road (from the Jiefang Road to the West Nanjing Road), the southern section of the Phoenix Road, the original big playground, the Shangba residential district (settlement area), the Wenhua Road, the East Ring Road, the east part of the North Riverside Road, the North Exit shale brick factory and the Bangfang Road of Jiaoli town.

2. 目的 Aims:

实现三都县整体打造。

Constructing the whole Sandu County.

分期规划图

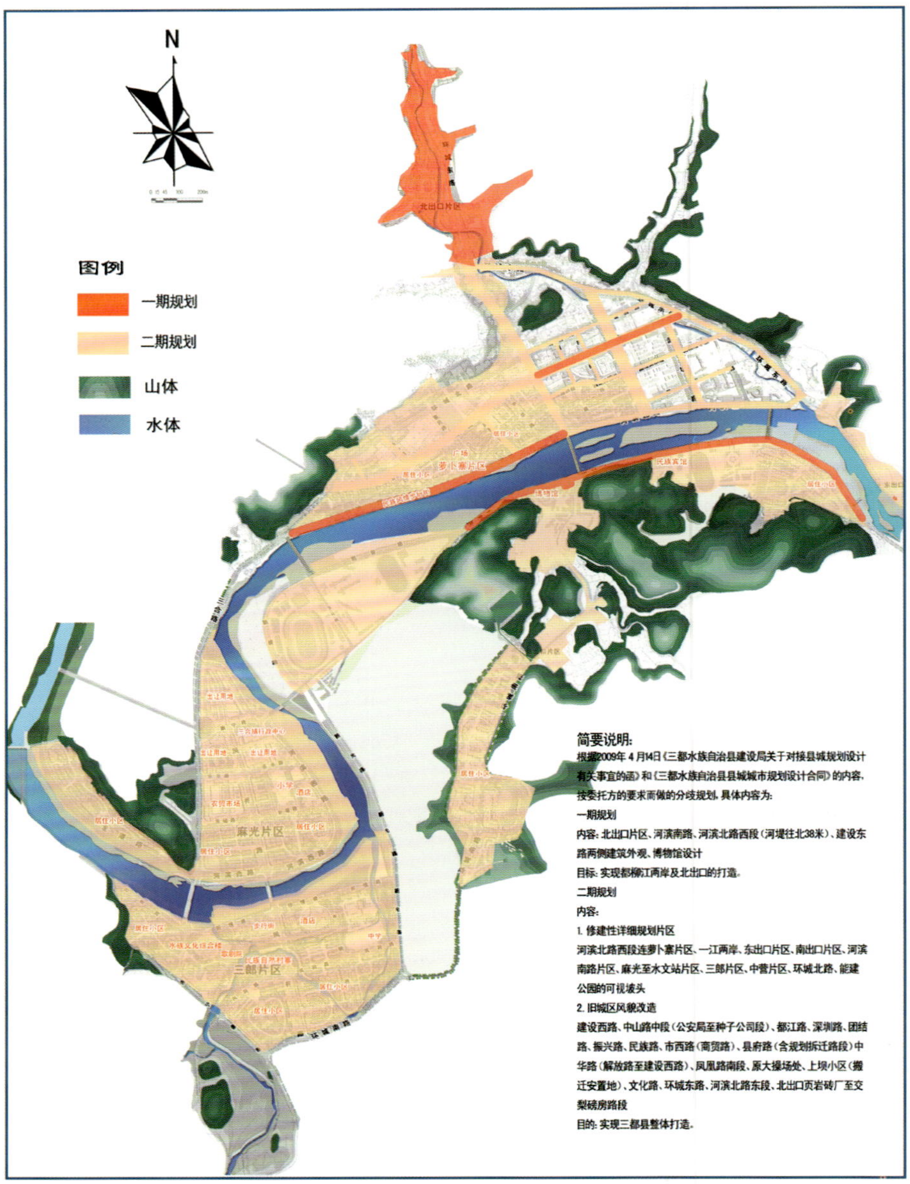

第六章 旅游线路规划

旅游线路能够将境内境外的旅游资源以及旅游服务串联起来，是连接旅游者与当地旅游的纽带，好的旅游线路的设计能够激发旅游者的游兴，带动旅游业的发展。

一、旅游线路规划原则

1. 特色原则

旅游线路的设计要体现特色。形象鲜明，富有特色的旅游线路会对旅游者构成较大的吸引力，因此要充分挖掘区域内文化内涵，分析区域内旅游特色，才能形成区域内的特色品牌。

2. 开放性原则

三都县范围较小，旅游景区景点，尤其是高品质的景区景点游线，在设计中要遵循开放性原则，在完善区内旅游线路的同时，设计出与区外旅游设计的接口，将县域旅游线路与跨县域背景下的旅游线路结合在一起。特别是要将县域内旅游线路纳入到相邻地区开发较为成熟、有一定知名度的旅游线路当中去，将会有利于三都县旅游业的发展。

3. 顺序与节奏原则

在旅游线路的设计中，必须充分考虑旅游者的心理和体力、精力状况，据此安排线路的结构顺序与节奏，做到动静结合、快慢结合。

4. 内容的互补原则

旅游线路的设计中，既要体现旅游主题的内容，又要做到在旅游活动项目上的多样化，以丰富旅游者的旅游生活和旅游体验。

二、区域旅游线路

1. 三都两日游

D1：早上由贵阳乘车前往三都县城，在三都享受午餐，下午参观三都博物馆，之后前往河滨北路步行街享受都柳江风光以及水族风情，在此享受水族特色餐饮；晚上前往三都歌剧院欣赏史诗晚会《远古走来的贵族》，宿三都五星级“葫芦酒店”。

D2：早上前往尧人山国家级森林公园游览，观赏尧人山原始森林的浩瀚壮阔，观赏勇猛彪悍的尧人山斗鱼，寻访神秘的“风流草” 之后前往水各大寨，在此享受原生态水族餐饮，体验水族歌舞、水族工艺、水族卯坡，享受“东方情人节——卯节”举办地——卯坡的浪漫气息，之后乘车返回贵阳。

Chapter VI Tourist Route Planning

Tourist routes can link the inside and outside tourism resources and services together. It is a tie to connect tourists and the local tourism. Good tourist routes are designed to stimulate the tourists' interests and promote the development of tourism.

I The principles of the Tourist Route Planning

1. The Principle of Characteristics

Tourist routes should reflect the characteristics of the county. Vivid and distinctive tourist routes will be more attractive to tourists, so the tourist brand with regional characteristics can be formed only by fully exploring the regional cultural connotations and analyzing the tourist attractions within the region.

2. The Principle of Openness

With small size of Sandu County and a few tourist scenic spots, especially high-quality ones, the design of tourist routes should follow the principle of openness. While the local tourist routes are improved, they are designed to link the routes outside, so the county's tourist routes and trans-county's ones are combined together. In particular, the tourist routes within the county should be incorporated into the mature and well-known tourist routes in the adjacent region, which will benefit the tourism of Sandu County.

3. The Principle of Sequence and Rhythm

During the design of the tourist routes, the travelers' status of psychological, physical and energetic strength must be taken full account of. The sequence and rhythm of the tourist routes are designed to be static and dynamic, fast and slow.

4. The Principle of Complementation

The tourist routes are designed to not only reflect the contents of the tourist theme but also diversify the tourist activities so as to enrich the tourists' life and experience.

II The Regional Tourist Routes

1. Two-day Tour of Sandu

D1: Travelling from Guiyang to the Sandu County by bus in the morning and having lunch in Sandu; visiting the Sandu Museum in the afternoon and then going to the Pedestrian Street of the North Riverside Road to visit the sceneries of the Duliu River, feel the customs of the Shui Nationality and enjoy the food and drink with the national characteristics; and watching the epic evening—"the Nobility from the Ancient Times" in the Sandu Opera House and then living in the Hulu Hotel, a five-star hotel in Sandu.

D2: Travelling to the Yaoren Mountain National Forest Park to visit the vast primeval forests of the Yaoren Mountain, watch the brave rumble fish fight, and seek the mysterious "merry grass"; going to the Shuige Village to enjoy the original ecological meals, dances and crafts of Shui nationality and the Merry charm of the Mao Slope, a place to celebrate the oriental Valentine's Day-Mao Festival; and then driving back to Guiyang.

2. 三都三日游

D1：早上由贵阳乘车前往三都县城，在三都享受午餐，下午乘船参观都柳江风景名胜区，晚上前往三都歌剧院欣赏史诗晚会《远古走来的贵族》，宿三都五星级“葫芦酒店”。

D2：早上乘车前往尧人山国家级森林公园游览，观赏尧人山原始森林的浩瀚壮阔，观赏勇猛彪悍的尧人山斗鱼，寻访神秘的“风流草”；中午在风景区享受午餐，下午参观三都县城，欣赏充满水族特色的沿街建筑；之后游览河滨北路步行街享受都柳江风光以及水族风情，在此享受水族特色餐饮；宿三都五星级“葫芦酒店”。

D3：早上乘车前往产蛋岩，之后前往水各大寨，在此享受原生态水族餐饮，体验水族歌舞、水族工艺、水族卯坡，享受“东方情人节——卯节”举办地——卯坡的浪漫气息，之后乘车返回贵阳。

三、跨区域旅游线路

贵阳—三都—榕江

贵阳—都匀—独山—三都

贵阳—丹寨—三都—榕江

贵阳—丹寨—三都—荔波—广西

四、主题旅游线路

1. 三都水族文化游

游线内容：参观三都县城，欣赏充满水族特色的沿街建筑；参观三都博物馆，了解三都人文风情；游览三都河滨北路步行街，品味原汁原味的三都生活；体验原生态水寨风情，欣赏水族歌舞表演，观赏体验水族工艺品的制作流程，切身感受三都水族的独特魅力。

2. 世界最长的节日——水族端节游

游线内容：每年农历 8 月的第一个“亥”日开始至 10 月的“亥”日结束，体验水族祭祀仪式，体验水族新年菜肴，体验水族端节赛马等水族新年活动。

3. 都柳风情游

游线内容：游览都柳江风景名胜区，沿岸观赏都柳江如画美景，观赏都柳江小溪摩崖，游览三都河滨北路步行街，品味原汁原味的三都生活。

4. 尧人山原始森林游

游线内容：观赏尧人山原始森林的浩瀚壮阔，观赏勇猛彪悍的尧人山斗鱼，寻访神秘的“风流草”、“产蛋岩”。

2. Three-day Tour of Sandu

D1: Travelling from Guiyang to the Sandu County by bus in the morning and having lunch in Sandu; visiting the sceneries of the Duliu River; and watching the epic evening—"the Nobility from the Ancient Times" in the Sandu Opera House and then living in the Hulu Hotel, a five-star hotel in Sandu.

D2: Travelling to the Yaoren Mountain National Forest Park to visit the vast primeval forests of the Yaoren Mountain, watch the brave rumble fish, and seek the mysterious "merry grass" in the morning; having lunch in the scenic spot; visiting the Sandu county town and the street buildings with the Shui national characteristics in the afternoon; going to the Pedestrian Street of the North Riverside Road to feel the customs of the Shui Nationality and enjoy the food and drink with the national characteristics; and living in the Hulu Hotel.

D3: Travelling to the Egg cliffs in the morning and going to the Shuige Village to enjoy the original ecological meals, dances and crafts of Shui nationality and the Merry charm of the Mao Slope, a place to celebrate the oriental Valentine's Day—Mao Festival; and then driving back to Guiyang.

Ⅲ Cross-Boundary Tourist Routes

Guiyang—Sandu—Rongjiang

Guiyang—Duyun—Dushan—Sandu

Guiyang—Danzhai—Sandu—Rongjiang

Guiyang—Danzhai—Sandu—Libo—Guangxi

Ⅳ Theme Tourist Routes

1. Sandu Shui National Cultural Tour

Tourist content: visiting Sandu county town to enjoy the street buildings with Shui national characteristics; visiting the Sandu museum to understand the local culture; walking on the Pedestrian Street of the North Riverside Road to enjoy the original taste of Sandu life; experiencing the original ecological lifestyle of the Shui villages, enjoying the dance performances, arts and crafts of the Shui nationality so that to feel the unique charm of the Sandu Shui nationality.

2. The Longest Festival in the World-Duan Festival Tour of the Shui Nationality

Tourist content: watching the New Year activities such as ritual events, horse racing, dishes making of the Shui nationality for the first "Hai" day in lunar August to the "Hai" day in October every year.

3. Duliu Tour

Tourist content: visiting the Duliu scenic spots, enjoying the picturesque coastal scenery of the Duliu River, and the Duliu creek and clifts, visiting the Pedestrian Street of the North Riverside Road, and feeling the authentic taste of Sandu life.

4. Yaoren Mountain Primeval Forests Tour

Tourist content: visiting the vast primeval forests of the Yaoren Mountain, watching the brave rumble fish, and seeking the mysterious "merry grass" and the "Egg cliffs".

第七章 旅游营销规划

围绕三都水族自治县定位“凤羽三都 心随水动”，打造“中国水族文化遗产旅游目的地”，我们建议三都实行分阶段、分地源市场进行推广。建议推广营销形式为：

1. 人员促销

派促销人员到重点的目标市场举办旅游展览会，使目标市场及时了解景区，增加信任感。促销人员与重点客源市场的代表性单位、团体、企业、家庭建立长期的联系关系。

2. 广告

这种促销方式可以比较快速和有效地将三都的旅游产品形象传播给大众，提高知名度。利用报业集团的优势，发挥新闻报道作用，加强舆论引导。

具体可以通过电视、广播、报纸、杂志、户外广告、网络等广告媒体，快速、有效地将三都旅游景区的形象传播给大众。

3. 公共关系

通过不同形式的公共手段，树立城市的品牌形象：

宣传性公关（针对新闻界）。向新闻界（报纸、电视、杂志、广播）提供宣传资料；接受记者采访；召开记者招待会、新闻发布会；进行专题报道；提供新闻性文章、新闻图片等。

社会型公关。依托于政府、行业，争取建立“老年健身基地”、“环保教育基地”等公益形象，深化品牌建设的内涵。

交际型公关。出版三都旅游景区宣传册；举办促销推介活动。

4. 营业推广

以优惠酬宾活动，价格折让优惠，吸引旅游者批量进入，加速提升旅游知名度，造成涉及面广、影响消费力的形象推广效应。

5. 旅游节事活动

邀请有关领导剪彩，举办庆祝活动。在每一年的恰当时间，举办富有特色的旅游节事活动。

6. 网络促销

加强网络建设，发挥网络优势，开展网上互动活动，开辟新的宣传渠道。通过网络向潜在的旅游者全方位地传播三都的旅游品牌形象。

7. 销售激励

(1) 向游客赠送旅游地纪念品；

(2) 旅游淡季时降价销售、奖励销售；

(3) 对旅游代理商和批发商进行销售奖励。

Chapter VII Tourist Marketing Planning

With the developmental orientation of the Sandu Shui Autonomous County, "Phoenix fluttering in Sandu, heart moving with the water", creating a cultural heritage tourist destination of Chinese Shui nationality, we propose to promote Sandu by different stages and resource markets. The marketing ways are as follows:

1. Staff Promotion: The promotional staff organizes tourism exhibitions in the key target markets to acquaint them with the scenic spots and increase their trust. The promotion staff builds long-term association with the representative units, organizations, businesses and families in the key target markets.

2. Advertisements: It is more rapid and effective to make the public know of the images of Sandu tourist products so as to raise public awareness by this way. We can take the advantages of newspaper groups to play the role of news reporting and strengthen public opinion guidance.

To be specific, the images of Sandu scenic spots can be rapidly and effectively transmitted to the public through the advertising media such as television, radio, newspapers, magazines, outdoor advertising and internet.

3. Public Relations: Establishing the urban brand image through various forms of public means:

Propaganda-based public relations (for mass communication): providing publicity material to the media (newspapers, television, magazines, and radio); accepting interviews; holding press conferences, news conferences; and special reports; providing news articles and news pictures; and so on.

Society-based public relations: relying on the governments and industries to establish the public images such as "old age fitness base", "environmental education base" and so on to deepen the connotations of the brand building.

Communication-based public relations: publishing Sandu tourist brochures and organizing promotional activities.

4. Sales Promotion: By the special offer and price discount activities, attract bulk tourists to flow in so as to speed up to enhance tourism awareness and create an image promotion effect widely influencing consumption.

5. Tourist Festival Activities

Invite the leaders to attend the ribbon-cutting ceremony during the celebrations. Organize distinctive tourist festival activities at the appropriate time every year.

6. Network Marketing: Strengthen the network construction and take the advantages of the network to hold on-line interactive activities and create new communication channels. All-dimensionally spread the Sandu tourist brand image to the potential tourists through the network.

7. Sales Incentive

(1) Present the Tourist souvenirs to the tourists;

(2) Price reduction and incentive sales during the off season;

(3) Give awards to the tourist agents and wholesalers.

第八章 项目可行性分析

第一节 效益分析

一、生态效益分析

进入 21 世纪，现代社会所有领域的发展几乎都与生态环境质量密不可分，生态环境对于旅游城市打造的重要性更是不言而喻。因此，三都旅游城市打造从一开始就是站在生态保护的基础上来进行的，而不是走传统城市先污染后治理的老路，生态效益将会收效明显。

1. 促进区域生态环境的进一步改善

为了保证城市有足够的旅游吸引力，在三都旅游城市打造中，政府、企业、市民等社会各界在他们活动的过程中，将会更加重视生态环境保护，并通过制定相应的地方法规，采取切实可行的措施来加大环保力度，这将对三都的生态环境产生正向效应，促进区域生态环境的进一步改善。

2. 加快城市生态文明建设的进程

在生态环保原则指导下的三都旅游城市打造，无论是建筑改造、景观塑造、城市功能布局还是旅游区规划设计，都把凸显其适宜的气候环境、原生态的自然资源、良好的人居环境作为重要任务，充分运用城市这一载体，强化生态文明的建设力度，使旅游发展和城市发展形成相互促进的合力，从而加快城市生态文明建设的进程。

二、经济效益分析

三都旅游城市打造是以旅游业的发展为先导的城市发展战略，目的是通过发挥旅游业的联动作用来推动城市社会经济的发展，其对城市经济发展的促进作用十分巨大。

1. 营造良好的招商引资环境，吸引投资

一个城市的区位是否有利、资源是否丰富、城市环境是否舒适、交通是否便利是衡量城市投资环境的重要指针。三都位于贵州南部综合旅游区的黄金地段，优良的区位优势、优质的自然资源、优美的山水风光和底蕴深厚的水族文化资源，本都是吸引投资者的有利条件，但落后的基础设施却让一些投资者望而却步，制约了其资源优势的发挥。目前三都对外交通主要依赖于 321 国道和 206 省道，路况较差，可进入性不强，城市内部公共交通系统也尚未完善；住宿设施总量不足，而且档次较低；城市建筑和景观都未能体现出独一无二的水族文化风貌，缺乏三都特色和持久的吸引力。鉴于此，三都旅游城市打造以改善城市形象为出发点，从确定城市发展方向、城市功能配套、城市景观重塑、城市街面改造入手，完善公共基础设施，提升城市的综合素质，创造良好的招商引资硬环境，必将吸引更多的投资者来三都投资兴业。

Chapter VIII Feasibility Analysis of the Project

Section Ⅰ Benefit Analysis

Ⅰ Ecological Benefit Analysis

In the 21st century, the development of all areas of modern society almost inextricably links with environmental quality, and ecological environment is of great importance to build a tourist city. Therefore, the Sandu tourist city is built on the basis of ecological protection, rather than the old path of pollution first and treatment later, from the start. The ecological benefit will be obviously high.

1. Promote Further Improvement of the Regional Ecological Environment

In order to ensure the city has enough tourist attraction, more attention to environmental protection will be paid in the activities of the government, enterprises, public and other community in various sectors during building Sandu tourist city. The related local laws and regulations are established and practical measures are taken to step up environmental protection efforts, which will produce positive effects on the ecological environment of Sandu and improve the regional ecological environment.

2. Speed up the Process of Urban Ecological Civilization Construction

With the principle of ecological environmental protection, Sandu tourist city construction, the building renovation, landscape shape, the layout of urban functions or the planning and design of tourist area, regards the highlight of the suitable climatic conditions, the original ecology of natural resources, good living environment as an important task. With full use of the city, the ecological civilization construction is strengthened to form joint forces between tourism and urban development so as to speed up the process of urban ecological civilization construction.

Ⅱ Economic Benefit Analysis

The Sandu tourist city construction is a tourism-led urban development strategy, with the intention to play the role of tourism in promoting the socio-economic development of the city, which is very significant to urban economic development.

1. Create a Good Investment Environment to Attract Investment

The condition of geography, resources, urban environment and transportation is the important indicator of the city's investment environment. Sandu County is located in golden section of the comprehensive tourist areas in southern Guizhou. Its excellent location advantage, high quality natural resources, beautiful landscapes and Shui national cultural resources should be the favorable conditions to attract investors, but poor infrastructure deters some investors and constraints to develop the resources advantages. Currently, the roads to outside world are mainly the 321 national highway and 206 provincial highway, poor road conditions and hard to access, and the public transport system inside is also to be improved; the accommodation facilities are insufficient and poor; the urban architectures and landscapes do not reflect the unique cultural style of Shui nationality and lack of Sandu characteristics and lasting appeal. So, the Sandu tourist city construction takes the improvement of urban tourist image as a starting point. The public infrastructure and the overall quality of the city will be improved to create a good hard investment environment through determining the direction of urban development, complementing the urban facilities, reshaping the urban landscapes, and reconstructing the urban street facade, which will attract more investors to Sandu.

2. 能促进旅游业的发展，带动区域产业结构优化升级

三都旅游城市打造是以旅游业发展为先导的城市发展战略，旅游业是旅游城市打造的核心，因此，必将促使当地下大力气塑造全新的旅游形象，提高其知名度和美誉度，增强旅游从业人员素质，改善旅游接待设施和服务，从而使当地旅游业得到大发展。"百业扶旅游，旅游促百业"。当旅游业得到大发展后，不仅能带动住宿业、餐饮业、交通运输业等相邻第三产业的发展，而且还能向第一产业和第二产业辐射，将大大加速各行各业的大交流、大协作，从而促进国民经济各部门的发展，带动区域产业发展由传统的一、二、三的结构向三、二、一的结构优化升级。

3. 增加就业机会，拉动城市内需

旅游业是一个劳动密集型的服务性产业，也是一个对劳动力素质要求多元化的产业。一方面，旅游业的许多岗位不需很高的技术，只需短期培训就可以很快胜任，因而为刚步入社会的青年人、下岗职工、进城务工农民提供了门槛较低的就业机会。另一方面，旅游业的就业层次较多，既有高水平科技人员发挥作用的空间，又有低层次人才展现才能的舞台。随着就业机会的增加，当地居民的工资收入不断提高，其消费需求也随之增加。由于部分工资收入用于购买本地生产的商品和提供的服务，进一步刺激本地经济活动的扩大，这又使得相关企业的营业量得以扩大，并导致收入和就业机会的进一步增加，不断创造出更多的经济效益，形成社会经济良性循环的发展趋势。

三、社会效益分析

1. 有助于民族文化的保护和传承

三都县作为全国唯一的水族自治县，其古朴神秘的水族文化在民族文化旅游资源中具有垄断性和独占性，早已被世人视为民族珍宝，但是它的开发与保护一直没有深度开展起来。三都旅游城市打造将会为水族文化的保护和传承发挥巨大的作用。随着旅游城市发展目标的确立和旅游业的发展，濒临湮灭的历史建筑又得到维护和管理，当地一些被人们遗忘的民族传统习俗和文化活动又得到开发和恢复，传统的手工技艺因市场需求的扩大又重新得到发展，传统的音乐、舞蹈、戏剧等非物质文化遗产又受到重视和发掘。城市成为展现民族文化的重要载体，将民族符号展现在建筑、街道、广场上，使原来几乎被遗忘的文化遗产获得了新生，有利于民族文化的保护和传承。

2. 有利于城镇化水平的提高

城镇化水平是反映区域社会发展水平的一个重要指标，用非农业人口（城镇人口）占总人口的比重来衡量。由于资源优势没有得到充分发挥，经济发展落后，三都城镇化水平一直处于全省末位。随着旅游城市规划的实施，就业机会增加，由于旅游业相对于农业的高回报率，越来越多的农业人口将会转而从事住宿、餐饮、导游、交通客运等与旅游相关的服务业，促使第一产业从业人数的比例随之下降，第三产业就业人数增加，这就意味着非农业人口比例将会迅速上升，三都城镇化水平将会得到大幅提高。

2. Promote the Development of Tourism and Regional Industrial Upgrading

The Sandu tourist city construction is a tourism-led urban development strategy, with the core of tourism. Therefore, it is necessary to make great efforts to shape the new tourist image, enhance the fame and reputation, improve the quality of tourist staff, and better the tourist reception facilities and services, which can make great development of local tourism. "All other industries promote tourism, and tourism promotes all other industries". the great development of tourism can propel not only the development of the related tertiary industries such as accommodation, catering, transportation and other industries but also the development of the first and secondary industries and accelerate the exchange and cooperation between various industries so as to promote the development of national economic sectors and the structural optimization and industrial upgrading from the traditional first-second-third structure to the third-second-first one.

3. Increase Employment Opportunities and Stimulate Urban Domestic Demands

Tourism is a labor-intensive service industry and an industry with plural requirements of labor quality. On the one hand, many tourism jobs do not need high technology and one can be a qualified worker after a short-term training. So, tourism offers employment opportunities with a low threshold for those young people just entering society, laid-off workers, and migrant farmers. On the other hand, there are many levels of tourism employment, with a space for high-level scientists and engineers to play great role and a stage for low-level talented people to show their talent. With the increased employment opportunities, the wages and consumption demands of the local residents are rising. As part of the wages for the purchase of locally produced goods and services, the local economic activities expand, which in turn enables the enterprises to expand trade volume and leads to a further increase in income and employment opportunities. So, more economic benefits continue to be created and a virtuous cycle of social and economic trends will be formed.

Ⅲ Social Benefit Analysis

1. Protect and Inherit the National Culture

As the only Shui Autonomous County in China, the mysterious Shui culture of Sandu County is exclusive and monopolistic. It has long been regarded as national treasures of the world, but its development and protection has not been done well. Sandu tourist city construction will play a major role in protecting and inheriting the Shui culture. With the establishment of the urban development goals and the development of tourism, the old historic buildings have been maintained and managed well, some local traditional practices and national cultural activities which were forgotten by people have been explored and restored, the traditional arts and crafts have been developed again because of the expansion of the demands, and the non-material cultural heritages such as traditional music, dance and drama have been explored and attached importance to. the city becomes an important carrier of national culture and the national symbols will be displayed in buildings, streets and squares, so that the almost forgotten cultural heritages have been rejuvenated which is conducive to protect and inherit the national culture.

2. Raise the Urbanization Level

The urbanization level is an important indicator to reflect regional social development, which is measured by the proportion of the non-agricultural population (urban population) to the total population. Owing to poor utilization of the resource advantages and lagged economic development, the urbanization level of Sandu has long been in the last place of the province. With the implementation of urban tourist planning, more employment opportunities have been created. Due to the high rate of return of tourism comparing to agriculture, more and more farmers worked in the tourism-related services such as accommodation, catering, tour guides, passenger transport and so on, and the proportion of employment in primary industry has declined and the employment in the tertiary industry increased, which means that the proportion of non-agricultural population will rapidly increase, the urbanization level of Sandu will be greatly enhanced.

3. 促进城市精神文明建设

三都旅游城市打造不是一个单纯地给城市“穿衣戴帽”的过程，而是一项“牵一发动全身”的社会工程，物质环境的改变带来的不仅仅是生活品质的改变，还有居民素质的提高乃至整个社会文明程度的提升。由于旅游业是一种外向型产业，以“人与人的交往”为主要活动，行业规范、服务标准、从业者的综合素质等都较其他行业的要求高，对于旅游从业者而言，这无疑是提高个人文化修养的良机；与此同时，外来旅游者在旅游目的地的活动过程中，由于同当地居民直接和间接交往的接触活动存在，也会以其有意无意地“示范”行为影响当地居民，促进居民提高自身的素质，特别是提高文明意识，从而加快城市精神文明建设的步伐。

3. Promote the City's Cultural and Ideological Progress

Sandu tourist city construction is not only a "Dressing Up" process for the city, but a "trigger" social engineering. The change of the physical environment not just follows the change of life quality but also improves the quality of some residents and even enhances the degree of civilization of the whole society. As tourism is an outward industry focusing on the communications between people, it needs higher requirements of industry standards, service standards, and the overall quality of practitioners than other industries. For tourist practitioners, this is undoubtedly a good opportunity to enhance personal culture. At the same time, because of the direct and indirect contacts with the local residents, tourists will affect the local residents to improve personal quality especially to raise awareness of civilization by their intentional or unintentional "model" behavior in the process of tourist activities so as to speed up the spiritual civilization construction of the city.

第二节 主要项目筹资分析

Section II Main Project Funding Analysis

一、主要项目筹资分析

三都县旅游城市打造规划设计方案，目的是为了通过对投资环境和城市形象的重塑，进而起到推动其他招商引资项目启动实施的作用。与此同时，我们也应清楚地认识到这一项目由于工程量大、涉及面广，资金筹措既成为城市改造项目的支点，同时也是项目得以实施的难点。可这并不等于说项目就缺乏可行性了，只要适当地变化思路，就能“化腐朽为神奇”，在促使城市改造项目得以实施的同时，也能早日实现其他项目的实施。以下是对启动项目筹资方式的几点思考：

第一，政府对投资商给予相应的倾向性政策，但附带的条件是投资商要对某一路段的建筑外立面改造进行承建。

这种方式，对政府而言，解决了城市改造的资金筹集问题；对投资商而言，投资项目获得批准，并得到政策性优惠，顺带投入的城市改造费用实际上已让其名利双收。投资要有眼光，它代表着未来的趋势；投资要抓机遇，它总是稍纵即逝；投资要冒风险，它的风险永远与利润相连。在如此良好的投资机遇面前，投资商的投资已成一种必然。

第二，政府拿出部分财政拨款，为城市主干道及“一江两岸”建筑外立面改造出资。

政府对城市改造需投入一定资金，但并非以政府投入为主。政府更应积极做好鼓励和引导工作，如：出台相关优惠政策、给予业主和投资商一定形式的补偿等方式，充分调动业主和投资商的积极性，以市场推动这一项目资金的获得。

第三，相关业主对三都城市主干道及“一江两岸”建筑外立面改造出资。

I Main Project Funding Analysis

The aim of the planning and design program of Sandu tourist city construction is to give impetus to the implementation of other investment projects by reshaping the investment environment and the city image. At the same time, we should clearly recognize that the financing issue is the fulcrum of the project and the difficulty of the implementation because of its large quantity and wide range. But this does not mean that the project is lack of feasibility. It will be a "miracle" with appropriate changes in thinking. While the urban renewal project is implemented, other projects can also be implemented at an early date. The following is some thoughts of financing of the start of the project:

First, the Government gives the investors the corresponding preferential policies, with the condition that the investors should construct the building facades of certain street.

For the government, it can solve the financial problem of the urban reconstruction; for investors, it is a good way to gain fame and fortune because the investment projects will be approved with preferential policies. Investment should be prescient and it represents the trend of the future; investment should be determined and it is always fleeting; investment has risks but its risks always combine with profits. With such a good investment opportunity, investors are bound to invest the project.

Second, the Government allocates some funding to reconstruct the urban main roads and the facades along the riverbanks.

The Government must invest in reconstructing the city but the fund offered by the government is not the main resource. The Government should actively encourage and guide well, such as make preferential policies to give owners and investors some compensation, fully mobilize the enthusiasm of owners and investors, and gain financial support by market.

Third, the owners concerned finance the reconstruction of the urban main roads and the facades along the riverbanks.

这一部分可通过两个渠道实现：其一，以办公和经营为主的企事业单位及政府机关建筑外立面改造费用的支出，通过引导、协商等方式，在平等自愿的前提下，以集体为单位予以出资；其二，以居住为主的商住建筑外立面改造费用，在平等自愿的前提下，以业主为基本单位出资。

This can be achieved through two channels: firstly, through guidance and consultation, the facade reconstruction of the office and business-oriented enterprises and government agencies can be funded by the collective units on an equal and voluntary basis; secondly, the facade reconstruction of residential buildings can be funded by respective owners on an equal and voluntary basis.

二、其他项目筹资分析

在三都城市主干道及“一江两岸”建筑的外立面改造和环境整治项目竣工后，三都县整个以旅游文化产业为龙头的投资环境将得到极大的改善。而以招商为最终目的的三都旅游城市打造规划设计方案中的各个项目，势必会吸引众多投资商的目光，持续地为三都城市发展带来惊喜。因此，三都旅游城市打造的基点就是解决城市主干道及“一江两岸”建筑的外立面改造和环境整治问题，而后再通过对其他主要项目的对外招商，以引凤凰筑巢（大财团的投资），从而达到发展三都的目的。

II Other Project Funding Analysis

When the reconstruction of urban main roads and the facades along the riverbanks and the environmental restoration project is completed, the investment environment in which the tourist cultural industry takes the leading role will be greatly improved. The projects of Sandu tourist city construction planning programs with the ultimate goal of foreign investment will be bound to attract the attention of many investors and there will more and more surprises in the process of Sandu urban development. Therefore, the basic point of Sandu tourist city construction is to solve the reconstruction problems of the urban main roads and the facades along the riverbanks and the environmental management problems, and then attract foreign investment (large consortia investment) by other major projects so as to achieve the purpose of development.

第九章 结论

依托于古朴神秘的水族文化和俊秀神奇的都柳江及尧人山国家森林公园等自然资源，将三都县打造成为独具民族特色和现代气息的旅游休闲度假城市，同时整合区内旅游资源、联合周边旅游景点，营造出辐射黔湘桂的区域旅游中心城市，使三都树立起“凤羽三都，心随水动”的旅游城市形象，引领并带动一方社会经济环境的持续发展。

这一目标愿景，在三都提出旅游城市打造战略思路的同时就已经走上了开始实现的道路，加之整个城市范围内高度统一、全面铺开的城市形象规划设计，使其更具备了发展的有力保证，成就三都明天的辉煌指日可待。

Chapter IX Conclusion

With the mysterious magic Shui national culture and the natural resources such as charming Duliu River and Yaoren Mountain National Forest Park, the Sandu County will be constructed to be a tourist city with unique national characteristics and modern flavor. At the same time, by integrating the local tourist resources and combining the adjacent tourist attractions, the regional tourist central city, radiating Guizhou, Hunan and Guangxi, will created, and the tourist city image of "Phoenix fluttering in Sandu, heart moving with the water" will be established to lead and promote the sustainable development of social and economic environment.

While the strategy of Sandu tourist city construction was made, this goal and vision has already on the way. In addition, with high degree of unity within the city and the comprehensive city image planning and design, the development of Sandu has been strongly guaranteed and the tomorrow's glory of Sandu is just around the corner.

①三都县城规划鸟瞰图 ②三都县城规划前鸟瞰图 1 ③三都县城规划前鸟瞰图 2 ④城区入口景观效果图

②

③

④

沿街建筑改造

Rebuilding Street–Buildings

Architectures, which are all modern buildings, along the main road in Sandu County have typical Chinese town style–monotony, lack of integrity. It is not enough for architectures with only utility function to leave people good impression, but cultural connotation is necessary.

In addition, part of the existing roads in Sandu County are quite crowded, mainly because the roads themselves are narrow, and road parking zones are divided on both sides, so most of the vehicles park there, causing road congested and cluttered.

一、概况 I Outline

三都县县城主干道沿街建筑均为现代建筑，具有典型的中国小城镇气息——千篇一律，缺乏整体性。建筑有魄无魂，具备实用功能，缺乏文化内涵，使建筑显得空洞，缺乏特色，难以让人产生美感，留下映像。

另外，三都现有部分道路较为拥挤，主要原因在于道路本身较为狭窄，还在道路两侧划分停车带，大多数车辆停靠于此，致使道路拥挤、凌乱。

二、规划思路

三都县城市建筑风貌塑造工程的重点在于沿街建筑的外立面改造和公共休闲设施的创意设计。通过城市建筑风貌的塑造，将文化气息之魂附于城市建筑之魄上，达到灵肉合一之效果。如此一来不仅能够提升三都县城人文内涵，使城市具有灵气，达到更高一层的境界，同时也能为三都县城增添新的景观，形成新的旅游资源，形成独具特色的三都城市标识，使得三都县旅游城市打造再添亮点。

文化是建筑的魂，要成功地塑造城市建筑形象就要从了解地区文化入手。在我国，悠久的历史铸就了灿烂的文化，许多城市都具有多重文化的共生性，致使城市在文化的定位上难以取舍，最后形成了多种文化的融合，好的融合能够兼容并包，形成自身特色，不好的融合却适得其反，使城市总体定位不清晰不明确。而三都却是一个特殊的城市，它是全国唯一的水族自治县，全县拥有60%以上的水族居民，水族人口占全国水族人口的52%以上，水族文化一枝独秀，毋庸置疑，古朴神秘的水族文化便是三都县沿街建筑外立面改造的不二主题，紧抓这个唯一性，不仅能形成自身特色，与周边县市的文化内涵区分开来，还与贵州省“大力发展民族旅游”的战略思想统一起来。从水族文化中挖掘具有代表性的元素构筑独具三都特色的建筑风格，定能形成代表三都的城市视觉景观。

II Alteration Layout for Street Buildings in Sandu County

The building engineering of Sandu County's architectural style focuses on the reform of the building facade and creative design of public recreational facilities. Through the creation of urban architectural style, the soul of culture is annexed to the city's architecture, and reaches the effect of combining soul and body. In this way not only to raise Sandu County's humane connotation, making the city has an anima, achieve higher realm, at the same time adding new landscape to Sandu County, forming a new tourism resource, forming a unique Sandu city identity, making the Sandu tourist city build another bright spot.

Culture is the soul of the construction, in order to shape the city's figure well we have to begin with its regional culture. In China, a long history makes the vast and splendid culture, many cities have symbiosis of multiple culture, making the city facing difficult choices in urban culture positioning, and finally form a multi–cultural fusion, a good integration can include all the culture and form their own characteristics, while a bad integration is just the opposite, so that the general position of the city is not clear. Sandu is a special city, it is the only municipality of the Shui nationality, the county has more than 60% inhabitants of Shui nationality, the population of Shui nationality account for more than 52% of the country's entire population, culture of Shui nationality thriving, and there is no doubt that the plain and mysterious Shui Culture is the only theme of reforming Sandu street building facade, stick to the uniqueness can not only form its own characteristics, and distinguish the cultural connotation from surrounding counties and cities, but also integrate with the "Great Development of National Tourism" strategic thinking in Guizhou Province. Excavated from the Shui culture, representative elements are built into unique characteristics of Sandu architectural styles, and will be able to form the city visual landscape on behalf of Sandu County.

三、设计理念

1. 总体风格

无处不在的体现三都水族风格，期间穿插布衣、苗族等其他生活在三都的少数民族风格，表现三都特色的同时也体现了民族大团结的融洽氛围。

2. 图腾标志

在了解三都文化习俗的基础上，将三都水族人民自古崇拜的图腾形象加以艺术的创意设计，形成形象鲜明，寓意深刻，独具特色的三都形象标志，大量运用于沿街建筑改造中，增添沿街建筑景观效果的同时，形成三都 logo 系统。

双鱼托葫芦：鱼在水族人民的日常生活中占据着重要地位，自古以来就受到水族人民的崇拜，因为水族人民认为鱼是水族先祖的救命恩人，传说在洪水泛滥中，是两条鱼托起了水族先祖兄妹避难的葫芦而幸免于难，才有了水族人民的繁衍生息。他们认为没有葫芦和鱼就没有水族。以双鱼托葫芦造型作为三都图腾标志不仅体现了三都水族文化，葫芦下方内方外圆的铜钱造型还寓意“福在眼前”。

还愿伞：水族人民有石崇拜的习俗，他们在祭石的时候会许下心愿，每当愿望达成，都会制作还愿伞祭石，将还愿伞运用于沿街建筑改造，设计为街道指示牌，向世人述说水族文化、水族人民知恩图报的淳朴性格。

3. 门头

屋檐：水族建筑的特色在于“双叠水”屋檐。因此在改造中部分建筑门头采用水族建筑特有“双叠水”屋檐，屋檐用具有民族特色的吊瓜进行装饰。局部而不是全部采用此屋檐形式主要有两方面的原因，一是受现有建筑构造的局限，无法改造为“双叠水”屋檐；二是局部使用能使建筑具有跳跃感，清新美观，而全部使用会使沿街建筑显得过于厚重、繁复，反而适得其反。

挑角：屋檐挑角采用“鱼尾巴”的形式，使沿街建筑具有民族特色的同时又能体现三都独特的水族韵味。

4. 材质

水族建筑多为干栏式木楼，根据这一水族民居特点，结合现代建筑风格，在三都沿街建筑外立面改造中主要使用外墙漆及文化石，局部采用喷砂装饰，使建筑风格保持三都特色的同时体现清新、典雅的气息。

外墙立柱通过包木处理及与外墙不同色彩的运用，使立柱更具立体感。

Ⅲ Alteration Design Concept of Street Architecture in Sandu County

1. Overall Style

Ubiquitous expression of Shui Style, adding styles of Buyi, Miao, and other minority living in all Sandu, thus showing the characteristics of Sandu as well as demonstrating harmonious atmosphere of the great national unity.

2. Totem Sign

Sandu logo system can be created by innovatively and artistically designing the ancient totem image of Sandu people to form vivid image and unique Sandu mark and greatly using in reformation of street buildings on the basis of understanding the culture and customs.

Two Fish Supporting the Luck: Fish occupies an important position in daily life of Shui people, and was worshiped by Shui Nationality since ancient time, because Shui people think that fish is the savior of Shui nationality's ancestor, the legend said that in the flooding, two fish supported cucurbit which survived the brother and sister who are the ancestor of the Shui nationality, thus allowed the Shui nationality live and multiply. They think that there will be no Shui nationality without cucurbit and fish. Pisces supporting the cucurbit shape not only reflects the Sandu Shui culture, the coin shape under the cucurbit also symbolizes "a blessing in the eyes".

Wish Umbrella: Shui people have the custom of stone worship, they will make a wish when they offer the stone, and when the wish is fulfilled, they will make the umbrella to offer the stone. The application of the umbrella of the design of direction board in transforming street architectures can tell the world the culture and grateful character of the Shui people.

3. Doors

Eave: The feature of shui architecture is the "Double Dip" roof. Therefore, in the reforming, part of the construction gate uses the unique 'Double Dip" roof of shui architecture; roof is decorated by Diaogua which has the national characteristics. Part but not all use this roof form has two main reasons, first, the limitations of the existing building structure, and can not be transformed into "double cascading" roof; second, using this partially can make the architecture have the sense of jumping, fresh and beautiful, while using in all will make the street architecture become too heavy and complex, leading to the opposite effect.

Tiaojiao: The Roof takes a form of "fish tail", so that the street buildings have features of the nationality while at the same time show the unique and charm of the the Shui nationality in Sandu.

4. Materials

Most of the Shui architecture are wooden house supported by column, according to this character of Shui house, combining the modern architectural style, using mainly external wall paint and culture stone in the reform of street building facades in Sandu, locally decorated by sandblasting, so that the architecture style can show fresh and elegant while at the same time keeping the characteristic of Sandu.

The Column of exterior wall use wood covering and different colors to make the column have a third dimension.

5. 色彩

色彩主要以暖色调,黄色色系为主,使建筑风格温馨、典雅。不同的路段采用不同的色彩。

北出口改造建筑采用仿木色，新建建筑采用暖黄色调；河滨北路、南京西路、南京东路、麻光三郎片区新建建筑采用暖黄色调；其余建筑采用白灰色调。

5. Color

Mainly in warm colors, yellow color takes the first place, making the architectural style warm and elegant.

Different road sections taking different colors.

Reform architecture in North exit use wood color, new buildings take warm yellow tone; new architecture in Hebing North Road, Nanjing West Road, Nanjing East Road, MaguangSanlang section use warm yellow tone; the other architecture using white and grey colors.

6. 窗

将窗作为装饰，使用民族木窗格的装饰效果。内窗采用铝合金玻璃窗，外窗在铝合金防盗窗的基础上喷涂仿木漆，形成木制花格窗罩效果，这样既满足了居民的使用要求，又使建筑具有简约、精美的装饰效果,使建筑风貌更具民族特色。

6. Window

Take the window as decoration, and use the decorative effect of ethnic wooden pane. The inner windows use aluminium windows, external windows spray paint wood paint on the basis of anti-theft aluminium window, forming the effect of wooden window hood, so that this not only satisfies the demands of residents, but also makes the architecture have a simple and elegant decorative effect, making the architecture style have more ethnical characteristic.

7. 细部装饰

三都沿街建筑外立面改造中，不时出现有别于建筑主题色调和材质的细部装饰，乍一看不明就里，这便是让水族闻名于世的水书文字。建筑外立面以水书文字作为装饰主要是从以下几个方面考虑的：第一，向世人展示水族神秘而精妙的水书文化；第二，装饰的水书文字均为“金银满屋”、“人寿年丰”、“吉祥如意”等具有良好寓意的文字，体现了三都对世人的美好祝愿；第三，奇特难懂的文字会最大限度地吸引游客的注意力；第四，水书文字与主色调的差异可以使外立面显得更加丰盈和生动。

7. Detail Decoration

During the reform of street architecture facade, often appears detailed decoration different from the theme color and material of architecture, people cannot understand this at first glance, and this is the famous Shui character that makes the Shui nationality famous in the world. The following aspects are considered when Sui character as decoration: first, to show the world the mysterious and sophisticated Shui culture; second, the decorative Shui words often have a good meaning such as "Gold and Silver Full House", "living a long and wealthy life", "good luck", showing Sandu people's good wishes to the world; third, strange words can best attract the attention of tourists; fourth, the difference between the Shui characters and main color can make facade more abundant and lively.

8. 街边广场

街边广场是沿街改造的重要组成部分，在现有街边广场或公共绿地的基础上，运用三度水族元素对广场进行定位及改造，形成沿街景观节点的同时为市民和游客提供独具三都特色的休憩场所。

8. Street Plaza

Street plaza is an important part of street building reconstruction, on the base of existing street plaza or public green belt, use the elements of Shui nationality in Sandu to locate and reform the plaza, forming the landscape nodes along the street and at the same time provide a relaxing place with Sandu characteristics for residents and visitors.

9. 景观小品

景观小品不是孤立存在的，而是要与城市沿街建筑整体风格相协调的，三都的各类街边雕塑、路灯、公车站台、广告牌、垃圾桶、指示牌都与沿街建筑改造的水族风格保持一致。

9. Landscape Sketch

Landscape sketch is not isolated, but in harmony with the overall style of urban street architecture, all types of street sculpture, street lamps, bus stations, billboards, garbage cans, street signs in Sandu are in correspondence with the Shui style of street architecture.

10. 入口牌坊

进入三都的主干道以及重要街口入口处都会设置牌坊，牌坊以区域文化为构成要素，以鲜明的个性特征体现区域特色，通过牌坊向游客传达属于三都、属于该街区的独特信息。

10. Entrance Arch

The arch will be set at the entrance of major street and street corner entering Sandu, the arch takes regional culture as structure element, and shows regional characteristic by distinctive personality characteristics, unique information of Sandu and the block is conveyed to visitors through the arch.

建设东路区位分析图

建设东路改造平面图

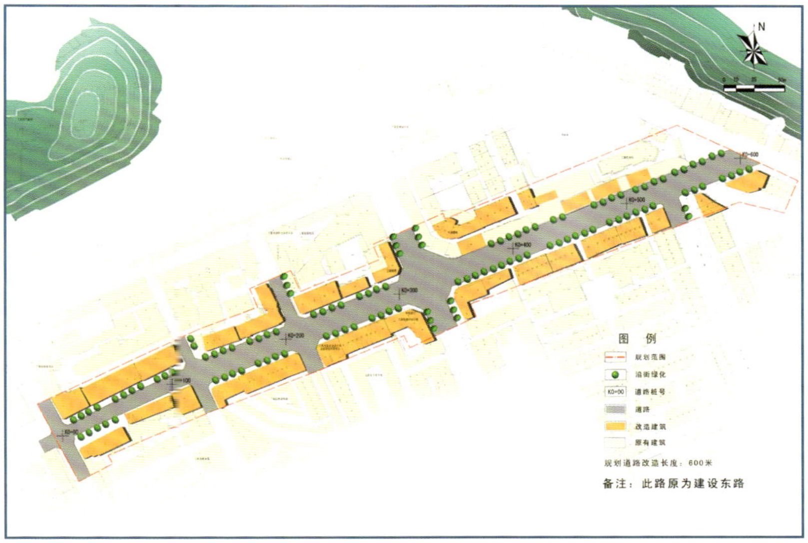

建设东路公共服务设施

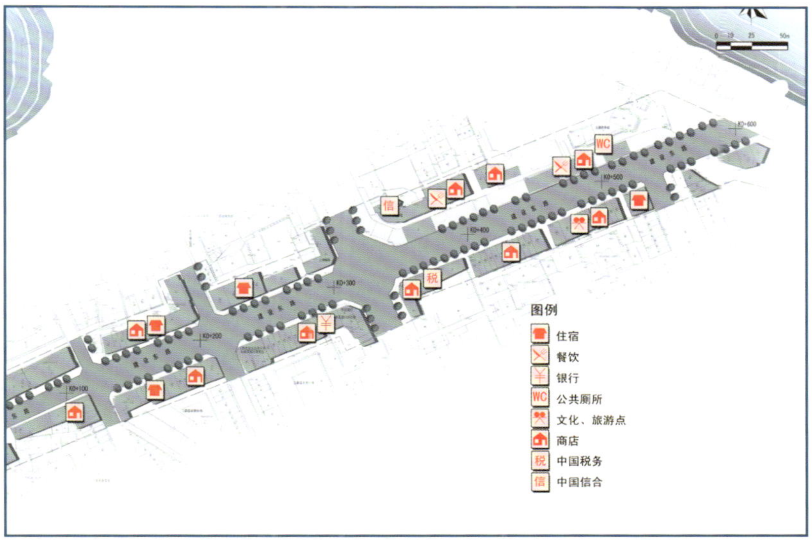

沿街建筑立面对照图

改造前

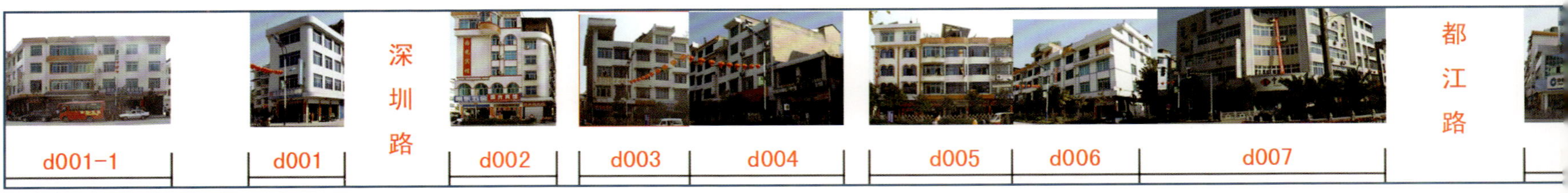

改造后

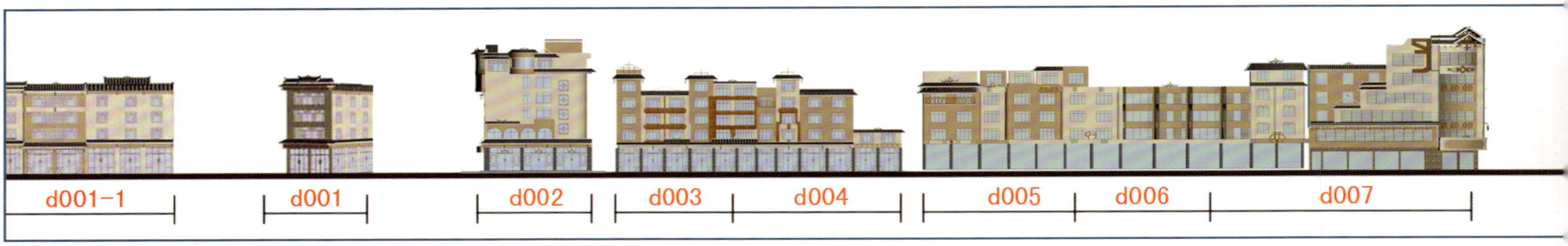

平面图

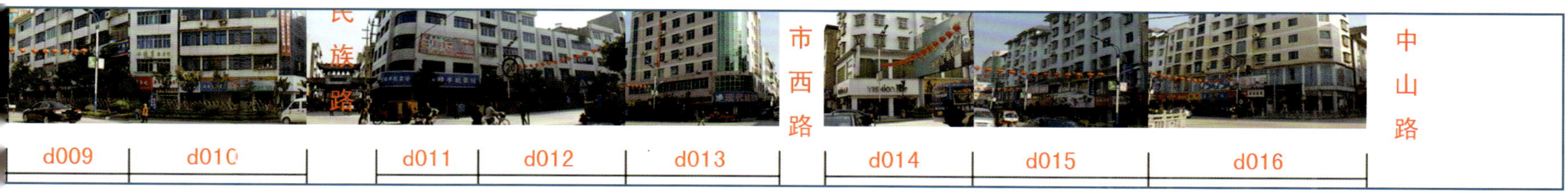
民族路
d009
d010
d011
d012
d013
市西路
d014
d015
d016
中山路

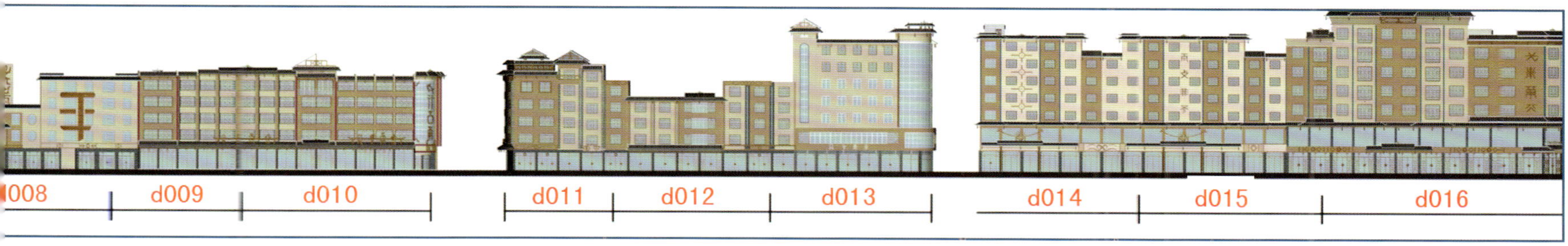
008
d009
d010
d011
d012
d013
d014
d015
d016

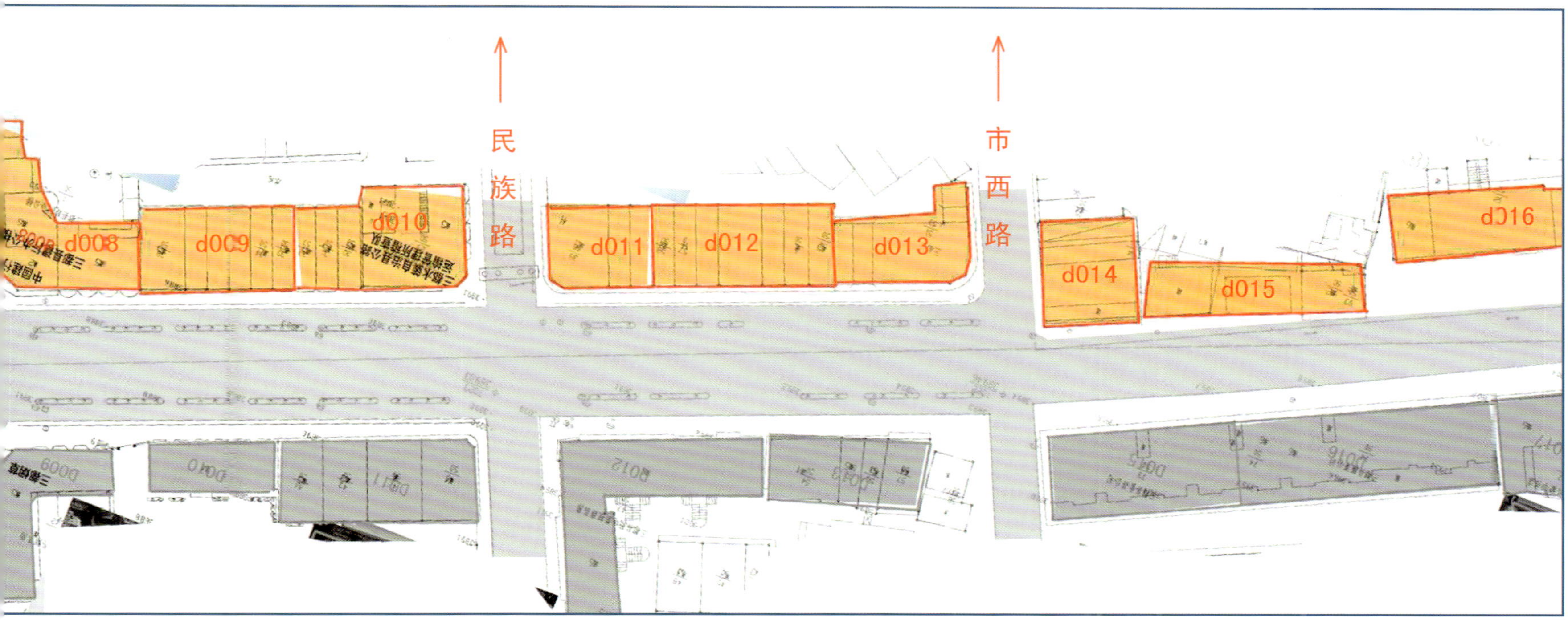
民族路
市西路
d008
d009
d010
d011
d012
d013
d014
d015
d016

①

①街道改造鸟瞰图 ②街道改造效果图 ③街道改造透视图

①

②

改造前

①～④街面建筑改造效果图

改造前

①街面建筑改造效果图

团结东路现状分析图

团结东路改造平面图

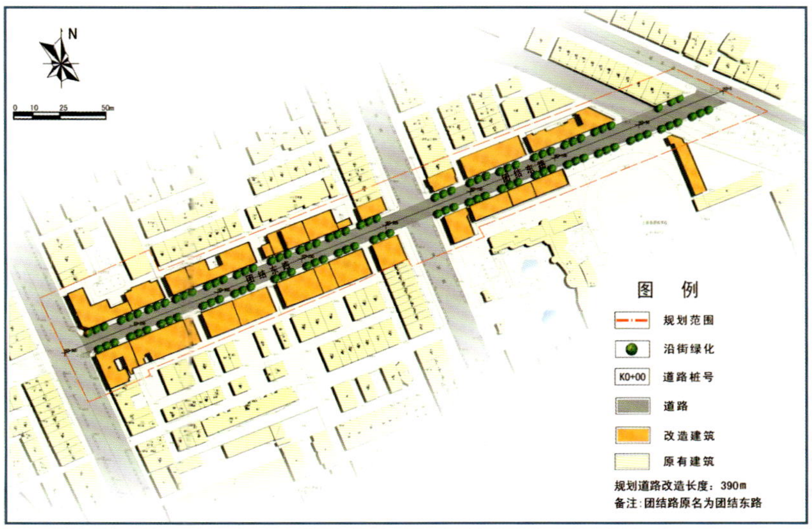

团结东路公共服务设施

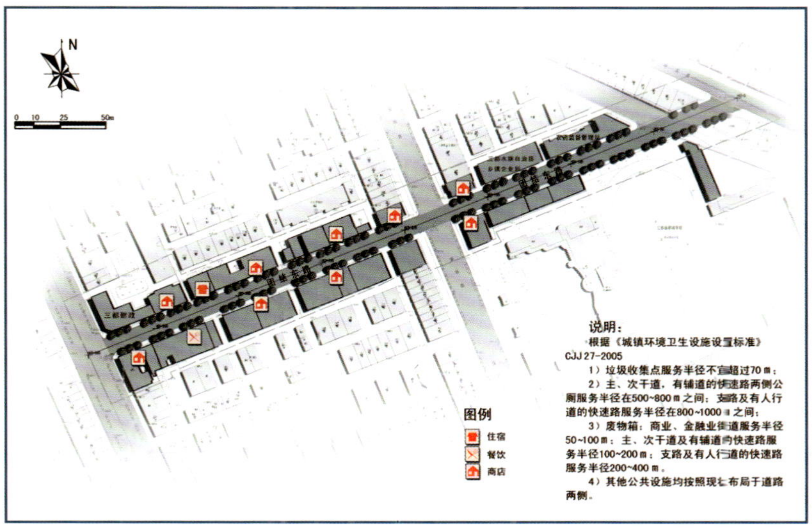

改造前

改造前

①～④街面建筑改造效果图

①

改造前

①～③街面建筑改造效果图

河滨北路（水族风情街）规划设计

Planning and Design of Hebin North Road (the Ethnic Street)

This land is situated on the bank of Duliu River, with beautiful landscapes and developed traffic and dense construction. Most buildings along both sides of the street are used as offices of enterprises. With good quality but different styles of buildings, the street is lack of unity, wholeness and identification.

一、现状 Ⅰ Current Situation

该地块处于都柳江沿岸，景观优美，交通发达，该片区道路两侧建筑密度较大，多为商住楼及事业单位建筑，整条街的建筑质量良好，但风格不统一，缺乏整体性和协调性以及可识别性。

二、规划思路

利用本地块的沿江景观条件，沿用现有道路及码头，将之规划为步行街，形成高低错落，具有民族特色的三都休闲娱乐亮点。

规划利用步行街及周边外围环境共同构筑片区旅游产业发展，用品位三都的六个主题去诠释一个心动的三都。六个主题分为品位三都之闲暇时光、品位三都之原汁原味、品位三都之精致生活、品位三都之激情浪漫、品位三都之水韵情怀、品位三都之悠情雅韵。

Ⅱ Planning Concept

The current roads and wharfs along with the riverbank landscapes of this land are designed to be a pedestrian street with different height constructions and ethnic properties.

In the plan, tourism industry of Sandu is developed relied on the pedestrian streets and its surrounding environment with six themes which are the theme of tasting leisure, origin, exquisite life, romance, water charm, and elegance of Sandu.

区位分析图

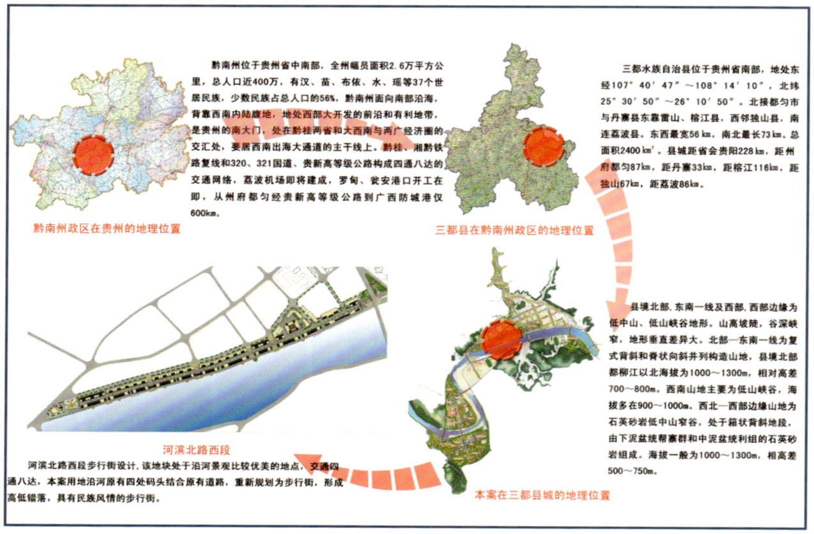

现状分析图

总体设计策略

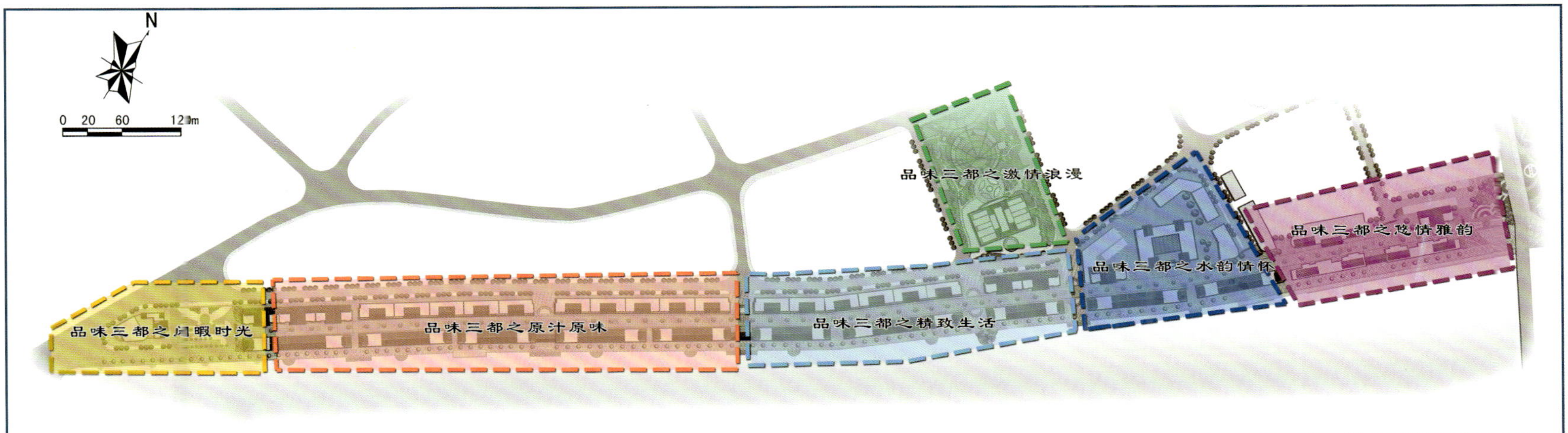

名称	品味三都之闲暇时光	品味三都之原汁原味	品味三都之精致生活	品味三都之激情浪漫	品味三都之水韵情怀	品味三都之悠情雅韵
功能	入口广场	土特产销售区	工艺品销售展示区	民族歌舞、戏剧展示区	休闲娱乐区	民族餐饮、客栈区
展示内容	民间健身活动：舞龙舞狮、斗角舞、斗鸡、斗鱼、斗狗、打苗毽、地下停车场	土特产：九阡酒、九阡李、甜茶、麻竹、中药材、野生绿色苦丁茶	民族工艺品：马尾绣、牛角雕、建筑、木刻、石刻、银饰、姊妹萧	民族歌舞：铜鼓斗角舞、铜鼓舞、双歌、单歌、长刷巴舞、短刷巴舞、古瓢舞；民族戏剧：花灯戏；民族乐器：长号、铜鼓、木鼓、芦笙、莽筒	滨江茶楼、室内歌舞剧表演、咖啡馆、KTV、SPA、游船、亲水平台	特色饮食：鱼包韭菜、风味酸［辣酸、毛辣酸（西红柿制成）、鱼酸、臭酸］、水族火锅、铁板烧、嗜鱼、香猪、香米；民族客栈：双叠水干栏式建筑

总平面图

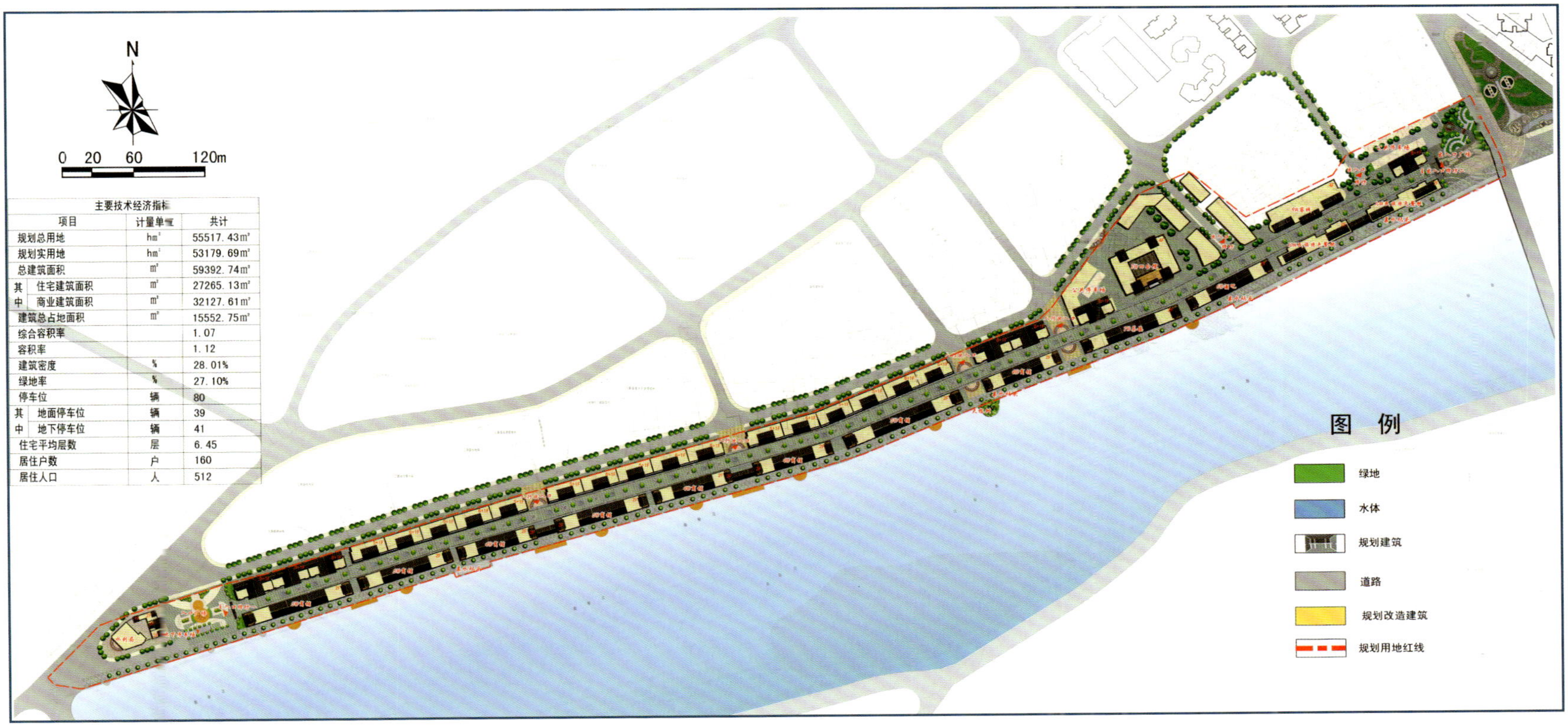

主要技术经济指标

项目		计量单位	共计
规划总用地		hm²	55517.43m²
规划实用地		hm²	53179.69m²
总建筑面积		m²	59392.74m²
其中	住宅建筑面积	m²	27265.13m²
	商业建筑面积	m²	32127.61m²
建筑总占地面积		m²	15552.75m²
综合容积率			1.07
容积率			1.12
建筑密度		%	28.01%
绿地率		%	27.10%
停车位		辆	80
其中	地面停车位	辆	39
	地下停车位	辆	41
住宅平均层数		层	6.45
居住户数		户	160
居住人口		人	512

三、功能分区

河滨北路西段步行街分为：入口景观展示区、商业节点展示区、民俗活动展示区、市民休闲活动区、滨河步行展示区、商业步行展示区。

整个功能布局沿河布置，两条主要步行街贯穿其中，步行街与其沿江亲水平台高低呼应，形成三都一道独特景观。

四、设计说明

1. 河滨北路步行街客栈

客栈无处不在地彰显了水族元素：客栈采用了水族建筑特有的“双叠水”屋檐，加上吊瓜装饰，更凸显出民族特色；客栈屋顶运用了水族形象标识“双鱼托葫芦”，与三都整体形象相呼应；客栈廊道采用镂空设计，使建筑整体显得简约灵动；客栈立柱运用国家级非物质文化遗产马尾绣的图案进行装饰，华美大方的同时呈现三都特色；客栈外立面采用水书“吉”、“福”进行装饰，传达了对客人的美好祝愿——“吉祥如意”、“福寿双全”。

Ⅲ Functional Division

The pedestrian on the west of North Hebin Road divides the whole pedestrian street into six areas including entrance landscape demonstration areas, business nodes demonstration areas, folk activities area, riverbank pedestrian demonstration areas business demonstration areas.

The whole functional layouts are set along river with two main pedestrian streets running through them. The different height between pedestrian streets and water balconies is a special landscape in Sandu.

Ⅳ Design Explanation

1. Pedestrian Street Inn of North Hebin Road

The architecture style of the inn is showing Shui elements in every detail. It adopts "double layers water" eaves of Shui people and hung melons decoration to highlight ethnic characteristics, the symbol of Shui people "two fishes carry gourd" on its roof to response the overall image of Sandu, hollow design on the corridor to make the whole building simple and smart, Mawei embroidered pattern to decorate the pillars beautifully and magnificently; and Shui characters of " 吉 " and " 福 " (these two characters have the meaning of wishes in Chinese) to ornament the facade expressing best wishes to guests.

功能结构分析

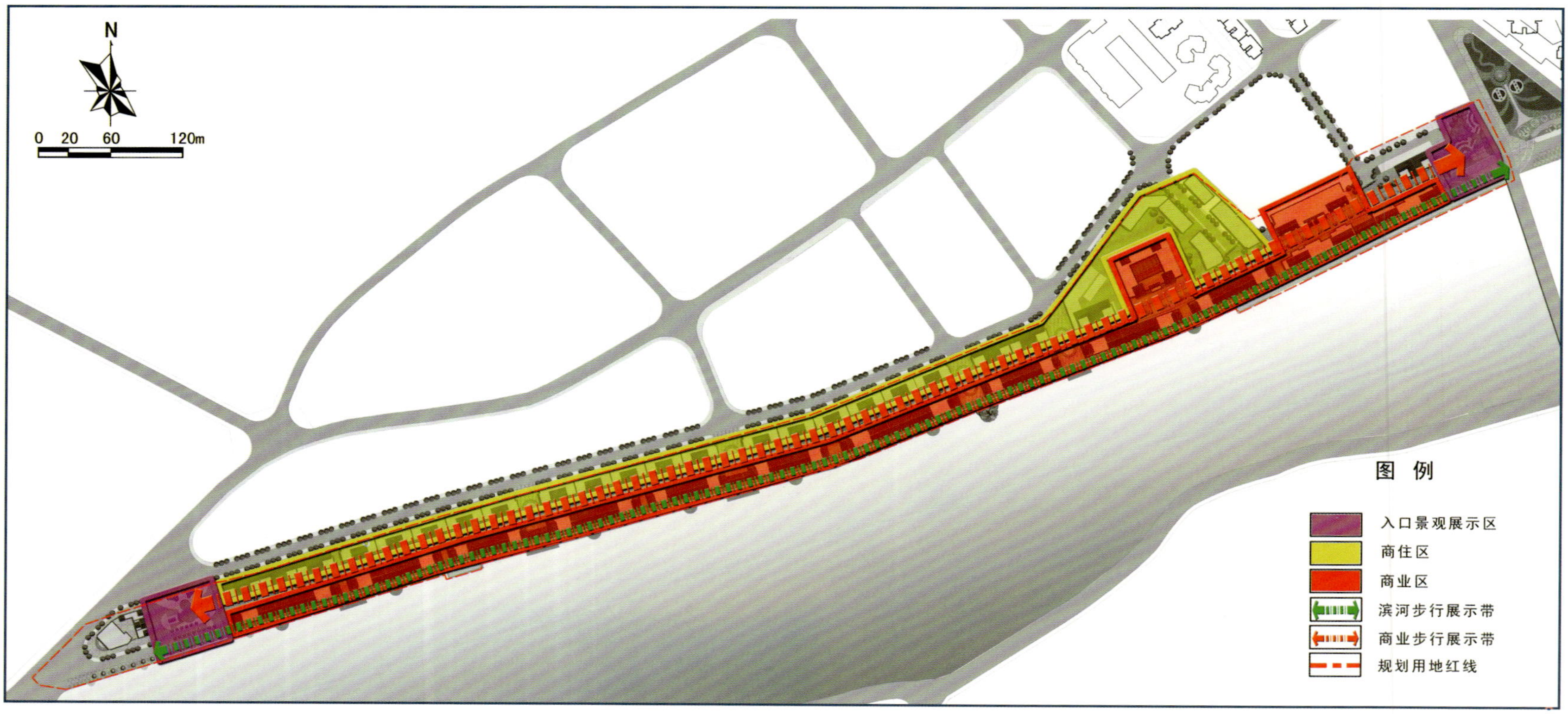

2. 入口牌坊

河滨北路民族风情街主入口牌坊，整体采用以“水”为主的思路。牌坊最上方用“双鱼托葫”去演绎古老的水族传说，牌坊屋檐采用递减式双叠水造型，它将民族的建筑具象化。加上左右两边的水书、吊瓜、牌坊立柱支撑面铜鼓的点缀，相信会给游人以全新的震撼，使人将三都铭记于心。

河滨北路民族风情街次入口牌坊方案一，采用抽象化的造型将水族的凤凰、还愿伞、铜鼓、卯、双鱼托葫融合为形象而简洁的次入口大门。

河滨北路民族风情街次入口牌坊方案二，此牌坊大气而富有美感。大门上用双鱼托葫造型贯穿其中，用“葫”、“福”的谐音，将三都的一份真诚的祝福送出。

2. Entrance Arch

The concept of "water" is adopted in the design of the memorial arch of the main entrance of North Hebin Road. The application of "Two Fishes Carry the Luck" above the arch expresses the ancient legend of Shui people; and "the double levels water" sculpture of eaves specializes the ethnic architecture. Along with the embellishment of the Shui Books, hanging melons and copper drums on the feet of the arch pillars on both sides of the arch, the entrance arch shall shock the visitors and leave them deep impression.

The signs of Shui people such as phoenix, wish umbrella, copper drum, the character " 卯 " and "Two Fishes Carry the Luck" are mingled to form the gate of the secondary entrance, vivid and simple, in the Proposal I of the secondary entrance arch of the Ethnic Street of North Hebin Road.

The arch is magnificent and full of aesthetics with the sculpture of "Two Fishes Carry the Luck" lying through it, which expresses sincere wishes to visitors of Sandu people in the Proposal II of the secondary entrance arch of the Ethnic Street of North Hebin Road.

公共服务设施

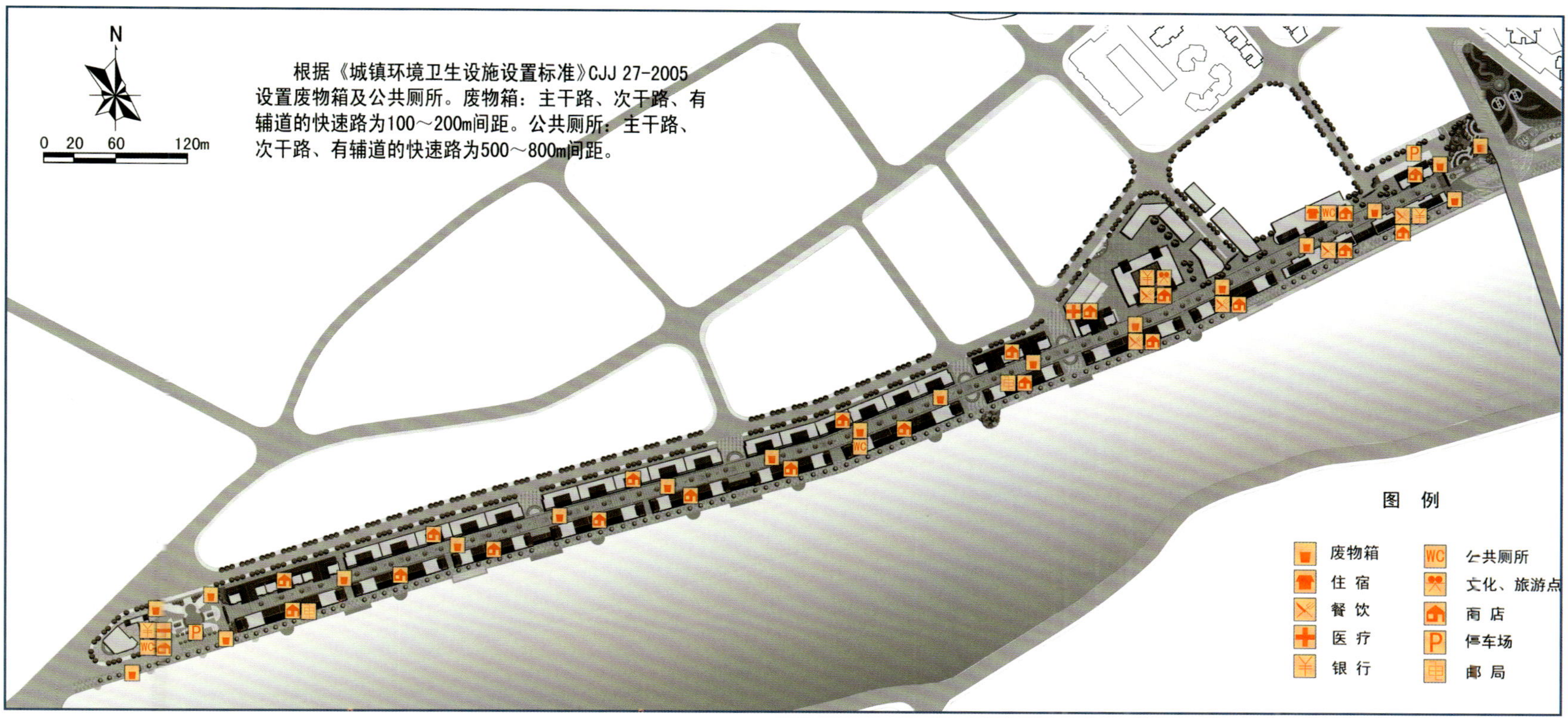

道路交通分析图

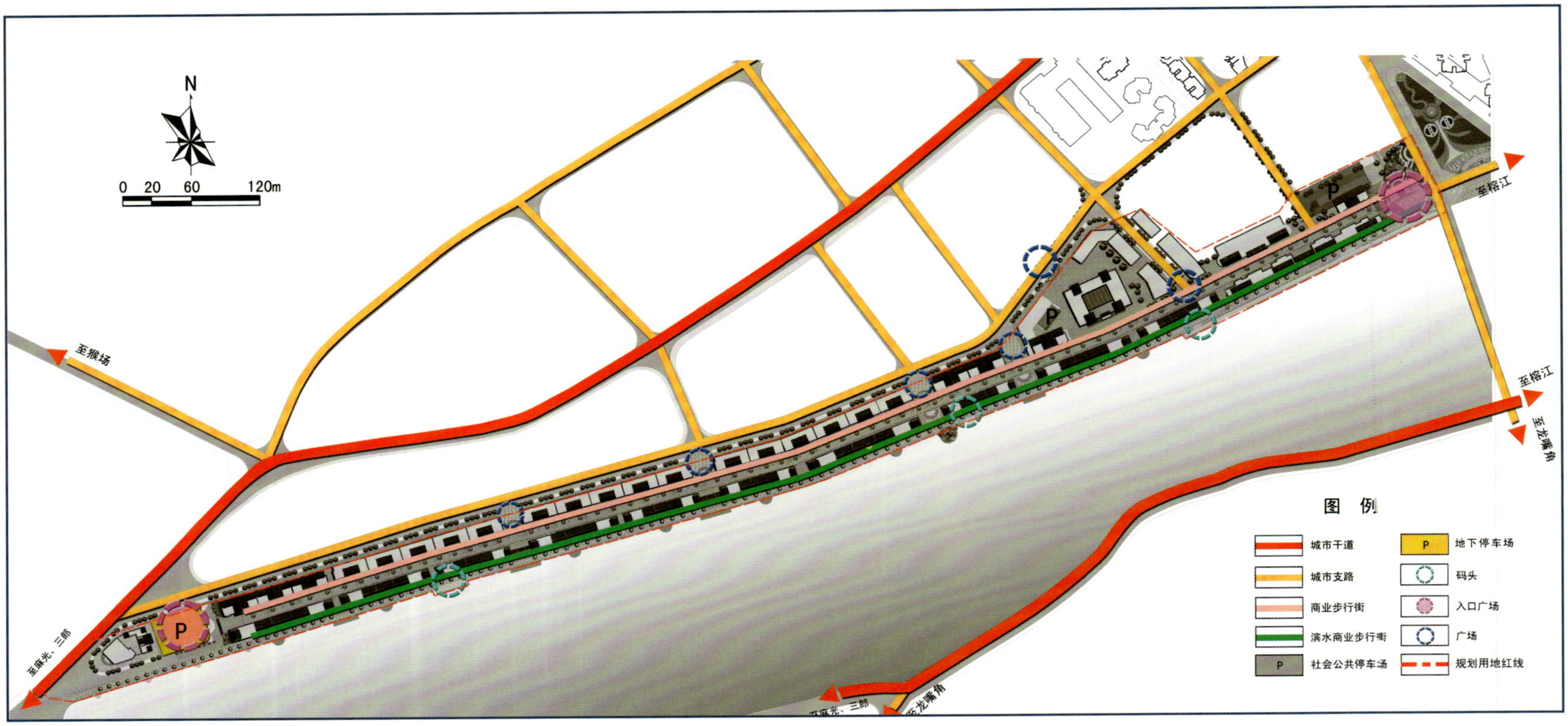

竖向设计图

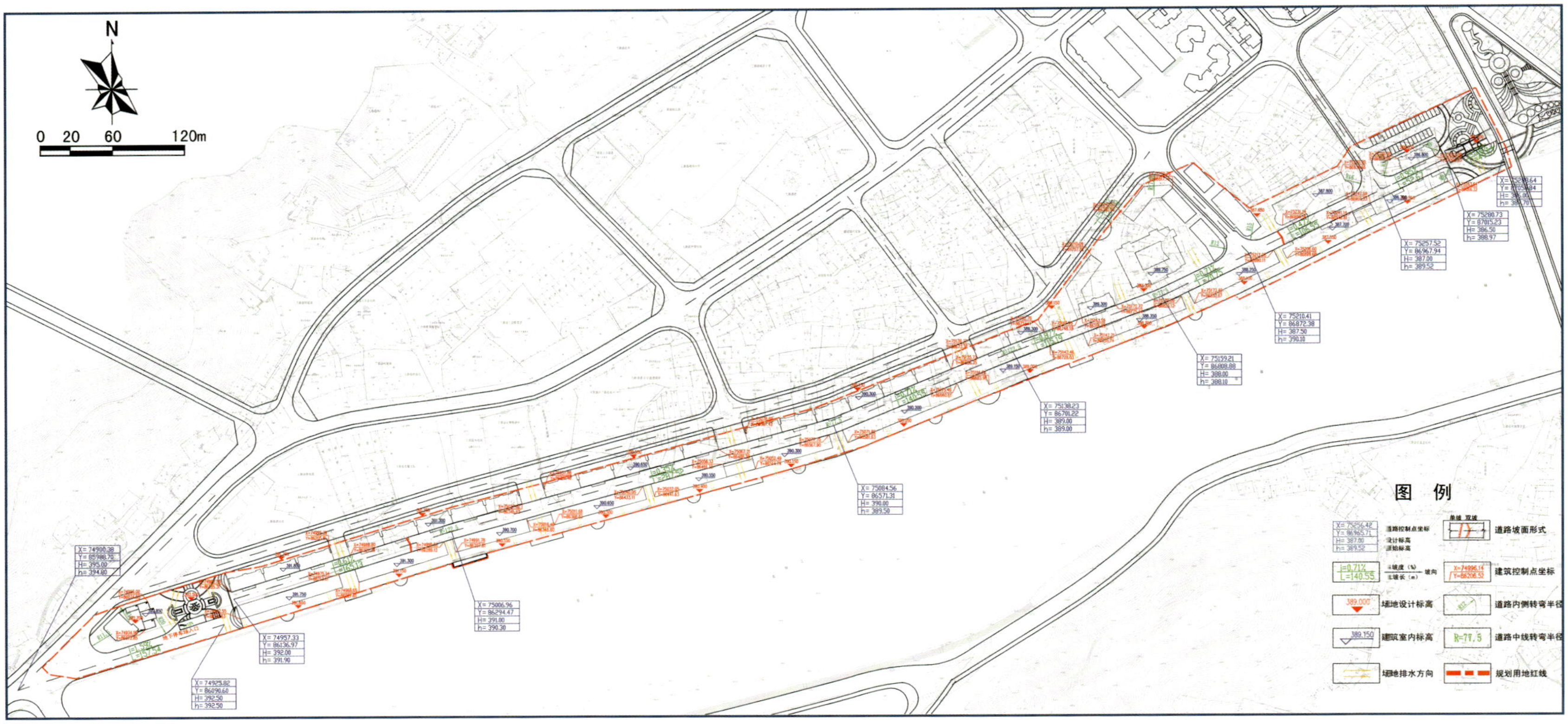

给水排水设计图

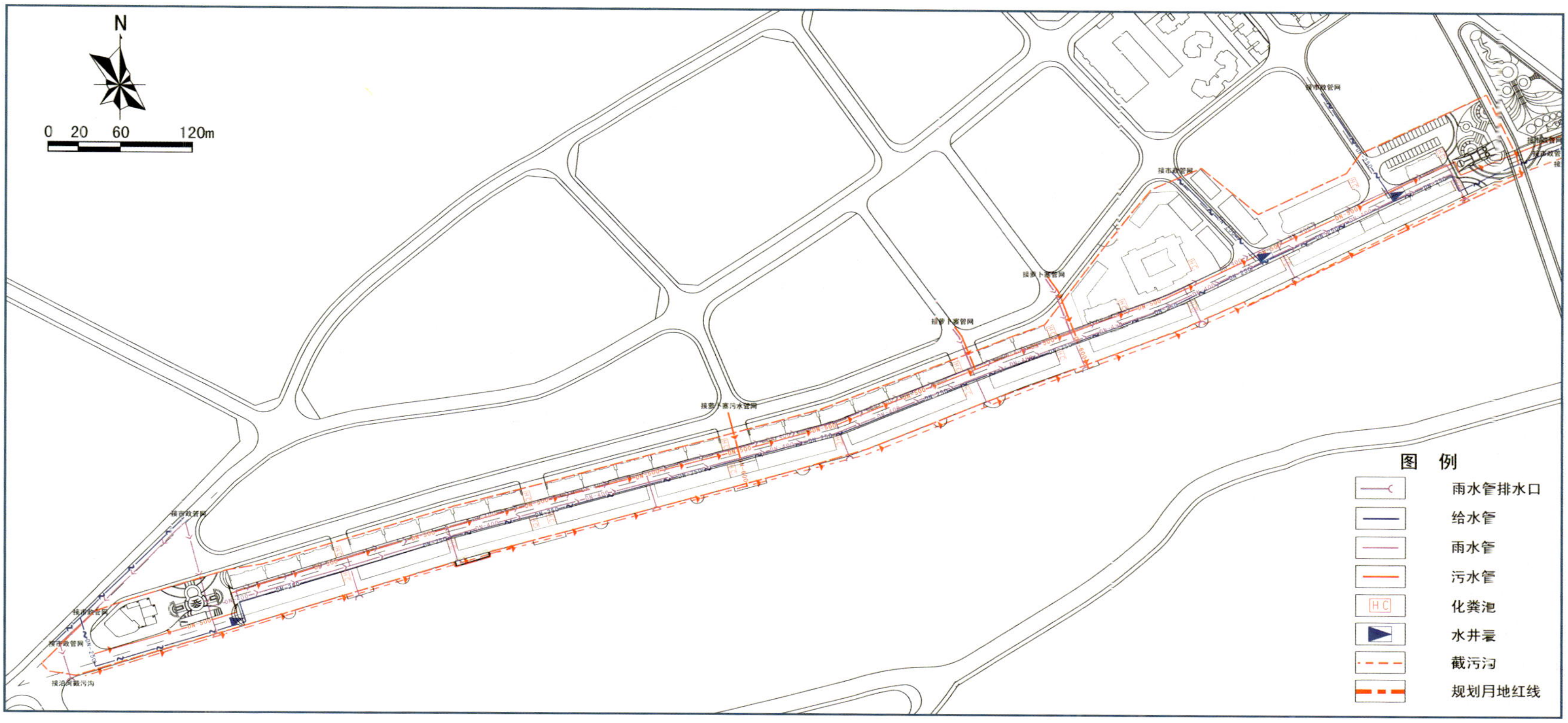

消防规划图

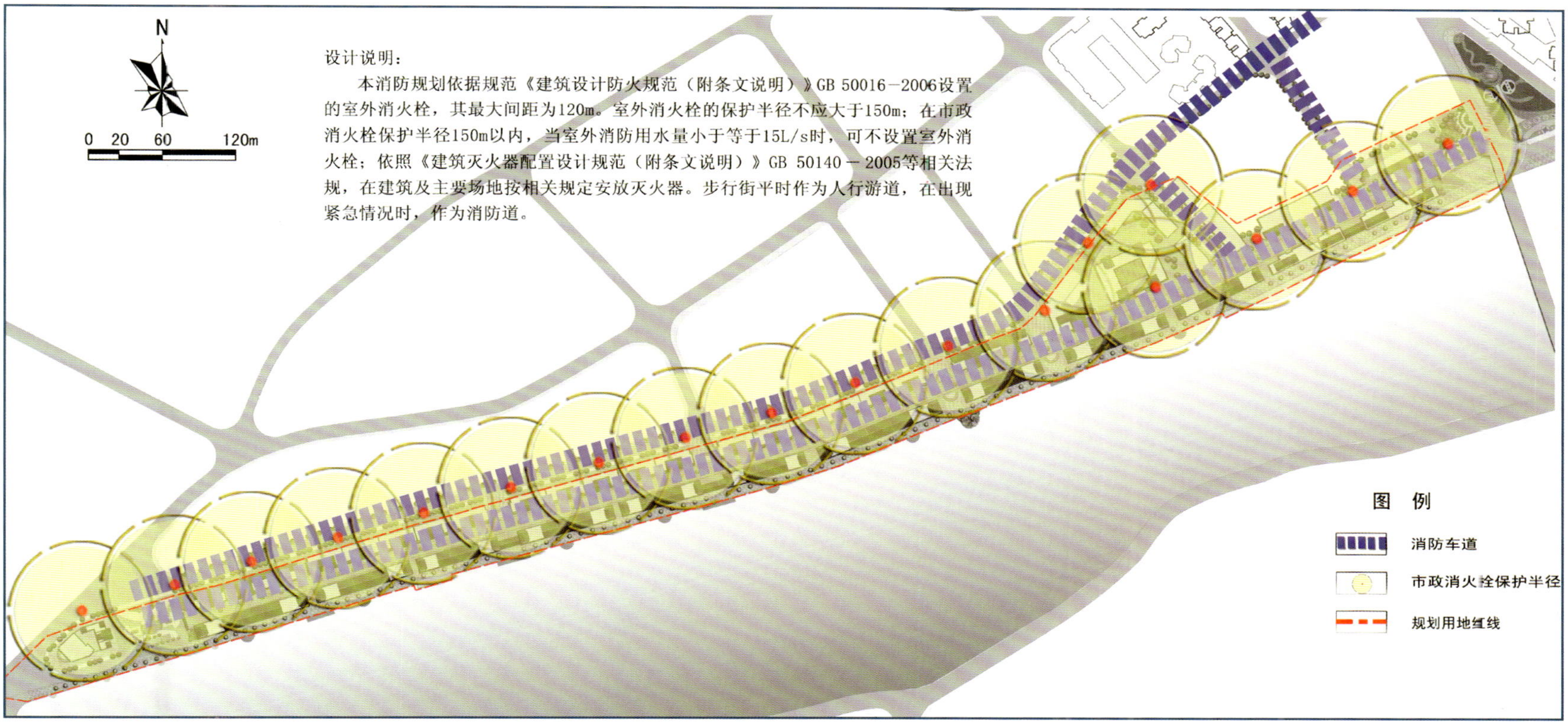

电力电信设计图

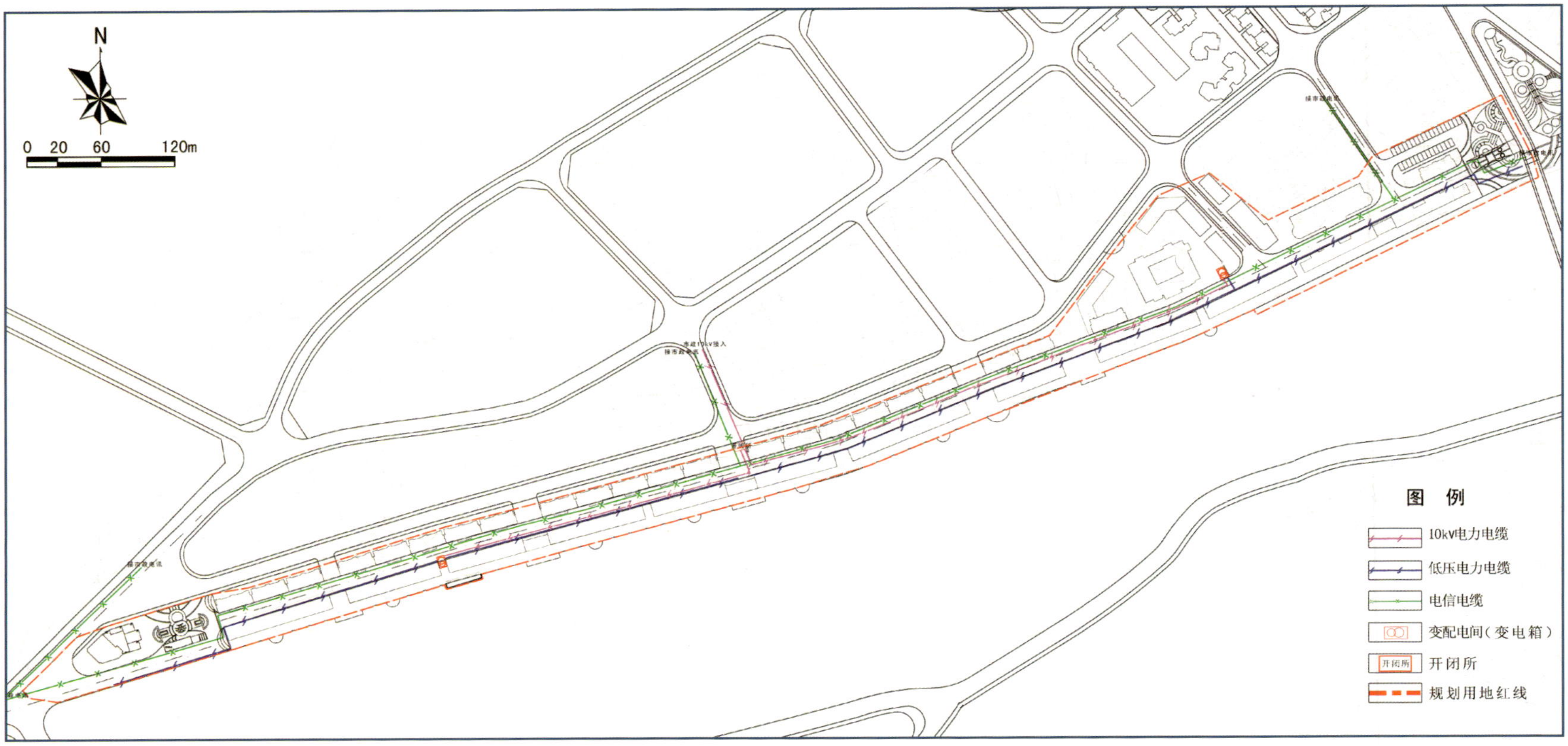

绿地系统分析图

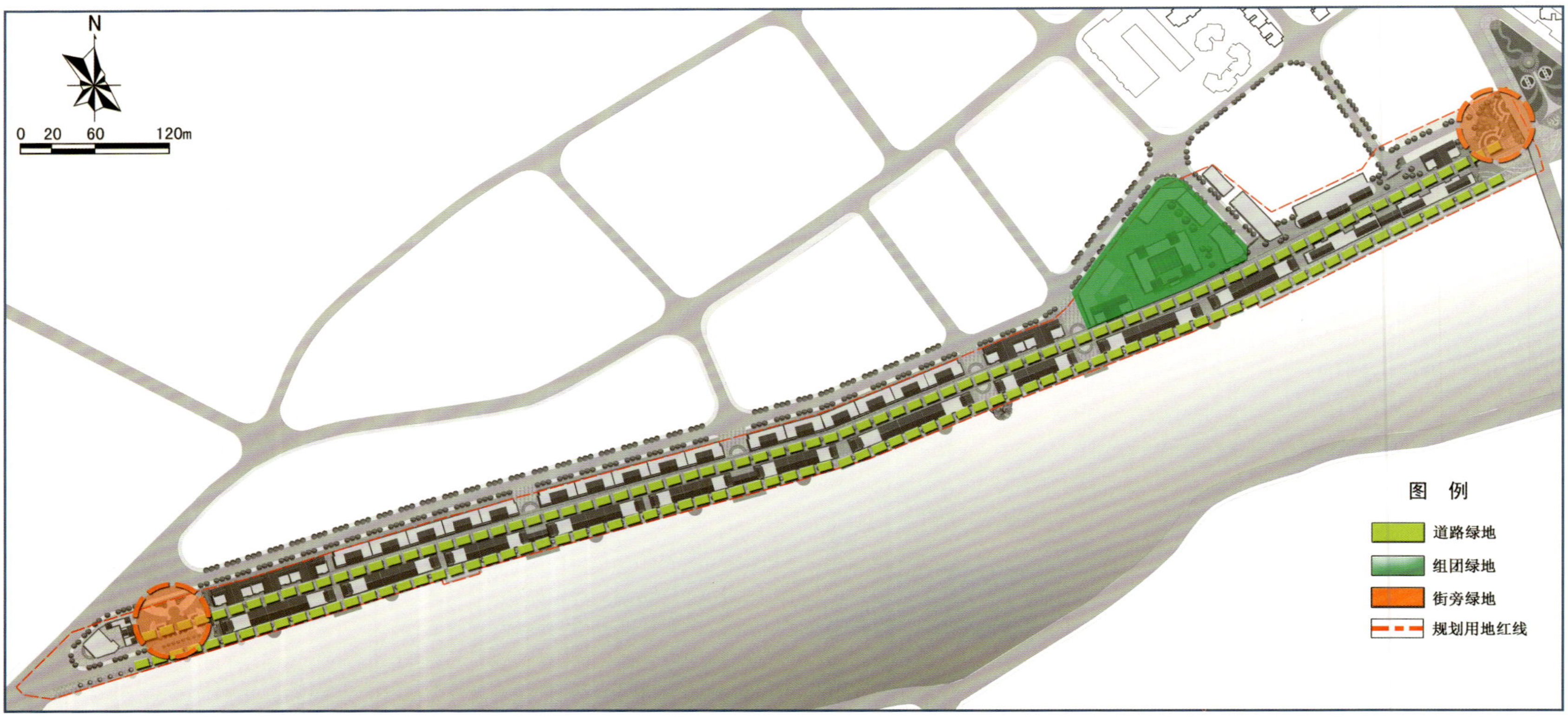

植物配置图

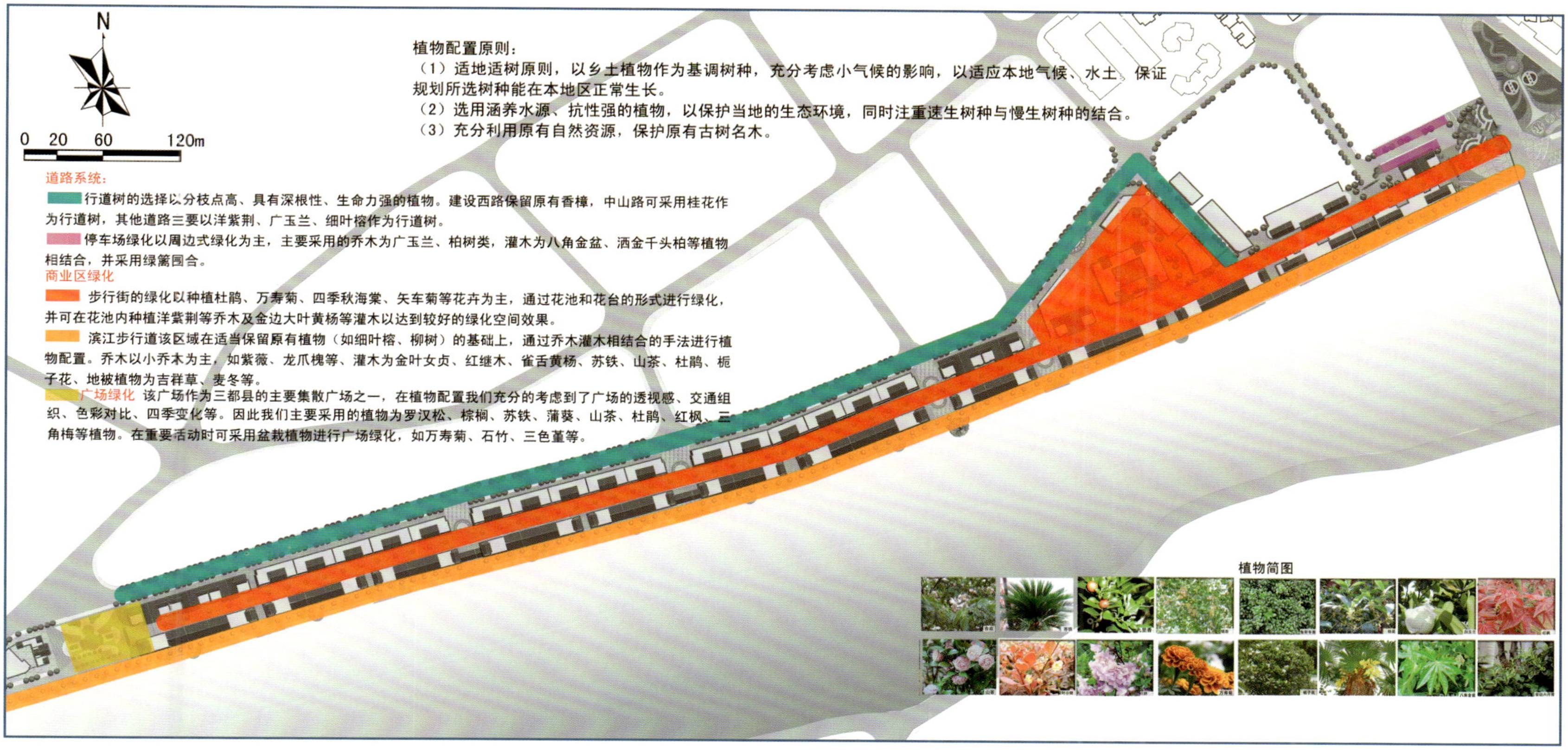

景点布置示意图

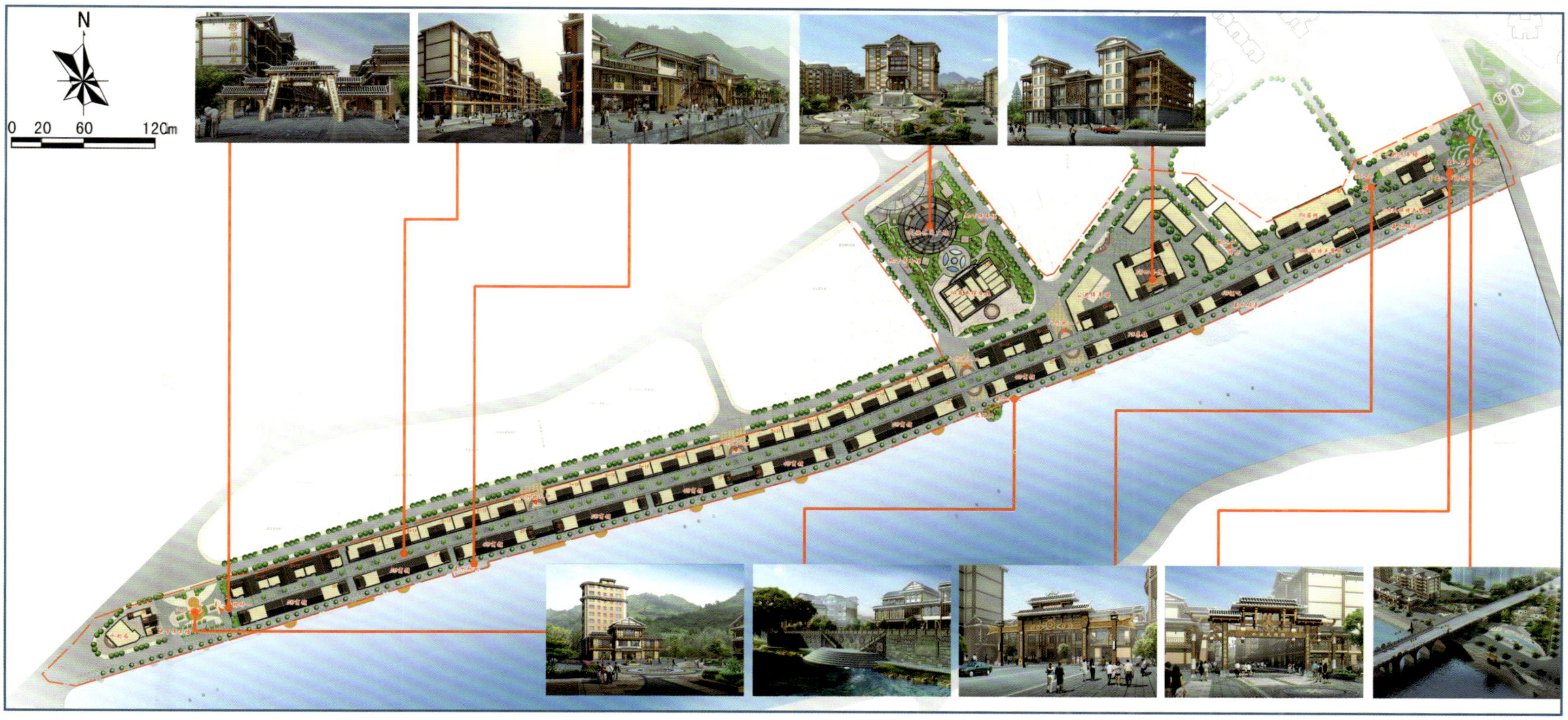

①

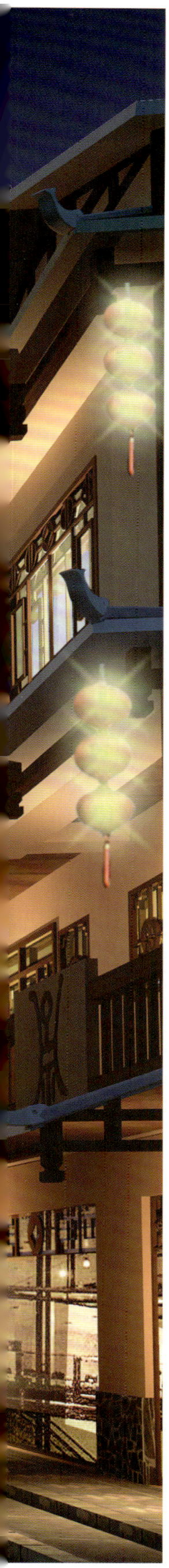

①风情街透视图 ②风情街入口效果图 ③风情街建筑效果图 ④风情街透视图 ⑤风情街建筑效果图 ⑥风情街建筑效果图 ⑦风情街透视图

①

①风情街滨江效果图 ②风情街入口效果图 ③风情街滨江局部效果图

三郎麻光新区规划设计

Planning and Design of the New Area of Maguang Sanlang

一、现状

麻光三郎片区位于三都老城区西南方向，是通往荔波的交通要道。都柳江从该片区中部穿过，将整个区域分为江北麻光片区和江南三郎片区两个部分。片区内建筑较少，多为农田，有待系统开发。其中麻光片区有办公楼、低层住宅等，建筑质量一般，约为 1 ～ 4 层，没有形成整体风格，更不能很好地反映当地的水族建筑特色。其余建筑大多为 1 ～ 2 层村寨民房，质量普遍较差。区域内道路质量较好的有三合路、北京路、普陀路。连接三都老城区的道路三合路为硬质路面，质量一般，其余村寨道路大多为黄泥路，路面较窄、质量较差，达不到市政道路基本要求。

I Current Situation

MaguangSanlang section, located at the Southwest direction in the Sandu old town, is the main road to Libo. Duliu River crosses the middle part of this section, and divided the section into two parts: Maguang section on the North side of the river and Sanlang section on the south side of the river. There aren't many buildings in the section, but mostly farmland, so it requires being developed systematically. In the Maguang section there are office building, low-rise dwelling and so on, the architectures have common quality, about 1~4 storeys, and haven't formed integral style, let alone reflecting the architectural characteristic of the local Shun nationality. The other buildings are mostly civilian building in the villiage, which are generally poor in quality. Roads with better quality in the section are Sanhe Road, Beijing Road, and Putuo Road. Sanhe Road, which is the access to old town, is hard surface road and has common quality, the other villiage roads are mostly yellow mud roads, which are narrow, poor in quality and cannot achieve the basic requirement of municipal roads.

区位分析图

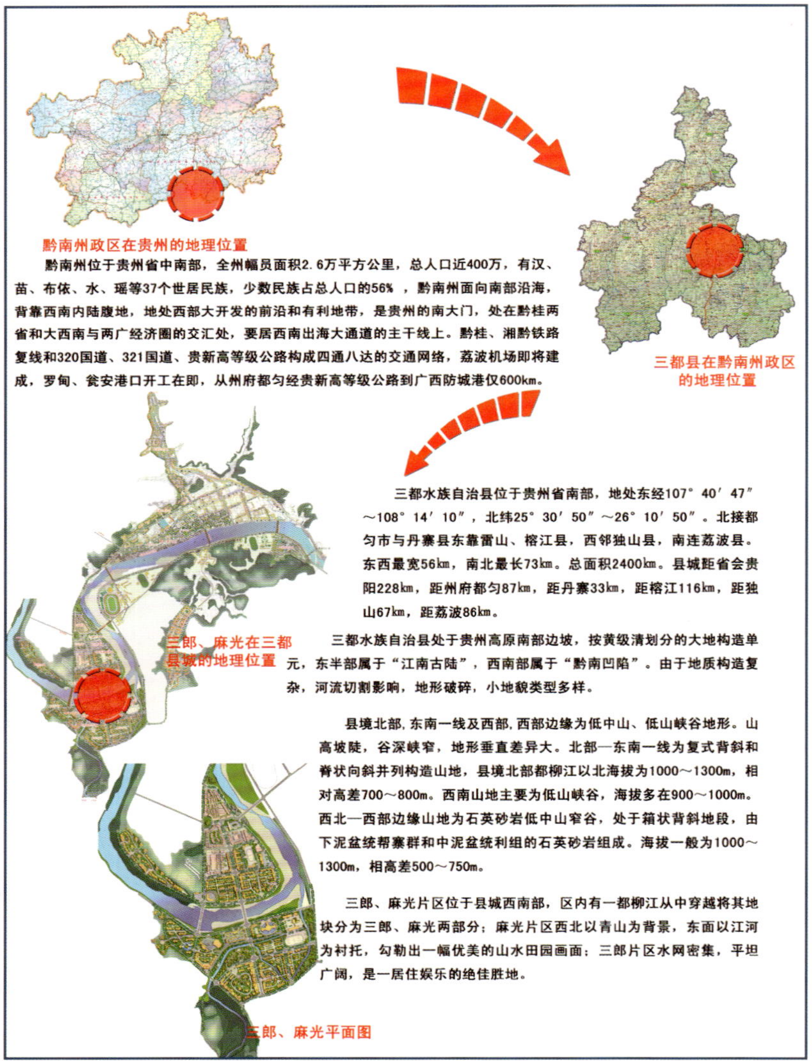

现状分析图

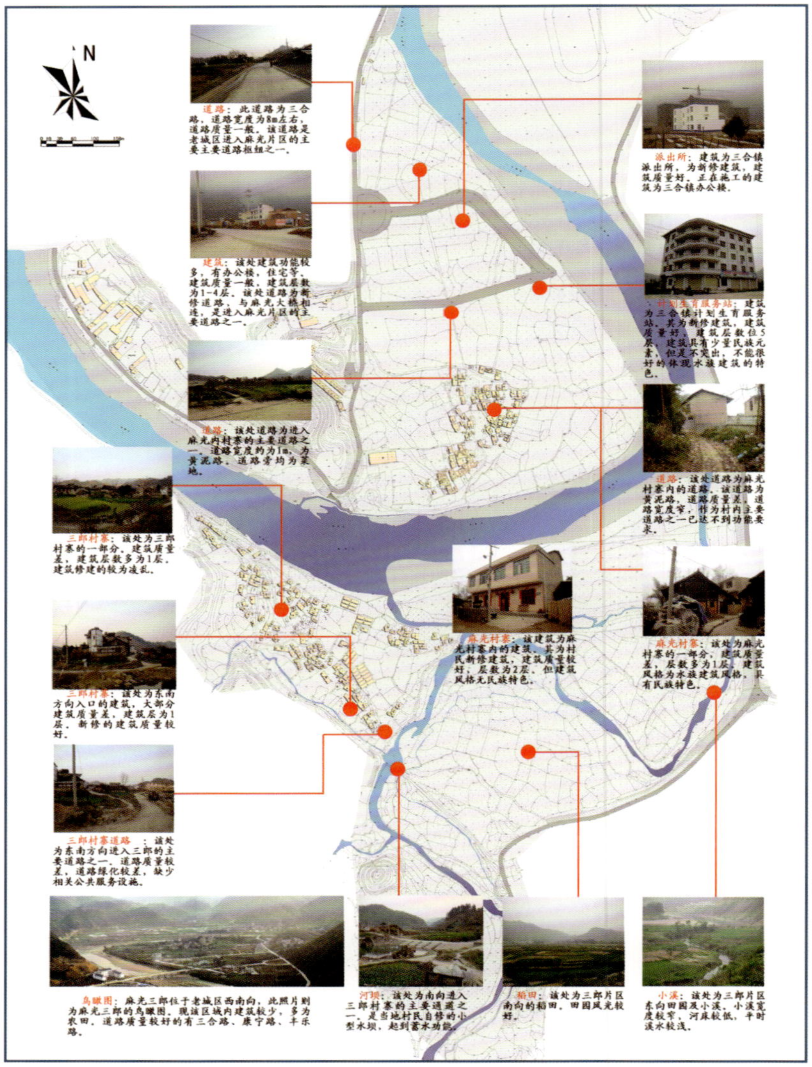

二、规划思路

麻光三郎片区是由南部进入三都县城的重要通道，也是三都建设中的新区。本区域规划用地总面积为 76.25hm^2。根据规划思路，将该规划区域定位为老城区的延续及水族文化的展示平台。在规划中按照道路划分，将规划区分为 15 个地块，主要通过建筑、景观等的创意设计，展现水族文化的风采，力求将其打造成为具有三都特色、水族风情、形象鲜明的县城新区。

II Planning Idea

Maguang Sanlang section is an important route of accessing Sandu county town from the South, and it is also a new section in the construction of Sandu. This section plans to use a gross area of 76.25 hectare. According to the idea of plan, this planning section is oriented as a continuation of old town and a show platform of the Shui culture. In the layout, the section is divided into 15 land parcels according to the road, mainly though creative design of architecture and landscape, to show the charm of the Shui culture, and strive to make it a lively new county section with Sandu characteristic and charm of the Shui nationality.

高程分析图

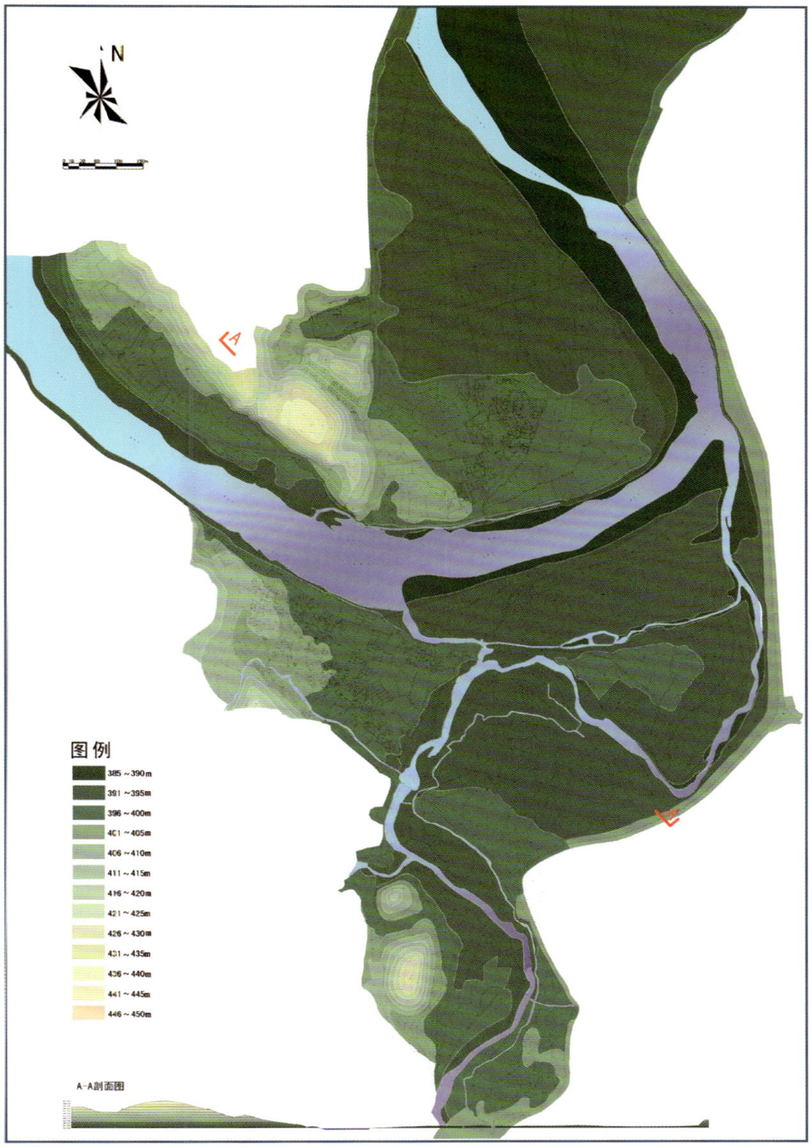

坡度分析图

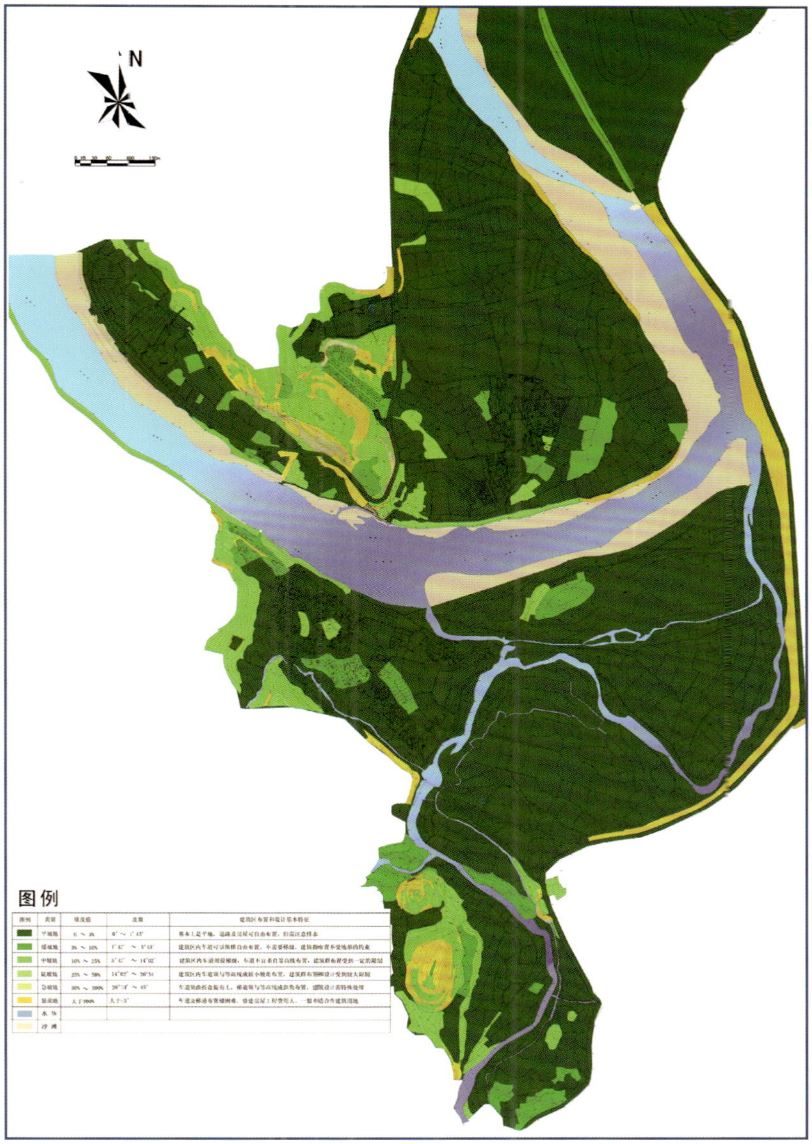

三、功能布局

本规划以《三都县城中南片区控制性详细规划》为依据，主要将该区域分为居住区、商业区、旅游度假区、文化办公区、幼儿园、小学、中学用地、医疗用地、农贸市场、民俗自然村寨、已出让地、滨水景观带、市政设施用地以及公共停车场等功能区。

Ⅲ Layout of Function

This plan is based on "detailed plan of central North section control in Sandu county", mainly divides this section into residential quarter, business quarter, tourism quarter, cultural and official quarter, kindergarten, elementary school, middle school, medical quarter, open fair, ethnic and natural village, sold land, landscape belt of Bing river, facility land of municipal administration, public parking lot and other functional section.

总平面图

功能分析图

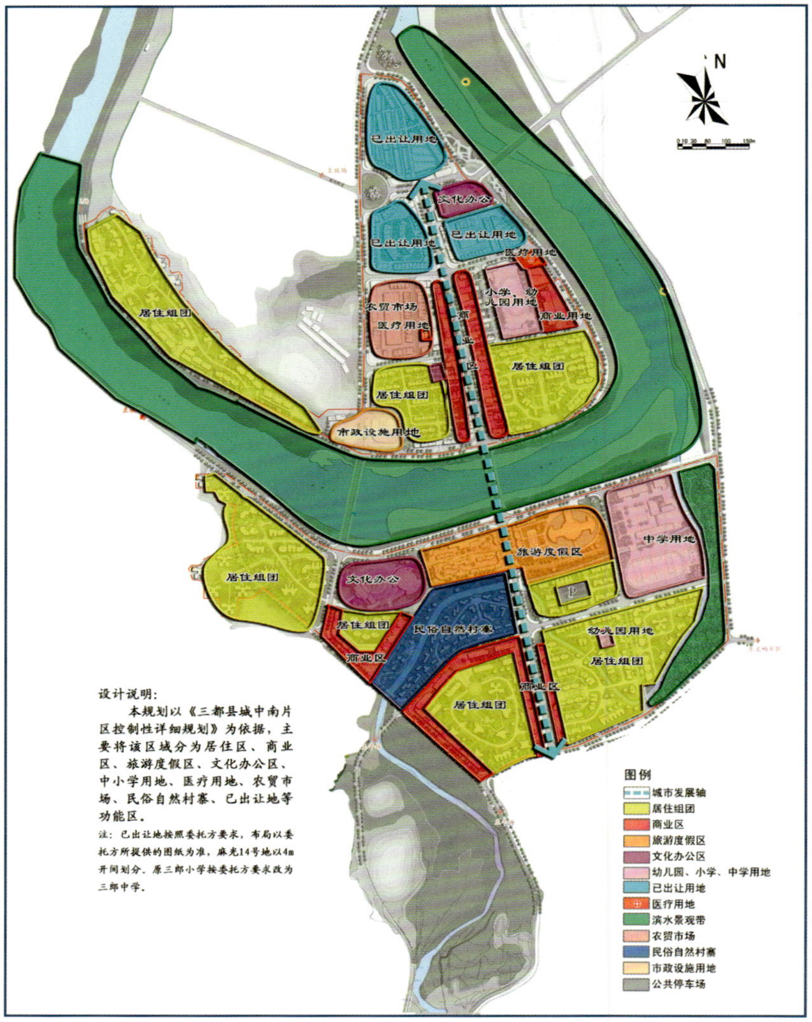

四、设计说明

1. 综合大楼

三都综合大楼是三都的新地标性建筑，大楼气势雄伟，但不失民族特征。用具象化的手段，彰显水族文化核心。通过建筑本身，去展现悠久的民族传说故事。用“水书”和“鱼”这样的特征符号，去揭开水族在人们心中留下的神秘色彩。该综合楼大门是水族的建筑元素“双鱼托福”，建筑主体部分由四把“许愿伞”构成，象征着水族人民对生活的美好祈愿。顶部中央的凤凰图腾和水族铜鼓将三都的民族特色表现得淋漓尽致。建筑上方用水书撰写的“谷米满仓、金银满屋，情满水乡、地久天长”十六个大字在建筑的宏伟气势下显得熠熠生辉，体现了水族人民对幸福生活的热爱和向往。

2. 歌剧院

歌剧院整体采用抽象凤凰造型，以现代化的玻璃与钢材结构，去演绎水族的民族精神。白色外墙面与玻璃幕墙互相映衬，使其独具文艺活动场所的艺术气息，给人俏丽生动的视觉享受。四只抽象的鱼托起了吉祥铜钱的大门造型，寓意着把祝福带给每一位前来的游客。两侧鱼鳞状的孔洞，增加了建筑的通透感，整个建筑因此而更具亲和力。建筑后部凤凰展尾的造型，不仅代表着“凤羽三都”的城市形象，还象征着城市蓄势待发、飞向未来的气势。

3. 酒店

五星级酒店的设计理念来源于水族“鱼托葫芦”的传说故事，恢宏大气的同时充满文化内涵。酒店主体建筑是葫芦的造型，酒店副楼及楼道由四条抽象的鱼组成，与主楼一起构成了“鱼托葫芦”的景观效果；主楼“葫芦”及副楼宝顶都采用了“双叠水”，体现了独具水族特色的建筑形式；酒店大门巧妙地运用了水书中的“吉”字造型，使整个酒店洋溢着浓郁的水族气息。酒店的点睛之笔在于酒店四周围绕着的水体景观，体现了鱼托葫芦于洪水中的现实景观之外，还增加了酒店的灵性及亲水性。

4. 沿街建筑

沿街建筑设计主要考虑建筑景观的连续性与整体协调性，设计通过层数变化形成有节奏的天际线，建筑立面适当运用当地民族元素以保持当地居民对自身文化的记忆性及对自身民族产生强烈的自豪感。因此，该区域沿街建筑风格与三都老城区沿街建筑的整体风格保持一致。

IV Design Specification

1. Integrated Building

The Sandu integrated building, which is a new landmark architecture in Sandu, is grand and also with ethnical characteristic. It shows the core of Shui culture by visualized method. Age old national legend is showed by the architecture itself. Characteristic symbol such as "Shui character" and "fish" are used to reveal the mystery of Shui nationality in people's mind. The gate of this integrated building uses the architectural element of the Shui nationality "two fish supporting the luck", the main body of the architecture consists of four "wish umbrella", which means Shui people's best wish for life. The phoenix totem and Shui bronze drum in the middle of roof describe fully the national characteristic of Sandu. Above the architecture, sixteen character written with Shui character "cereal and rice fills the granary, gold and silver fills the house, heart in water town and will last forever" shines in the grand architecture, and reflects Shui people's passion and longing for a happy life.

2. Opera House

The whole opera house uses the shape of an abstract phoenix, and uses the modern glass and steel structure to represent the ethnical spirit of the Shui nationality. The white exterior wall and glass curtain wall relieve against each other, and makes the opera house have the unique artistic style of literature and art activity place, bring people beautiful and vivid visual enjoyment. Four abstract fish support the gate which has the shape of copper cash, implying the meaning of bringing blessing to every visitor here. The aperture of fish scale on both sides adds sense of vision to the architecture, thus making the whole architecture more amiable. The back of the architecture has the shape of a phoenix stretching its tail, which not only represents the city's image "Sandu the feather of phoenix", but also symbolizes the city vigor of getting ready and fly to the future

3. Hotel

The design conception of the five-star hotel comes from the legend of "fish supporting cucurbit" in the Shui nationality, magnificent and at the same time full of cultural connotation. The main body of hotel is the shape of a cucurbit, the annex building of hotel and corridor are made of four abstract fish, and forms the scenery effect of "fish supporting the cucurbit" with the main building. The roof of main building "cucurbit" and annex building use the effect of "double water", which shows the architectural form with unique Shui feature; the gate of hotel cleverly used the shape of "luck" in Shui character, making the whole hotel filled with the Shui national atmosphere. The finishing touch of the hotel is the water scenery around the hotel, which shows the real scenery of fish supporting cucurbit in the flood and also adds spirituality and hydrophilic to the hotel.

4. Street Architecture

The design of street architecture mainly considers the continuity and whole harmony of architectural scenery, it forms a rhythmic skyline through the changing number of piles, facade of architecture use local national elements properly–to reserve local people's memory of their own culture and strongly feel proud of their nation. So the street building in this section keeps consistent with the street building style in Sandu's old town.

公共服务设施

道路交通分析图

竖向设计图

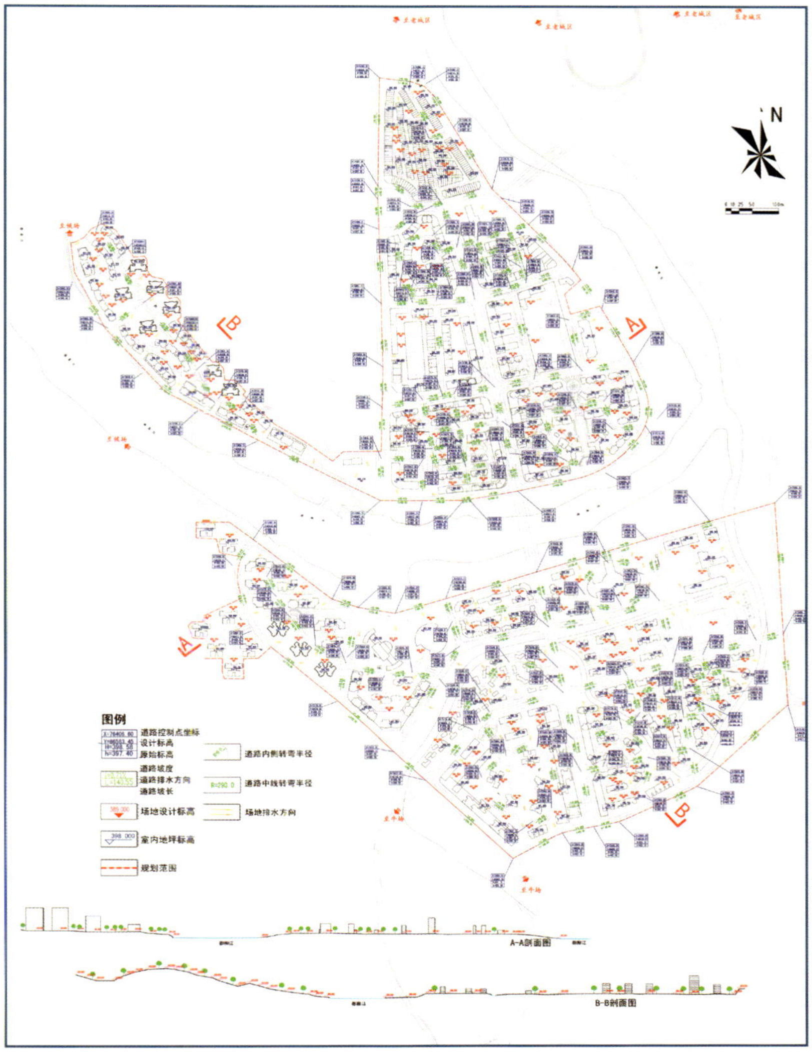

给水排水规划

消防规划图

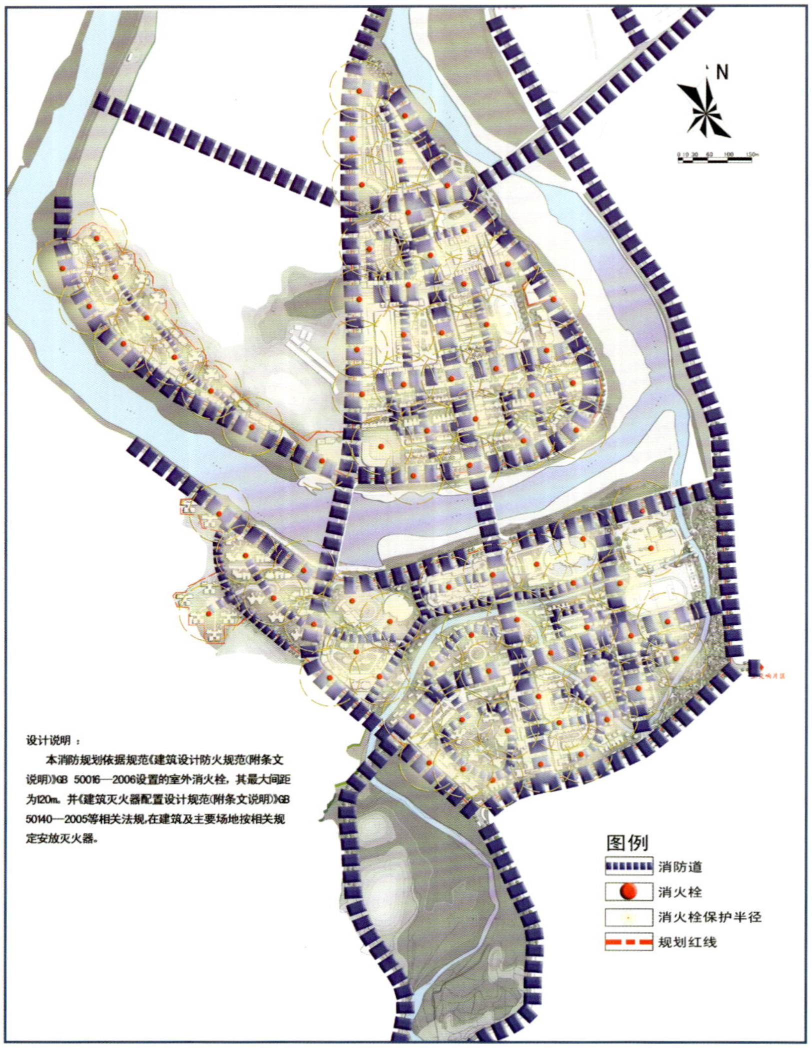

电力电信规划

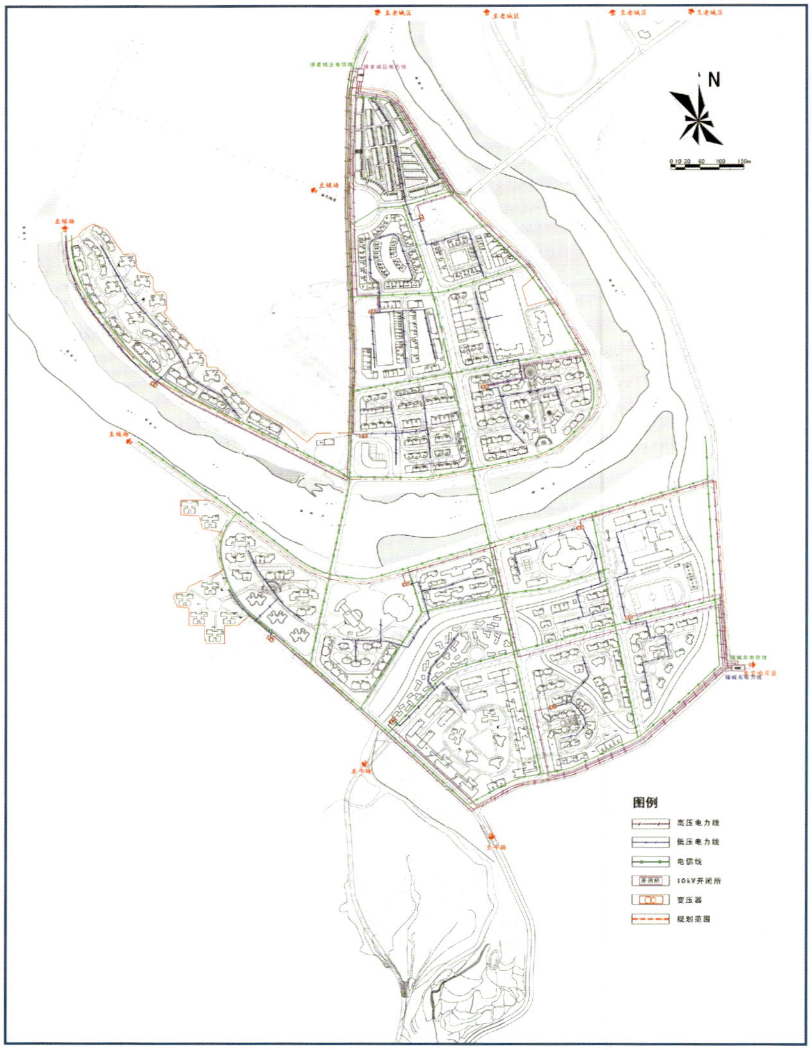

植物配置图

N
植物配置原则：
（1）适地适树原则，以乡土植物作为基调树种，充分考虑小气候的影响，以适应本地气候、水土，保证规划所选树种能在本地区正常生长。
（2）选用涵养水源、抗性强的植物，以保护当地的生态环境，同时注重速生树种与慢生树种的结合。
（3）营造自然山林带、景观林带、果园农田种植带，为游人活动提供丰富的空间体验。
滨江绿化 河流内的植物以挺水植物为主，如芦苇等，种植在水深50CM的区域内。河岸边种植一些具有较高观赏价值及保护功能的植物为
主，如柳树、紫薇以及抗风能力较
强的圆柏、竹类等。
等。
山体绿化 在原有油桐、马尾松等树种的基础上，增种一些果树如桃子、杨梅等，将该区域上的荒地开发出来作为经济果园。
小、中学及幼儿园绿化
该区域的绿化忌用有毒、有刺、有过多飞絮的植物，该区应选择些叶、花、果形状奇特、色彩鲜艳 的植物。如鹅掌楸、白玉兰、扶桑、木槿、美人蕉、杜鹃、佛肚竹等植物。
道路系统：
车行道植物的选择以香樟为主要行道树，采用乔木与灌木相结合并点缀花草的方式，营造一种生态的氛围。
停车场的植物为对称的布局手法，力求在对称中有变化。主要采用具有较强抗污染气体的植物。如广玉兰、侧柏、小叶黄杨等。
居住区道路绿化树种采用银杏、桂花、红继木、金叶女贞、杜鹃、麦冬等，营造一种欣欣向荣的氛围。
居住区： 居住区内以色彩的变化如花色、叶色果色、树干色让人产生一种赏心悦目的感觉。主要以片群植为主，采用的植物有广玉兰、合欢、紫薇、木槿、腊梅、栀子、红背桂、苏铁、边六月雪、山茶、九里香、紫叶小檗、番红花等。
民族酒店区绿化 该区域的植物宜采用一些名贵树种，如银杏、鹅掌楸、海藻、腊梅等，并可适当的配置桩景及多年生草本植物，如紫薇桩景、黑松桩景和菊花、金莲花、白芨等，注重植物色彩的搭配及季向变化。
商业区绿化 该区域主要运用对称的手法主，通过已修剪的灌木为主乔木为辅的形式，以提高景观平台绿化档次，体现一种现代感及欣欣向荣的氛围。主要种植棕榈、苏铁、蒲葵、山茶、杜鹃、万寿菊等植物。
自然村寨绿化 该区域主要运用自然式种植手法为主，主要采用的树种为楠竹、泡竹、苦竹、柏树、油茶、桑树、榕树、构树、女贞、金银花、铁扫帚、五倍子等当地树种，营造一种自然生态的环境。
广场绿化 该区域主要运用对称的手法主，主要种植棕榈、苏铁、蒲葵、山茶、杜鹃、万寿菊、红枫、龟甲冬青等植物。通过对称中求自然变化的方式，追求多样化的绿化形式。
设施绿地：该区域包括公共设施和市政设施两个区域，可根据不同的区域采用不同的植物居住区内以色彩的变化如花色、叶色果色、树干色让人产生一种赏心悦目的感觉。主要以片群植为主，采用的植物有广玉兰、合欢、紫薇、木槿、腊梅、苏铁、边六边月雪、山茶、紫叶小檗等。
植物简图

①

①麻光三郎新区规划鸟瞰 ②河道景观效果图 ③河道景观效果图 ④街道透视图

①景观广场效果图 ②~⑧街面建筑效果图 ⑨星级酒店效果图

①

①综合大楼效果图 ②歌剧院效果图 ③街面建筑效果图 ④街面建筑效果图

②

③

④

①

②

③

④

①小区建筑效果图 ②~⑧街面建筑效果图

黔西篇 Qianxi Profile

城市设计篇（第二版）

变城记

City Changing

前期策划
Prophase Planning

Introduction

In the recent years, Qianxi has relied on its resource advantages to accelerate its industrialization represented by source, so that the local economic energy and fundament have been constantly stable, which has provided the succession development with experience and fund.

At the new time and beginning, how to implement and fulfill Scientific Development View, and stress its own advantages by combining its own conditions with the principles of "men–oriented", "comprehensive coordination and succession" and lead the development of surround region even the whole trial district of Bijie prefecture?

We believe that Qianxi should press ahead with urbanization and improve urban function and public related facilities to find out urban properties, and excavate cultural elements to construct urban connotation and form unique humanistic landscape resources with the fundament of primitive industrialization, so that it can make good use of its own advantages to strengthen urban competition.

前言

近年来，黔西依托其资源优势，大力推进以能源为代表的工业化进程，使得地方经济活力与根基不断牢固，为后继的可持续发展提供了借鉴的经验和物质的保障。

在新时期、新起点上，黔西将如何落实和实践科学发展观呢？如何将发展、以人为本、全面协调可持续性、统筹兼顾与黔西自身条件相结合，从而突出自身优势呢？如何抓住贵州百年难遇的发展良机谋发展，走协调可持续之路、抓好统筹兼顾之道，力拔头筹，做带动周边区域发展乃至整个毕节试验区发展的领头羊呢？

我们认为在拥有了一个初具规模的工业化平台上，未来黔西县发展应加快城镇化水平，完善城市功能，健全公共配套设施，寻找城市特质，挖掘文化元素，构筑城市元素，形成独特的人文景观资源。以城带县，以城带产业发展，用自己的优势，提升城市的竞争力。

一、概述

黔西是一块有着悠久历史文化和优美自然风光的宝地，吸引着众多的创业者和观光客。但是由于一些历史原因，黔西县城规划建设缺乏本地特色，城市吸引力不强，这与黔西快速发展的社会经济现状十分不协调。所以，要在周边众多的城镇中脱颖而出，在激烈的竞争中立于不败之地，就必须对县城的发展方向进行重新定位，并对城镇空间进行重新打造，从而提升城市品质和县城人居环境，形成更加强有力的核心竞争力。

二、定位

充分挖掘黔西文化内涵，以黔西杜鹃花品牌为核心，以黔西水西文化、苗族文化为补充，全方位实施城市改造，将黔西打造为宜居、宜游、宜投资，唯美浪漫的杜鹃花都。

三、策划思路

1. 利用品牌优势，塑造城市形象。黔西杜鹃花经过数年的发展已经具有相当的国内影响力及一定的国际知名度，是黔西最知名的城市名片，以此为核心元素，进行城市形象的塑造，避免与周边县市产生同质性，体现黔西特色。

2. 利用民族文化，丰盈城市内涵。打破城市有魄无魂的现状，将文化之魂注入城市之魄，为城市带来生命力。

3. 利用城市设计，凸显城市张力。将黔西的自然特色和人文风貌通过城市改造进行具象化的表现，将之变成城市的元素符合实体，以此展现城市魅力。

四、打造策略

1. 完善城市功能布局，新建图书馆、博物馆、体育馆等设施，提升城市品质和创造良好的都市氛围。

2. 黔西需加强城市硬环境建设，完善的城市配套设施特别是公共设施，改善投资环境和人居环境。

3. 增加城市景观作为城市活力亮点，打造河道景观，并合理规划城市公共绿地景观，形成令人印象深刻的、具有黔西特色的城市空间，与宜居、宜游、宜投资，唯美浪漫的杜鹃花都的定位，遥相呼应。

4. 以杜鹃花为核心元素，来打造黔西城市形象，用杜鹃花来装扮城市、塑造城市景观，让来到黔西的游客处处都能看到花的身影，闻到花的芳香，感觉就像来到花的海洋，从而使黔西从周边众多城市中脱颖而出，树立起自己鲜明的形象，成为有识之士投资的热土和游客流连忘返的胜地。

Ⅰ Overall View

Qianxi, with profound historic culture and beautiful natural view, attracts numerous enterprises and visitors. Nevertheless, Qianxi is lack of local characteristics in its planning and construction due to some historic reasons that its attraction has been not enough, which can not match with its rapid development of social and economic situation. Therefore, Qianxi can not stand out from the numerous cities and towns and keep its winning position unless redesign the development position of the city and recreate urban space to promote urban quality and citizens' life and form stronger core competition.

Ⅱ Orientation

To fulfill overall urban reformation by excavating completely cultural connotation of Qianxi and with the core of azalea brand and the complementation of the culture of Shuixi and Miao ethnic group, so as to create Qianxi to be a beautiful and romantic azalea capital comfortable for living, travelling and investment.

Ⅲ Scheme Idea

1. To create urban image with brand advantages: Azalea of Qianxi has had certain impact and fame in the world, and become a well-known name card of Qianxi, which can be used as core element to create urban image in order to avoid similarity with surrounding counties and cities, and embody specificity of Qianxi.

2. To enrich urban connotation with national culture: The city will be injected with cultural soul to change the current situation of the city without soul, and to bring life to the city.

3.To highlight urban attraction with city reformation: The natural characteristics and humanistic appearances can be showed in the constructions through city reformation, and become reality in accordance with urban elements to demonstrate urban charm.

Ⅳ Creation Strategy

1. To improve urban functional layout by new constructions of library, museum and stadium, promote urban quality and create good urban atmosphere.

2. To strengthen construction of hardware facilities of urban environment, and improve related facilities especially infrastructures and investment and inhabitant environment.

3. To increase urban landscape as shining point, create river landscape and reasonably design and scheme urban public green lands, so as to form impressive and special urban space of Qianxi, to response to the orientation of romantic and beautiful azalea capital comfortable for living, travelling and investment.

4. Azalea, as the core element and ornament, is adopted to form urban image and create urban landscape for Qianxi, so that visitors can get the sense of travelling in the sea of azalea with blossom and fragrance of azalea around them, and Qianxi can stand out from many of the surrounding cities with its obvious image and become the hot place for knowledgeable people to invest and the attractive resort for tourists to travel.

设计元素（非物质文化）
Design Element

设计元素（彝族建筑）
Design Element

设计元素（明清建筑）
Design Element

城市总说

Outline

第一章 总则

第一节 规划编制背景

城市与旅游资源具有很紧密的关联性，旅游资源随着城市的发展而沉淀。在当代，旅游将成为 21 世纪的主要产业，城市与旅游资源的相关性又出现了新的特点和变化。在一些旅游资源丰富和品位高的地方，旅游资源的开发，反过来又促进了城市的发展，甚至由此产生了新兴的旅游城市。因旅游资源的开发，推动城市发展和新型旅游城市诞生的新趋势，在全国许多旅游城市中也得到了佐证。认识城市与旅游资源的相关性，及其在当代演变特点，对我们自觉地进行城市规划建设、职能分工，合理地开发利用城市旅游资源，保护城市的生态环境等各个方面，都具有极其重要的意义。

我们在对黔西进行旅游城市建设规划和风貌改造的过程中与其旅游总体规划结合起来。一方面，城市建设布局形成城市的面貌和特色，这本身就是一种旅游资源和人造环境，有利于发展城市旅游业。同时，在城市建设布局中，还十分注意妥善保护古建筑物、古遗址等人文旅游资源和生态环境。另一方面，旅游总体布局也成为城市建设布局的一个有机部分，使景区和景点的建设纳入城市建设布局规划，浑然一体。这样，景区和景点的建设才能得到城市各方面建设的配合。在这方面许多古城、名城的建设，累积了宝贵的经验。

现在，有的城市在进行城市设计和建设中不注意与本地旅游资源的特点和风格相一致，城市建筑物千篇一律，像堆积着一块块火柴盒式的积木。这样，城市本身就没有自己的面貌了，更谈不上形成吸引旅游者的独特的美好形象。这个问题在民族地区特别重要。因此，黔西县充分依托其特色鲜明的自然旅游资源及底蕴深厚的人文旅游资源来进行城市风貌打造，使黔西县城成为具有独特自然风光和多元文化的旅游服务城市，是黔西县城发展的当务之急。

第二节 规划依据

一、国家、省、市、县相关法律法规

二、项目委托书

第三节 规划范围及年限

一、规划范围

1. 旧城主干道沿街建筑外立面改造片区

莲城大道、水西大道、文峰路、公园路、水西公园周边建筑、16m 街（恭勤街）、文化路、迎宾路、迎宾北路、迎宾西路、老城区。

Chapter I General Provisions

Section I Planning Background

City and tourist resources have a very close correlation while the tourist resources accumulate with the urban development. In modern times, tourism will become a major industry in the 21st century, and the new features and changes of the relevance of cities and tourist resources have appeared. In some areas with rich and high-grade tourist resources, the development of tourism resources in turn promotes the development of the city, and even achieves the emerging tourist city. Due to the development of tourist resources, the new trend of promoting urban development and creating new tourist cities has been proved in many tourist cities in China. It is of extreme significance for us to understand the relevance of city and tourism resources and its characteristics of evolution in contemporary world in our works of consciously scheming and constructing cities and dividing functions, rationally utilizing and developing urban tourism resources, protecting urban ecological environment, and so on.

We combine the overall tourism planning with the planning and construction of the tourist city and the features transformation of Qianxi City. On the one hand, urban construction layout forms the city's appearance and feature, which itself is a kind of tourist resources and man-made environment conducive to the development of urban tourism. Also, in the process of the layout of urban construction, great attention is paid to properly protect the ancient buildings, archaeological sites and other human tourist resources and ecological environment. On the other hand, the general tourist layout tourism has become an integral part of the layout of urban construction so that the construction of scenic spots and attractions seamlessly is integrated into the layout of urban construction planning. Thus, the construction of scenic spots and attractions can adapt to all aspects of urban construction. In this regard the construction of many ancient cities has created valuable experiences.

Now, some cities pay no attention to be consistent with the characteristics and style of local tourist resources during urban planning and building. The urban buildings are monotonous and look like box-like building blocks. In this way, the city itself does not have its own style, not to mention to build a good and unique image to attract tourists. This issue is particularly important in minority areas. Therefore, through fully relying on its natural tourist resources with distinctive characteristics and human tourist resources with deep foundation, Qianxi County constructs the urban style to be a tourist service city with unique natural sceneries and pluralistic cultures which is the priority of the development of Qianxi County.

Section II Planning Basis

Ⅰ The Related National, Provincial, Prefectural and County Laws and Regulations

Ⅱ Letter of Authorization of the Project

Section III Scope and Period of the Planning

Ⅰ Scope of the Planning

1. The Reconstruction Area of the Street Facade along the Main Road of the Old Downtown

Liancheng Avenue, Shuixi Avenue, Wenfeng Road, Park Road, the surrounding buildings of Shuixi Park, 16m Street (Gongqin Street), Wenhua Road, Yingbin Road, North Yingbin Road, West Yingbin Road, and the old downtown.

2. 规划设计片区

(1) 贯城河（含博物馆、图书馆、体育馆、游泳馆）

(2) 风情街（水西广场至老县政府）

(3) 八大场馆建筑

(4) 古街新建建筑

3. 创意设计部分

(1) 少年宫、歌影剧院、体育馆、高端酒店

(2) 东西南北四大入口景观设计（前后 50m）

(3) 城区可视坡头景观

二、规划年限

根据《黔西县城市发展总体规划》，规划期限调整为：

1. 规划期：2009 ~ 2025 年，17 年。

2. 近　期：2009 ~ 2014 年，6 年。

3. 远　期：2015 ~ 2025 年，11 年。

第四节 城市规划规模

根据《黔西县城市总体规划》预测，2020 年黔西县：

一、人口规模

全县总人口达 96 万人左右，县城人口 18 万。

二、用地规模

县城范围控制在 39.6km^2，县城人均城市建设用地 81.2m^2，城市建设用地 14.6km^2。

第五节 规划指导思想

一、全面协调原则

二、可持续发展原则

三、地域个性原则

四、以人为本原则

五、历史文脉原则

第六节 基础概况

一、自然地理

1. 地理位置

黔西县位于贵州中部偏西北，乌江中游鸭池河、六广河及六冲河下游左岸，介于东经 105°47′16″ ~ 105°26′40″，北纬 26°45′49″ ~ 27°20′57″ 之间。东邻修文县，南界清镇市、织金县，西北连大方县，东北接金沙县。县域总面积为 2554.1km^2，占全省总面积的 1.46%。县城距省会贵阳市 117、118km，距地区驻地毕节市 115、102km（沿贵毕公路），行车时间 1.5h。黔西县地处省会贵阳市、地区驻地毕节市的经济辐射圈交叉部位。

2. Areas of the Planning and Design

(1) Guancheng River (including museum, library, gymnasium and swimming pool)

(2) Ethnic Street (from Shuixi Plaza to the old county government)

(3) Eight Venues

(4) New Buildings in the Ancient Street

3. The Creative Design Part

(1) Children's Palace, Opera House and Theater, stadium, high-end hotels

(2) The landscape design of four entrances on the east, west, south and north (50m wide)

(3) Visual slope landscapes of the downtown

Ⅱ Period of the Planning

According to "Qianxi Urban Development Master Plan" for the planning process be adjusted to:

1. Of the planning period: 2009~2025, 17 years.

2. Recent: 2009~2014, 6 years.

3. Forward: 2015 to 2025, 11 years.

Section IV Scale Urban Planning

According to "the overall urban planning of Qianxi County", Qianxi County in 2020:

Ⅰ Population Size

The total population will be 960 000, with 180 000 in county town.

Ⅱ Land Size

The area of county town will be controlled in 39.6 square kilometers, the area of per capita urban construction land 81.2 square meters, and the area of urban construction land 14.6 square kilometers.

Section V Planning Guidelines

Ⅰ The Principle of Overall Coordination

Ⅱ The Principle of Sustainable Development

Ⅲ The Principle of Geographical Character

Ⅳ The Principle of People-oriented

Ⅴ The Principle of Historical Context

Section V I Basic Overview

Ⅰ Physical Geography

1.Location

Qianxi County is located by Yachi River, Liuguang River of the middle part of Wujiang River and the left bank of Liuguang River in northwest-central Guizhou, between 105° 47'16" to 105° 26'40"E, 26° 45'49" and 27° 20'57"N, bordering Xiuwen County in the east, Qingzhen City and Zhijin County in the south, Dafang County in the northwest, and Jinsha County in the northeast. The total area is 2 554.1 square kilometers, accounting for 1.46% of the total area of the province. The county town is 117 or 118km far from the provincial capital Guiyang, 115 or 102km far from Bijie city (along GuiBi Road), with travel time of 1.5 hours. Qianxi County is at the junction of economic circle of the provincial capital Guiyang city and Bijie City.

2. 地质地貌

黔西县位于黔中丘原向黔西北山原过渡的斜坡地段。全县地势西北高，东南低，最高点是西北部金坡乡八抱山，海拔 1821.2m，最低点是东部太来乡野济河口，海拔 704.4m，相对高差 1116.8m。西、南、东三面为连续的深切峡谷，西北、西南和东北为山地，中部广大地区，分布着起伏平缓的丘陵洼地。

3. 气候条件

(1) 气候特征——春迟夏短，秋早冬长，四季分明；夏季热而不酷，冬季寒而不严；水热同季，干湿分明；小气候差异显著，气候类型复杂多样。

(2) 光照——百分率日照 30%，属于我国日照低值区。

(3) 气温——县城年平均气温 13.2 ~ 14.5℃，七月的月均气温为 23℃，一月的月均气温为 3.3℃；县境东南部河谷地带气温稍高，其余地区相差不大；全县气温年际变化不大。

(4) 降水——多年平均降水量为 1005.2mm，县境西南部、南部降水最多，县境中部降水量最少。

4. 水文状况

乌江干流左岸支流，横贯县境西部、中部和东北部，是本县最大的河流。野济河全长 106km。

二、人文环境

1. 政区人口

人口以汉族为主，有彝、苗、布依、仡佬、白、满等 17 个少数民族，少数民族占 24.7%。2001 年全县总人口 80.07 万，其中非农业人口 5.67 万人，城镇人口约 14 万，城镇化水平 17.5%。

2. 历史沿革

据沙井乡“观音洞”出土文物考证，黔西县是人类发祥地之一，属长江以南旧石器时代早期文化的代表，早在五六十万年前就有人类在黔西这片土地上繁衍生息。夏、商时为梁州南徼外“荒裔”。商末周初属鬼方联盟的卢夷国。春秋、战国时，先后为蜀国和鳖国地。秦嬴政二十七年入版图，属夜郎县。汉元光五年属鳖县。齐、梁之际，昆明夷闽支卢鹿部占据鳖县等地，实行部族统治。唐初，卢鹿部遥附。贞观中期，鸭池河、六归河沿岸入琰川县。武后万岁通天二年，基本以今县境置龚州。北宋沿置。皇祐至嘉祐年间，卢鹿部建立“罗氏鬼国”，州名消逝。元初，置亦溪不薛总管府和宣慰司。大德七年废亦溪不薛府、司。洪武十五年，筑水西城（今县城）；次年，置“龙九驿”中的谷里、水西、奢香（西溪）三驿站。崇祯三年属水西宣慰司，水西城为设治地，分置则窝，以著则溪。

2. Geological Features

Qianxi County is located in slopes of the central Guizhou downy area towards northwestern Guizhou highland. The land of the county is higher in the northwest, and lower in the southeast. The highest point is on the Babao Hill of Jinpo Town in the northwest, with the elevation of 1 821.2m; the lowest point is located at the outlet of Yeji River of Tailai Town, with the elevation of 704.4m. The relative height difference is 1 116.8m. there is a row of deep gorges in the west, south and east, mountainous land in the northwest, southwest and northeast, and a large area of gentle hilly and mash land distributed in the central region.

3. Climatic Conditions

(1) Climatic Characteristics—late spring and short summer, early autumn and long winter, four distinct seasons; hot summer with extreme heat, cold winter but not severe cold; rain and heat at the same season, distinct wet and dry; significantly different microclimate, complex and diverse climates.

(2) Light—30% of the percentage of sunshine, the low sunshine area in China.

(3) Temperature—annual average temperature is 13.2~14.5℃ in the county town, the average temperature in July 23℃, and 3.3℃ in January; the temperature is much higher in the river valley area in the southeast and little difference in the rest; the annual temperature changes slightly in the whole county.

(4) Rainfall—The average annual rainfall is 1,005.2mm, with most rainfall in the southwest and south, and least in the middle part of the county.

4. Hydrological Situation

The left bank branch of the Wujiang River runs through the western, central and northeastern county, which is the largest river in the county. And the Yeji River is 106km long.

Ⅱ Humanistic Environment

1. Population of the Administrative Region

Most of the people belong to Han nationality and there are 17 other ethnic minorities such as Yi, Miao, Bouyei, Gelao, Bai and Man, accounting for 24.7%. In 2001, the population of the whole county is 800,700, of which 56,700 non-agricultural population and 140,000 of urban population. The level of urbanization is 17.5%.

2. History

According to the unearthed relics from the Avalokitesvara Cave of Shajing Township, Qianxi is one of the cradles of Chinese people and the representative of early Paleolithic culture in the south of the Yangtzes. Human beings lived and multiplied on this land of Qianxi as early as 500~600 thousand years ago. It was a desert community outside the south Jiaowai of the Liang state in Xia and Shang dynasty and belonged to the Luyi country of the Guifang alliance at the end of Shang dynasty and the beginning of Zhou dynasty. In the periods of the Spring and Autumn and Warring States, it was the land of Shu and Bie countries successively. It was Yelang County in the twenty-seventh year of Ying Zheng in Qin dynasty and Bi County in the fifth year of Yuanguang in Han dynasty. The branch of Lulu tribe of Yimin Kunming occupied Bi County to implement tribal rules during Qi and Liang. At the beginning of Tang dynasty, Lulu tribe attached remotely. The region along Yachi River and Liugui River was incorporated into Yanchuan County in the mid-Zhenguan period. Gongzhou was built mainly in the territory of today's county in the second year of Tongtian of the Heavenly Empress Wu and Song dynasty as the same. Lulu tribe built Luoshi ghost country between the periods of Huangyou and Jiayou and the name of the state was abolished. The Yixibuxue general government and the Xuanwei office were established at the beginning of Yuan dynasty. The Yixibuxue government and office were abolished in the seventh year of Dade. The Shuixi wall (now the county town) was built in the fifteen year of Hongwu; the three posts of Guli, Shuixi and Shexiang (Xixi) of Jiulong Post were established in the following year. It was affiliated to Shuixi Xuanwei office in the third year of Chongzhen. Shuixi city was the administrative site and include two Zexi of Zewo and Yizhu.

清康熙五年，以水西城置黔西府，辖则窝、以著、雄所三则溪。从此改土归流，结束昆明夷闽支卢鹿部在水西地区1100多年的世袭统治。康熙二十二年，降府为州。乾隆元年，上拨四川叙永厅岩上、岩下地入黔西州，全州划9个里、51甲、113寨。

民国3年元月，改称黔西县，县下行政区划为区、乡镇、保、甲。民国30年6月前，全县分9个区，有4887km^{2}面积，7月，上划北部4个区2400.53km^{2}面积入新置的金沙县。余下境域划为24乡、4镇，共218保2667甲；后又撤并23乡、3镇，直至解放。

1949年11月25日，中国人民解放军解放黔西。12月3日，中国共产党黔西县委员会和黔西县人民政府正式宣布成立。当月，全县划5个区。1950年划10个区、231个村，1953年划12个区、131个乡镇。1958年9月，各乡镇并改为81个人民公社。当年，12月下旬，将各区合并为5个大公社；1959年2月下旬，以区建公社，乡镇改称管理区。1961年，上级划大方县的青冈公社入黔西县，次年恢复各区，管理区改为公社，全县为11个区，1个区级镇，90个公社，754个生产大队、3874个生产队。1984年夏，改革政社合一体制，恢复各乡、镇、村组建置。1991年建并撤后，全县分为19乡、9镇、741个行政村，33个居委会。2001年机构改革把741个行政村调整为382个行政村。

3. 社会经济

2005年，全县生产总值（GDP）达264814万元，比上年增长20.9%。其中：第一产业98434万元，比上年增长7.7%；第二产业87415万元，比上年增长45.6%；第三产业78965万元，比上年增长15%。人均生产总值（GDP）达3151元，比上年增长20%。工业增加值53780万元，比上年增长93%。其中较大规模工业增加值33659万元，比上年增长163.1%。全社会固定资产投资达210715万元。财政总收入达22092万元，比上年增长29.8%，支出41863万元，比上年增长35.8%。社会消费品零售总额达43877万元，比上年增长16.2%。金融机构各项存款201152万元，比年初增长34%，贷款余额为160673万元，比上年增长174.8%。城镇居民人均可支配收入达6237元，比上年增长13.8%，人均增长837元。农民人均纯收入达1900元，比上年增长10.4%，人均增长296元。人口自然增长率为6.45%。

4. 基础设施

黔西地处贵阳至毕节的中间站，区位优势明显，交通便捷，贵毕高等级公路、321国道横贯全境，县内60%的乡镇通了柏油路。农业农村基础设施条件明显改善，全县通车里程达2920.2km，实现了“乡乡通”和90%以上的行政村通公路。

In the fifth year of Kangxi in Qing dynasty, the Qianxi Fu was established in Shuixi city to administrate the three Zexis of Zewo, Yizhu and Xiongsuo. Since then, the branch of the Lulu tribe of Yimin Kunming ended the over 1100-year hereditary rule in Shuixi region. The Qianxi Fu was reduced to Qianxi Zhou in the twenty-second year of Kangxi. In the first year of Qianlong, the Yanshang and Yanxia of Xuyongting Sichuan were allocated to Qianxi state which included 9 Li, 51 Jia and 113 Zhai.

In January of the third year of the Republic of China, it renamed Qianxi County with the administrative division of the district, town, Bao, and Jia under the county. The county was divided into 9 districts with total area of 4,887 square kilometers before June of the thirtieth year of the Republic of China. In July, the Jinsha County was established with an area of 2 400.53 square kilometers of 4 districts in the north. The rest territory was divided into 24 townships and 4 towns, accounting for 218 Bao and 2667 Jia; then merged into 23 townships and 3 towns until the liberation.

On November 25 1949, Qianxi was liberated by the People's Liberation Army. On December 3, the Qianxi Committee of the Chinese Communist Party and the People's Government of Qianxi were proclaimed. The county was designated as 5 districts at the same month, 10 districts and 231 villages in 1950, and 12 districts and 131 townships in 1953. All districts and towns were changed to be 81 communes in September 1958. All districts were amalgamated into 5 major communes in late December of this year; the communes were built on the basis of districts and towns renamed administrative districts in late February 1959. The superior government allocated the Qinggang commune of Dafang County into Qianxi County, resumed all the districts in the following year, and changed administrative districts into communes in 1961 when the county included 11 districts, a district-level town, 90 communes, 754 production brigades, and 3 874 production teams. The system of the unity of government and commune was reformed to restore the township-town-village-group system in summer of 1984. After the establishment, incorporation and abolishment in 1991, the county was divided into 19 townships, 9 towns, 741 administrative villages, and 33 neighborhood committees. 741 administrative villages were adjusted to be 382 ones during the institutional reform in 2001.

3. Social Economy

In 2005, the county's gross domestic product (GDP) reached 2.64814 billion Yuan, up 20.9%. Of which: primary industry 984.34 million Yuan, an increase of 7.7%; secondary industry 874.15 million Yuan, up 45.6%; the tertiary industry 789.65 million Yuan, an increase of 15%. Per capita gross domestic product (GDP) reached 3,151 Yuan, an increase of 20%. Industrial added value reached 537.8 million Yuan, an increase of 93%, of which industrial added value above designated size 336.59 million Yuan, up 163.1%. Total fixed asset investment was 2.10715 billion Yuan. Total fiscal revenue was 220.92 million yuan, up 29.8%, with expenditure of 418.63 million yuan, an increase of 35.8%. Total retail sales of social consumer goods reached 438.77 million yuan, an increase of 16.2%. The deposits in financial institutions were 2.01152 billion yuan, up 34% over the beginning of this year, and the loan balance was 1.60673 billion yuan, up 174.8% over the previous year. Disposable income of urban residents reached 6 237 yuan, up 13.8%, with an increase of 837 yuan per capita. Per capita income of farmers reached 1 900 yuan, up 10.4%, with an increase of 296 yuan per capita. The natural population growth rate was 6.45%.

4. Infrastructure

Qianxi is located at the junction of Guiyang and Bijie, with advantageous location and convenient traffic. The Gui—Bi high-grade highway and No. 321 National Road run through the county. 60% of the townships were connected by asphalt roads. Agricultural and rural infrastructure has improved well. The mileage of road in the whole county is up to 2 920.2km. all townships and 90% of villages can be accessible by roads.

第二章 SWOT 分析

一、优势

1. 区位优势

从交通区位上看，黔西是毕节的东大门，贵阳“1 小时经济圈”辐射范围内，贵毕高等级公路穿越县域全境，县城东距贵阳 117km，西离毕节 115km，交通便利。

从旅游区位来说，黔西是贵州省规划的全省 6 个特色旅游区涉及县之一，处于黔西——织金特色旅游区中。

从经济区位来看，黔西县地处省会贵阳市、地区驻地毕节市的经济辐射圈交叉部位并且和贵阳——清镇区域、开阳——息烽区域、金沙——仁怀区域、毕节——大方区域、织金——安顺区域共五大区域交汇地，有利于资源互补以及旅游客源市场的拓展。

2. 资源优势

旅游资源丰富，资源品味较高，垄断性较强，为旅游业的快速发展提供前提。特别是“百里杜鹃”已在国内具有一定的影响和知名度，对旅游产品的打造和旅游产业的发展十分有利。另外，黔西的观音洞文化、象祠文化、水西文化等人文资源在全国范围来说具有很强的垄断性，假以时日，通过适当的开发必将成为黔西旅游的又一知名品牌。

能源资源丰富，为经济发展提供基础。

3. 经济优势

经过数年的发展，旅游业界已经认识到旅游业并非“投资小”的产业，要实现旅游业的全面发展势必需要强大的经济实力作为支持，这样既能为旅游业发展直接投入资金，又能创造良好的投资环境吸引投资。近年来，黔西能源产业发展迅速，使得黔西经济发展势头迅猛，为黔西城市的建设、旅游的发展提供了良好的经济基础。

二、劣势

1. 城市缺乏特色

近年来黔西发展势头迅猛，城市建设速度惊人，但是仍然走进了城镇建设的一大误区——只注重城镇建设的功能，从而出现了千城一面的城镇建设现状，缺乏自身独有的风格，难以形成视觉上的冲击，难以使旅游者留下深刻的印象。

2. 基础配套设施不完善

首先，全县仅有 3 家旅行社营业部，从业人员不足百人，懂经营、会管理、高素质的旅游人才队伍更是少之又少，远远不能满足旅游产业发展的需求。一方面，黔西县部分景区还处于原始状态或虽已开发但却处在停滞不前的初期状态，旅游景区景点基础设施建设滞后。另一方面，由于全县旅游资源分布在 15 个乡镇，景点分布广、间距大，甚至一些景点地处偏僻，水、电、路、通信等基础设施不能满足旅游发展的需要。

Chapter Ⅱ SWOT Analysis

Ⅰ Advantage

1. Location Advantage

In terms of Traffic Location, Qianxi is the east door of Bijie, and is located in the "one hour economic circle" of Guiyang city; the GuiBi high standard highway passes through the whole city, the county town is 117 kilometers east of Guiyang, and 115 kilometers west of Bijie, the transportation is convenient.

In terms of Tourism Location, Qianxi is the involving county of one of the six characteristic tourist district planned by the Guizhou Province; it is located in the Qianxi—Zhijin characteristic tourism district.

In terms of Economic Location, Qianxi county is located in the cross position of two economic radiation circle: the provincial capital—Guiyang city and regional resident—Bijie city, and it is also the joint of the five areas: Guiyang—Qingzhen Area, Kaiyang—Xifeng Area, Jinsha—Renhuai Area, Bijie—Dafang Area and Zhijin —Anshun Area, so it has the advantages in resource offsetting and the development of tourism source market.

2. Resource Advantage

The rich tourism resource, high resource taste as well as strong monopolization has offered the prerequisite for the rapid development of the tourism industry. In particular, the "Hundred—Mile Azalea" already has its reputation and influence in the country, which is very good for the tourism products building and the tourism industry development. What's more, the cultural tourism resource in Qianxi, such as the Avalokitesvara Cave culture, Xiang Ancestral Hall culture, and Shuixi Culture, has the strong monopolization compared with other cities in the country. given some time it would be another famous brand of the Qianxi tourism through appropriate development.

Qianxi is rich in energy resource, which provides the basis for economic development.

3. Economic Advantage

Over the years of development, the tourism industry has realized that tourism is not the "small investment" industry; strong economic strength is needed to support the overall development of tourism industry, which will not only invest funds into the tourism development directly, but also create a favorable investment environment to attract investment. In the recent years, the energy industry in Qianxi developed rapidly, made Qianxi experience rapid economic growth, and provided the good economic foundation for the city construction and tourism development for Qianxi.

Ⅱ Disadvantages

1. City Lacks Characteristic

Qianxi enjoys the rapid economic growth these years, and the city construction is at the incredible speed, however it still walk into the major error of the city construction—only pays attention to the functions of the city construction, then the current situation of "1000 cities, one appearance" appears, the city lacks its own characteristics, and it's hard to have visual impact, so it is hard to be impressive to the tourists.

2. Lacks of Infrastructures

First, there are only three Travel Agencies with no more than 100 employees, yet the number of tourist talents who has the knowledge of operation, management and with high quality is much less, which is far away from meeting the demands of the tourism industry development. On the one hand, some tourist spots in Qianxi remain in the natural condition, or may have started to develop but come to a standstill in the initial state; the infrastructure construction in the tourist spot is lag. On the other hand, as the tourism resource is spread in the 15 towns, so the tourist spots are widespread with long distance, some tourism spots are in poor neighborhood, the infrastructure such as water, electricity, highway, communication cannot meet the demand of tourism development.

其次，县城缺乏必要的相关配套设施、旅游标识系统还不完善，使得县城缺乏打造旅游城市的基本条件。

第三，缺乏大型的公共场馆满足日益增长的城市人口需求。根据马斯洛需求层次理论，当人们的生活温饱得到满足后就会追求更高层次的需求，黔西社会经济条件较好，温饱问题已经解决了，但是却缺乏满足居民更高层次精神需求的场所，一些供居民使用的大型公共场馆，如体育场、歌剧院、游泳馆等便是现阶段满足居民精神需求的物质载体。

3. 城镇建设与旅游发展脱节

相关部门尚未认识到城镇建设与旅游发展之间的联系，事实上二者之间是相辅相成的，是可以相互促进的。然而目前黔西的城镇建设与旅游开发却无法有机地结合在一起，更有甚者，就连旅游资源都分属城建、林业、水利、宗教、文化等多部门管理，各自为政，各行其是，尚未形成合力。这样做的后果便是重复建设、不必要的建设大规模的产生，造成时间、金钱以及城市形象的多重损失。

三、机遇

1. 政策导向明显

首先，《中共中央关于制定国民经济和社会发展“九五”计划和 2010 年远景目标的建议》首次把旅游业排在第三产业中“积极发展”的新兴产业序列的第一位。旅游业的特殊功能和无可替代的作用已受到世界各国和中国各级政府的高度重视，“旅游促进发展”已成为世界大多数国家的共识。目前，我国已有 23 个省、市、自治区将旅游业列为支柱产业、先导产业或龙头产业，并对旅游业的发展给予前所未有的政策性扶持。旅游业在国民经济和社会发展中的作用日益受到政府重视。

其次，近年来，贵州省委、省政府越发重视旅游业的发展，出台了《中共贵州省委、省政府关于加快旅游业发展的意见》，提出了“加快把旅游业培育为我省新的支柱产业，逐步建成自然风光与民族文化相结合的旅游大省”的战略思路；制定了建设旅游大省并向旅游强省转变的战略目标……这一系列政策的出台增强了各级政府部门加大加快旅游业发展的决心，增强了省内外企业投资旅游业的信心。

第三，中央经济工作会议提出了“以发展城郊农业、旅游观光业促进小城镇建设”的战略方针。在 2009 年的中央经济工作会议上提出立足当地实际，挖掘生态资源、旅游资源，依靠中心城市辐射和带动小城镇发展，也是加快小城镇建设的一条有效途径。这样发展起来的小城镇，集生产、服务、观光、旅游、休闲、娱乐为一体，可以形成城乡联动发展的新格局。

Second, the county town is in lack of necessary relevant supporting facilities, and the tourism sign system is not perfect, so the county town lacks the basic requirements to build a tourism city.

Third, large public building is in need to meet the ever increasing needs of the urban population. According to the Maslow-Need-hierarchy theory, people will seek to achieve nonmaterial or spiritual goals after their basic material needs are satisfied, the social and economic conditions in Qianxi is good, and it has already resolved the problem of feeding the people, but it lacks the places to meet people's higher spiritual demands. The major public buildings provided for the residents, such as stadium, opera and swimming poor is the substance carrier of the citizen's spiritual demands.

3. Gap Between City Construction and Tourism Development

The relevant government has not yet come to realize the connection between the city construction and tourism development, but in fact these two aspects are complementary to each other, and can promote mutually. However, at present the city construction and tourism development cannot be combined organically, what's worse, the tourism resource is in charge of different departments such as city construction, forestry, water conservancy, religion, culture separately, each department goes its own way, so they can not form resultant force. The consequence of this is the frequent appearance of repeated construction as well as unnecessary construction, which causes multiple losses in time, money as well as city identity.

Ⅲ Opportunities

1. Obvious Government Guidance

First, the tourism industry was placed top in the series of the "actively developed" emerging industry of the tertiary industry for the first time, according to the "the Central Committee of the Communist Party of China's suggestions for making the plans of the national economy and social development and 2010 long term vision". The special function as well as irreplaceable role of tourism has made the countries all over the world and Chinese government at all levels pay high attention to this industry, "tourism promotes economic growth" has been recognized by most countries of the world. At present, tourism has been listed as pillar industry, forerunner industry or leading industry in 23 provinces, municipality or autonomous region of our country, and has been given the unprecedented policy support for its development. The role of tourism in the national economic and social development has received increasing attention.

Second, Guizhou provincial Party committee and provincial government has paid more and more attention to the development of tourism industry in the recent years, they issued "Guizhou provincial Party committee and provincial government's suggestion about speeding up the development of tourism industry", put forward the strategy of "accelerate the step of cultivating tourism industry as the new pillar industry in our province, build up the major tourists' destination province step by step that combine natural scene with national culture"; set the goals of building big travel province and then changing to strong travel province… The appearance of these policies increase all levels of government department's determination to speed up the development of tourism industry, and enhance the confidence of enterprises' both in and out of province to invest in the tourism industry.

Third, the Central Economic Working Conference put forward the strategic concept of "promote the construction of small towns through the development of Suburban Agriculture and tourism industry". In the year of 2009, during the Central Economic Working Conference, it was put forward that it is an effective way to speed up the construction of small towns to take advantage of ecological resource, tourism resource based on local conditions, and promote the development of small towns depend on the radiation of the central cities. Under the circumstances, the small town will be a collection of production, service, sightseeing, tourism, leisure and entertainment, and the new pattern of urban and rural integrated development will appear.

第四，“旅游活县、城建带县”战略方针向纵深发展。自黔西县提出“旅游活县、城建带县”战略发展方针以来，黔西各级政府领导将之提上议事日程，经过数年来的推进证明这样的战略发展方针是切实可行，卓有成效的，因此在今后的发展过程中势必会将这一战略方针朝向纵深推进，为黔西县旅游发展及城镇建设带来更大的机遇。

2. 交通区位越发明显

成贵铁路在黔西境内设站，届时从黔西到贵阳只需要20min车程，而从成都到黔西的车程也将大大缩短。另外，成贵快速铁路是与贵广快速铁路连接的，届时黔西通往珠三角的交通也将被打通，为黔西开拓省外市场带来了巨大优势。

黔西至织金、清镇的快速公路将于今年内开工建设，建成通车后，黔西将与贵州旅游的传统精品线路——“贵州西线旅游”串联起来，届时将与天龙屯堡、织金洞等国内知名景区直接连接、并将进一步辐射到黄果树瀑布、龙宫等贵州精品景区，形成强有力的联合。与此同时，黔西旅游将形成多重通道，区位优势越发显现。

Fourth, to develop the strategic "promote the development of county through tourism, and support the development by city construction" in both scope and depth. Since this strategic was put forward, the leaders at all government levels in Qianxi have put it on their agenda, and it has been proved to be practical, effective through these years' practical experience, so this strategic will continue to be advanced to depth in the future development of the process, and bring greater opportunities to the tourism development and city construction in Qianxi.

2. Increasing Traffic Location Advantage

The ChengGui Railway will establish its station in Qianxi County, by the time it would cost only 20 minutes between Qianxi and Guiyang, and the time between Chengdu and Qianxi will also be greatly reduced. What's more, the ChengGui Railway is connected with GuiGuang express railway, so the transportation line to Pearl River Delta will be open to traffic, which will bring huge advantages for Qianxi to explore markets in other provinces.

The freeway between Qianxi and Zhijin, Qingzhen will begin its construction this year, when it open to traffic, Qianxi will then connect with the traditional fine line of Guizhou tourism industry–Guizhou west line traveling, at that time it will connect directly with the famous tourist spots such as Tianlong Tunpu, Zhijin hole, and will further radiate to the major tourist spots in Guizhou such as Huangguoshu Falls and the Dragon King's palace, and form a strong coalition. At that time, the tourism in Qianxi will have multipath, and its location advantage will be more obvious.

四、挑战

1. 城镇建设是否能落实到位

近年来，各地政府越来越认识到了规划编制的重要性，各级各类规划层出不穷，这本来是好事，但是许多地方政府的实践经验却显示规划对城镇建设的指导作用不尽如人意，多数规划都逃不出“墙上挂挂”的尴尬境地，规划与实施的脱节成为城镇建设中的普遍现象，黔西县的城镇建设能否摆脱这一怪圈，落到实地是当前面临的一大挑战。但是只要黔西县县委县政府能充分认识到这一点，理顺体制，下定决心将之落实到底，势必就能将此威胁转化为机遇。

2. 周边景观的威胁

黔西旅游资源的亮点之一“百里杜鹃”处于黔西、大方交界处，两个县对资源共享；“支嘎阿努”湖处于黔西、大方两县的争夺中；“东风湖——乌江源百里画廊”处于黔西、织金、清镇的争夺中；索营峡谷处于黔西与修文的争夺中……黔西资源与周边县市存在一定的同质性，为其旅游业的发展带来一定的威胁。但是只要黔西县抢占先机，做到“快人一步，抢先开发”，形成先入为主的态势，势必能将威胁转化为机遇。

Ⅳ Challenges

1. Whether the City Construction Can be Put in Place

Local governments have increasingly recognized the importance of planning compilation these years, and various and all kinds of planning emerge one after another, this is a favor originally, but the practical experience of many local government shows the guidance role of planning to the city construction is not satisfactory, most planning cannot escape from the awkward situation of "hanging on the wall". It is common that the planning is separated from implementation, so it's a big challenge for Qianxi to shake off this weird cycle and implement the city construction in place. But as long as the Qianxi government can fully realize this problem, rationalize institutional issues, and make up its mind to carry it out, this threat will certainly changes to be opportunity.

2. Threat of the nearbyTourism Spot

One highlight of the Qianxi tourism resource "Rhododendron thymifolium Maxim" is located in junction of Qianxi and Dafang, the two cities share this tourism resource; "Zhi Ga A Nu Lake" is in the contention between Qianxi and Dafang county; "DongFeng Lake—Wujiang source Barry Gallery" is in the contention between Qianxi, Zhijin and Qing Town; "Suofengying Canyon" is in contention between Qianxi and Xiuwen. There is homogeneity in the tourist resource between Qianxi and other cities nearby, which will bring certain threat to its tourism development. But as long as Qianxi can seize the opportunity, develop his tourism industry before the others, and form the preoccupation situation, the threat will be changed to be opportunity.

第三章 形象定位及总体目标

一、战略定位

充分挖掘黔西文化内涵，以黔西水西古彝族文化为核心、杜鹃花文化、明清文化为补充的宜居、宜游、宜投资，唯美浪漫的旅游城市。

Chapter Ⅲ Image Positioning and Overall Objective

Ⅰ Strategic Positioning

To fully explore the cultural connotation of Qianxi, and become the beautiful romantic tourism city suitable for living, sightseeing and investing, with Qianxi Shuixi Ancient Yi culture as its core, and azalea culture as well as Ming and Qing culture supplemented.

二、形象定位

水西胜境，杜鹃花都——中国黔西

黔西，是人类发祥地之一，早在五六十万年前就有人类繁衍生息，因此在这块土地上留下了丰厚的文化成果，是水西文化的中心地带：从远古观音洞文化到象祠文化到地方少数民族文化再到水西彝族文化都在黔西得到了发扬光大，使得黔西绽放异彩，向世人呈现出了“水西胜境”的意境，但是随着时代的推移，水西文化也慢慢淡化，水西文化是黔西的根基，是黔西的精髓，重塑水西文化，再现“水西胜境”便成为了城市形象当仁不让的选择。

杜鹃花是传统十大名花之一，唐代大诗人白居易就有“花中此物是西施”的美誉。黔西境内的国家级森林公园“百里杜鹃”是地球同纬度地区保持最完好、规模最大、品种最齐全的杜鹃花带，这里的杜鹃花汇聚了众多优良品种，每到花期千山锦绣，灿若云霞，让游客留下了“花繁已临危世少，品珍更叹绝代奇”的诗句，因此黔西被称为“中国杜鹃花都”，在此基础上，黔西以花为媒，举办了“中国贵州国际百里杜鹃花节”提升了黔西旅游形象的同时，也搭建了良好的投资平台。综上，可以说黔西是“杜鹃花都”实至名归，更重要的是“杜鹃花都”这一形象已经具有了一定的知名度和美誉度，因此保留“杜鹃花都”这一主题形象便顺理成章。

三、目标定位

通过此规划实现总规关于黔西“多元文化的旅游服务城市”的定位。

通过旅游城市打造对黔西县城主体风貌的塑造，提升城市的人文内涵，增添城市的游憩景观，从内容到形式上丰富县城的文化生活。

通过旅游城市打造，全方位构筑自身特色，提升城市品位，形成宜居、宜游、宜投资的特色城镇，从而带动黔西经济社会的全面发展。

第四章 功能布局

第一节 老城区功能分区

临街商铺区、商业中心区、停车场及广场区、古街区、文化娱乐区、商住混合区、教学科研区、住宅区

第二节 贯城河功能分区

一、度假休闲区

理念：感受黔西之悠然自得

内容：天然游泳池、会所、茶楼、玲珑洞、游泳馆、儿童戏水池、青少年宫

Ⅱ Image Positioning

Shuixi Tourism Resort, Capital of Azalea—China Qianxi

Qianxi is one of the human birthplaces; people lived and multiplied here as long as five to six hundred thousand years ago, and left rich cultural fruit in this land. It is the heartland of the Shuixi cultural: from ancient Avalokitesvara Cave culture to Xiang Ancestral Hall culture, from local minority culture to Shuixi Yi culture, they are carried forward in Qianxi, boomed Qianxi with civilizations light, and presented people with the artistic conception of "Shuixi attraction". However, the Shuixi culture is fading as time goes by. Shuixi culture is the basis as well as quintessence of Qianxi, so it is the solid choice to rebuild Shuixi culture and re-present the "Shuixi attraction".

Azalea is one of the ten traditional famous flowers, it was granted the reputation as "Xishi of the flowers" by the well-known Tang Dynasty poet Bai Juyi. The national forest park "Rhododendron thymifolium Maxim"in Qianxi is the most well-preserved, largest azalea flower belt with most complete variety in the same latitudes of the earth. It has many quality varieties of azalea, during the flower season, the mountains are as beautiful as brocade and looks like rosy clouds, the tourists left the poem "It is very difficult to see so many flowers, especially those rare high quality ones" to describe this scene. So Qianxi is called as the "Capital of azalea". On this basis, Qianxi took flowers as media and held the "Rhododendron thymifolium Maxim Festival of Guizhou China", which not only enhanced the image of cultural tourism in Qianxi but also set up the good platform of investment. To sum up, Qianxi deserves the name of "Capital of azalea", what's more, this name already has certain brand awareness and reputation, so it all make sense to keep the thematic image of "capital of azalea".

Ⅲ Objective Positioning

To realize the objective positioning of "Multi-culture Tourism Service City" in the overall regulatory through this plan.

To enhance the city's cultural connotations, increase the recreation Landscape, and enrich the cultural life of the county town, by building the major character of Qianxi County in the process of developing the tourism city.

To build the city's own character, lift the taste of the city; make it a characteristic town suitable for living, sightseeing and investment, and drive the full-scale development of economy and society in Qianxi, through the development of tourism city.

Chapter 4 Functional Layout

Section Ⅰ Old Town Function Zone

Street shops area, central business district, parking lot and square area, Heritage Streets, cultural entertainment district, mixed commercial and residential area, teaching and research area, residential area

Section Ⅱ Guancheng river function zone

Ⅰ Holiday Resort

Idea: feel leisure and relaxation of Qianxi

Content: Natural Swimming Pools, Club, Teahouse, Delicate Cave, Swimming Hall, Children's Wading Pool, the Children's Palace

二、健身休闲区

理念：感受黔西之笑傲风月

内容：健身广场

三、滨河景观休闲区

理念：感受黔西之闲情逸致

内容：三角健身广场、沿河景观步道、山顶景观休憩茶楼和景观塔楼

四、娱乐文化休闲区

理念：感受黔西之恬淡静谧

内容：四星级酒店、水幕表演广场、休闲广场

五、生态湿地休闲区

理念：感受黔西之绿景怡情

内容：民族食府、钓鱼岛、梯田景观

第三节 风情街功能分区

一、入口景观区

功能内容：入口广场、牌坊、社会停车场、服务中心

二、工艺品展示区

功能内容：扬琴、蜡染、刺绣、根雕、花卉盆景、盆景、观景石、贴画、木刻、字画、杜鹃标本、影册

三、土特产购物区

功能内容：麻辣脆、糍粑包豆干、茶食（清宫御点）、擂茶、窝丝糖、核桃糖、油豆干、黄粑、玉米制品、辣椒制品、中草药（杜仲、天麻、厚朴、黄柏、白果、首乌等）

四、文化历史展示区

功能内容：文琴戏（黔剧前身）、春节灯谜、征联、苗族多声部民歌："倮朵"、"朵逗"、芦笙舞、布依族歌舞、花灯戏、傩戏、贵州梆子

五、住宿餐饮娱乐区

功能内容：茶楼、客栈、酒店、会所

Ⅱ Fitness Areas

Idea: feel nature and taste of Qianxi

Content: Exercise Plaza

Ⅲ Riverside landscape leisure areas

Idea: feel elegance and comfort of Qianxi

Content: Triangle Exercise Plaza, Sightseeing Pavement along the Rver, Sightseeing Teahouse in the Mountaintop, Sightseeing Tower

Ⅳ Cultural Recreation Areas

Idea: feel tranquillization and beauty of Qianxi

Content: Four-Star Hotel, Water Curtain Performances Square, Entertainment Plaza

Ⅴ Ecological Wetland Areas

Idea: feel green and scene of Qianxi

Content: Ethnic Restaurant. Fishing Island, Terraced Rice Fields

Section Ⅲ Folk Custom Street Function Zones

Ⅰ Entrance Scenic Area

Functional content: entrance plaza, Memorial arch, public parking area, service center

Ⅱ Handiwork Exhibition Area

Functional content: dulcimer, batik, embroidery, root carving, ornamental plant bonsai, bonsai, viewing stone, stickers, woodcut, azalea specimen, film list

Ⅲ Local Specialty Shopping Area

Functional content: spicy crisp, dried tofu in glutinous rice cake, cakes and sweetmeats (served the Qing emperor), mashed tea, wosi sugar, walnut candy, Oil dried tofu, yellow rice cake, Corn products, pepper products, Chinese herbal medicine (Eucommia Bark, Gastrodia elata Blume, bark of magnolia, golden cypress, Ginkgo Seed, fleece-flower root, etc.)

Ⅳ Cultural & Historical Exhibition Area

Functional content: Wen Qin Opera (predecessor of Qian Opera), lantern riddles in the Chinese New Year, matching antithetical couplet, Multi-tone folk song of the Miao nationality: "luo duo", "duo dou", reed pipe dance, Buyi dance, lantern show, nuo opera, Guizhou clapper

Ⅴ Accommodation Catering Area

Functional content: teahouse, inn, hotel, club

黔西县全城鸟瞰图

沿街建筑改造

Rebuilding Street–Buildings

Planning of Architecture

First of all, it is because the noble–minded ancient Yi people's contribution on maintaining social stability, national solidarity and national unity and their indelible contribution on promoting the economic development of Shuixi district that makes Shuixi Culture still be loved after the changes of dynasties. Qianxi, as the core of Suixi culture, should take responsibility of rebuilding the glory of the Culture. Thus, with the breakthrough point of ancient Yi culture which is the essence of Shuixi culture, we will excavate Shuixi culture at its root to form its features and rebuild its historic glory, and also we would like to share the ancient Yi civilization with citizens and visitors through specific sceneries.

一、规划思路

黔西沿街建筑外立面改造主要从三个方面进行考量：

首先，水西文化历经朝代的更替还能为世人所津津乐道，在于古彝人民积极维护社会稳定、民族团结以及国家统一的高尚情操，在于古彝人民为推动水西地区经济社会的发展所作出的不可磨灭的贡献，黔西，作为水西文化的中心地带理应肩负起再塑水西文化的辉煌重任。因此，在这里，我们以水西文化精髓——古彝文化作为突破口，以此形成自身特色的同时从根源上挖掘水西文化，进行历史辉煌的再塑，并通过具象的实物与市民、游客一起分享水西古彝文明。

这里需要说明的是黔西运用的彝族元素不仅局限于水西彝族，而是放眼整个彝族，力求抢占贵州彝族文化运用的先机，实现先入为主的战略发展局面。

It should be pointed out that Yi ethnic elements we will adopt are not only the one in Shuixi Qianxi, but of the whole so as to firstly make use of Yi culture of Guizhou and achieve the advantageous strategic developing situation of "first impressions are firmly entrenched".

其次，黔西作为“中国杜鹃花都”，在城镇建设中理应体现这一大自然赋予的礼物，因此，我们在这次城镇改造中也充分运用了杜鹃花元素。这一元素的运用主要体现在外立面的图案装饰上以及城市景观的营造上。

Second, Qianxi is regarded as "the Azalea Capital of China" which should be embodied in its construction. Therefore, we make fully use of azalea elements in the reconstruction of cities and towns, which are mainly demonstrated on pattern decoration of facades and creation of urban landscape.

第三，水西公园是黔西重要的标志性区域，该区域不仅承载着黔西的厚重历史，见证了黔西古彝文明，也是当地的重要休憩场所，而水西公园保留了浓郁的民清建筑风格，因此在沿街建筑改造的过程中，水西公园周边建筑沿用民清建筑风格，与水西公园保持协调。

Third, Shuixi Park, a symbolic area and an important rest place of Qianxi, has been recording the long history and witnessing the ancient Yi civilization of Qianxi. As Shuixi Park has been remained strong architectural style of Ming and Qing Dynasties, its surrounding buildings should be designed with the style of Ming and Qing Dynasties to coordinate with Shuixi Park.

综上，我们在进行沿街建筑立面改造的过程中主要运用了这三种元素，形成了风格统一，定位明确，形式各异的沿街建筑，形成视觉上的差异，向世人展示黔西的丰富文化内涵。

In all, three elements are used in street architectures reconstruction to achieve uniformities in style, clear and exact positioning, diverse flavors of street architectures and diverse vision to show people the abundant culture of Qianxi.

二、设计理念

1. 总体风格

充分挖掘黔西文化内涵，以黔西水西古彝文化为核心、杜鹃花文化、明清文化为补充，结合现代手法，全方位实施城市改造，将黔西打造成文化底蕴深厚、唯美浪漫的旅游城市。

2. 材质

彝族建筑多为干栏式木楼，根据这一彝族民居特点，结合现代建筑风格，在黔西沿街建筑外立面改造中主要使用外墙涂料、装饰木纹、深浅不一的文化石、盲道石及青石板以及青瓦，使建筑风格保持黔西特色的同时体现清新、典雅的气息。

3. 色彩

主要以黄色色系和灰白色为主，使建筑风格典雅；暖色调体现了旅游城市浪漫温馨的特点和彝族人民热情好客的风格。

不同的路段采用不同的色彩：迎宾路、莲城大道、水西大道以及民安路沿街建筑采用暖黄色调；恭勤路、黔金路、文峰路、公园路及水西公园周边建筑采用白灰色调。

4. 翘角

彝族建筑的特色在于古朴粗犷的斗拱，采用这种形式的翘角，使沿街建筑具有黔西独特的彝族韵味。

5. 屋檐

在现有建筑的基础上加上马头墙、民族风格的屋檐，与古彝文化、民清文化相呼应，美观大方的同时，形成自身特色。

6. 门、窗

沿街商业店铺的门采用简洁的民族木框加以修饰，与沿街建筑整体风格保持一致的同时能够起到点缀及装饰的效果。

多数沿街建筑保持现有的铝合金造型，与黔西简洁大方的整体造型相协调；部分路段将窗作为装饰，在原有铝合金窗的外围使用简洁的民族木窗格的装饰效果，起到点缀整体风格的效果。

7. 细部装饰

黔西沿街建筑外立面改造中，局部使用木质条纹装饰；同时，于建筑屋顶及建筑物外阳台安装木质护栏，美观实用。

Chapter Ⅱ Design Concept of Reconstructing Builcings along Streets of Qianxi

1. General Style

With the core of ancient Yi culture of Qianxi and the complement of Azalea culture and culture of Ming and Qing Dynasties, Qianxi will be constructed as a tourism city with thick cultural connotation, beauty and romance through fully excavating cultural connotation of Qianxi and with modern technique.

2. Material

Most of Yi constructions are wooden building built on stilts. According to this feature of Yi constructions and combining with modern architectural style, we mostly use exterior wall coating material, ornate wood grain, cultural stones and blind stones, blue stone plate and grey tile when reconstruct architecture facades along streets in Qianxi to retain the special style of Qianxi construction, and also display fresh, elegant flavor.

3. Color

Yellow and white will be adopted to make architectural style elegant, and warm colors can embody romance and warmth of tourism city and hospitality of Yi people.

Different color is adopted in different part of the roads: warm yellow is used on buildings along Yingbin Avenue, Liancheng Avenue, Shuixi Avenue and Minan Road; and hoar tone is given to buildings along Qingong Road, Qianjin Road, Wenfeng Road, Gongyuan Road and around Shuixi Park.

4. Rake Angle

Architectures of Yi ethnics are with the feature of rough and ancient corbel bracket with which the buildings along the streets can be of the unique Yi flavor.

5. Eaves

Horse head walls and eaves with national feature are added in existing buildings in order to response to ancient Yi culture and Ming and Qing Culture, elegant and unique.

6. Doors and Windows

Shops along streets are decorated and embellished with simple wooden frame in order to coordinate with the overall style of the buildings.

Street architectures keep the existing aluminum alloy style to consistent with simple but elegant general style of Qianxi; Windows are used as ornament in some roads. To outstand the general style by using simple ethnic wooden panel around the existing aluminum alloy window.

7. Detail Ornament

Wooden stripe will be used to decorate some of the buildings along streets of Qianxi when they will be reconstructed. Meanwhile, wooden guard bars will be installed on the roofs and the outside veranda of the buildings, which are good looking and practical.

四大入口设计方案

Design Proposal of the Four Main Entrance

改造前

迎宾路
Design Explanation for Street Scenery Planning

主要街道特色打造
Transformation of Main Streets in Special Style

黔西古城（老城区）规划设计

Plannig and Design of Qianxi City（the Old City）

一、现状

老城区沿水西路两侧呈紧凑布局，城市依山而建。贯城河由城西北向东南穿过，犹如盘龙环抱黔西县城，县城的景观资源丰富。

规划区主要包括城东路、城西路、城北路及城南路，四条道路分别位于老城区的东西南北方向，现有道路路面情况一般，沿街建筑质量一般。

I Current Situation

The layout of the old city along Shuixi road presents compactness. The city was built by the mountain with Guancheng River running through from northwest to southeast as if pleycosaurs embracing Qianxi County where there are abundant landscape resources.

The planning area mainly includes Chengdong road, Chengxi Road, Chengbei Road and Chengnan Road with each of them located in four directions of old city respectively. The existing road conditions and the quality of buildings along the street are general.

区位分析图

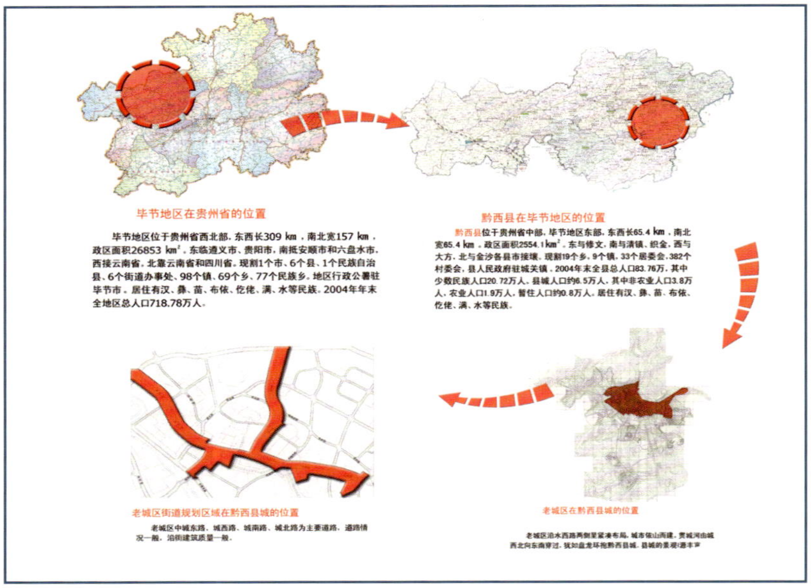

现状分析图

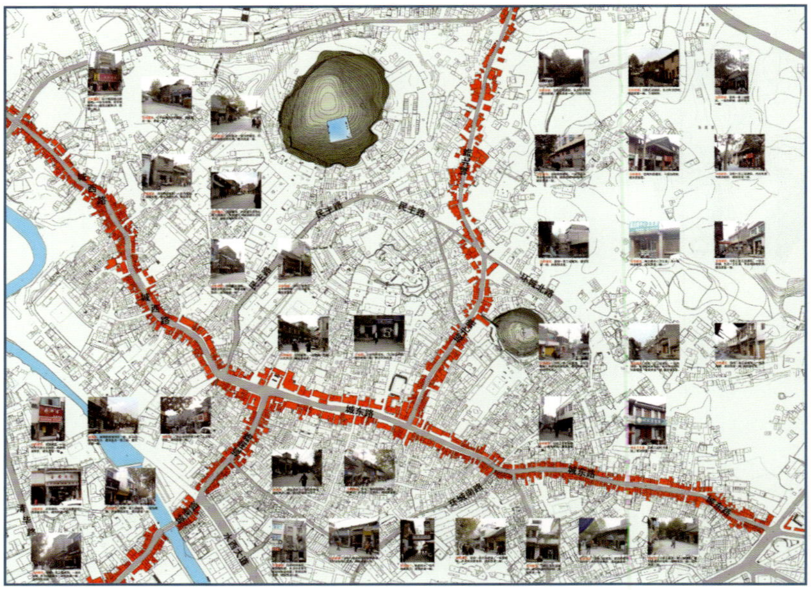

二、规划思路

根据规划局提供正在修编中的空位道路宽度及沿街用地性质为依据，对老城区东南西北四条道路进行规划，对道路周边建筑进行重建，以黔西沿街建筑整体风格为依据，控制沿街建筑的风格走向。

II Planning Idea

According to vacant road width and along the street land usage which are being revised and proviced by planning bureau, the road heading all directions in the old city will be designed and buildings around the road reconstructed. Based on the overall style of buildings along Qianxi Street, the direction of style along the street will be controlled.

总平面图

主要经济技术指标一览表

名称	单位	数量
规划总用地面积	m²	315842.25
规划实用地面积	m²	230195.71
总建筑面积	m²	323528.19
建筑占地面积	m²	83230.78
建筑密度	%	26.35
容积率	——	1.03
净容积率	——	1.41
绿地率	%	31.65
停车位	个	520

三、功能分区

根据对老城区的功能定位及其特殊性，将老城区划分为临街商铺区、商业中心区、停车场及广场区、古街区、文化娱乐区、商住混合区、教学科研区、住宅区等功能区。（详见功能分区图）

四、设计说明

1. 沿街建筑

该规划区四条道路与风情街相互交织，因此，该区域沿街建筑风格与风情街沿街建筑的整体风格基本保持一致却又有所区别。老城区沿街建筑沿用了风情街明清建筑风格与现代建筑风格相融合的理念，但是在色彩上使用了木质廊柱的原木色以此与风情街的红漆廊柱相互呼应。

沿街建筑设计还要考虑建筑景观的连续性与整体协调性，建筑立面适当运用当地民族元素以保持当地居民对自身文化的记忆性及对自身民族产生强烈的自豪感。

2. 酒店

极具彝族风格和古典韵味的设计，让游客有一种恍若隔世的感觉；民族与时尚的完美交融，宣告了古老与现代同在。将彝族元素——分解，再融合进这座建筑当中，合理穿插现代元素，力求凸显民族格调和地域风情，特别讲究简约、实用、美观。

青色瓦片的屋顶，翘角飞檐，采用彝族斗拱、彝族花纹、羊头等图纹作为装饰，使建筑更加别致、古朴；木质门窗框以及青石板墙的大量使用，更把原生态的风格体现得淋漓尽致，与其他彝族风格的建筑相互呼应。

Ⅲ Sectorization

According to functional localization and particularity of old city, it can be divided into street shop area, commercial centre zone, parking lot and plaza zone, old city street zone, cultural and recreational zone, teaching and research zone as well as residential areas. (for details please refer to sectorization plan)

Ⅳ Design Explanation

1. Buildings along the Street

As four roads in the planning area are intertwined with Ethnic Street, the street buildings are basically consistent with the overall style of the buildings along Ethnic Street while reserving differences. The street buildings of the old city inherit the idea of fusing building styles in Ming and Qing dynasties in Ethnic Street and modern building styles. However, the color of the buildings is burly wood of wood pillars in order to fit red varnish pillars in the Ethnic Street.

The street building design has also to take into account continuity and overall coordination of building landscapes. The elevation of the building appropriately uses local ethnic elements to maintain cultural memorability of local people and strong pride towards their own nation.

2. Hotels

The design of hotels possesses the style of Yi nationality and classic tastes, giving tourists a feeling as if they were living in another world. The perfect blendness of nation and fashion announces the presence of both the ancient and the modern, decomposing ethnic elements of Yi people and fusing into the building. The hotels reasonably intersperse modern elements in an effort to highlight national connotations and local customs, especially simplicity, practicalness and beautifulness.

The roof of hotel is cyan tiles with projecting eaves and cornices, using corbel bracket and pattern of Yi ethnic groups and sheep head pattern as its decoration, which makes the buildings more delicate and simple. A massive use of wooden window frame and gray stone wall demonstrate the original eco-style to its fullest and interact with other styles of Yi people.

功能分区图

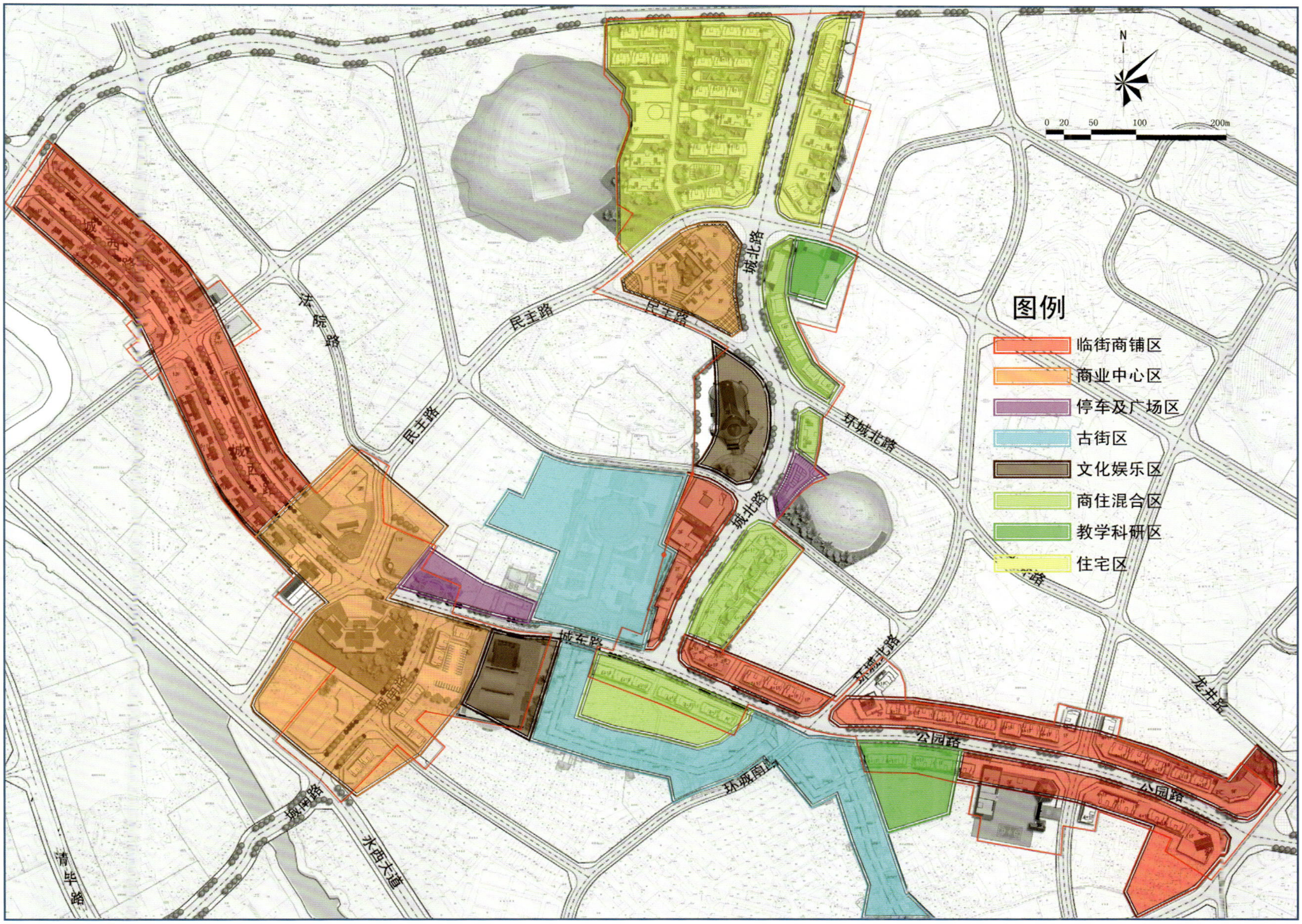
图例
临街商铺区
商业中心区
停车及广场区
古街区
文化娱乐区
商住混合区
教学科研区
住宅区
0 20 50 100 200m
N
民主路
城北路
环城北路
城东路
公园路
环城南路
水西大道
清毕路
龙井路

公共服务设施

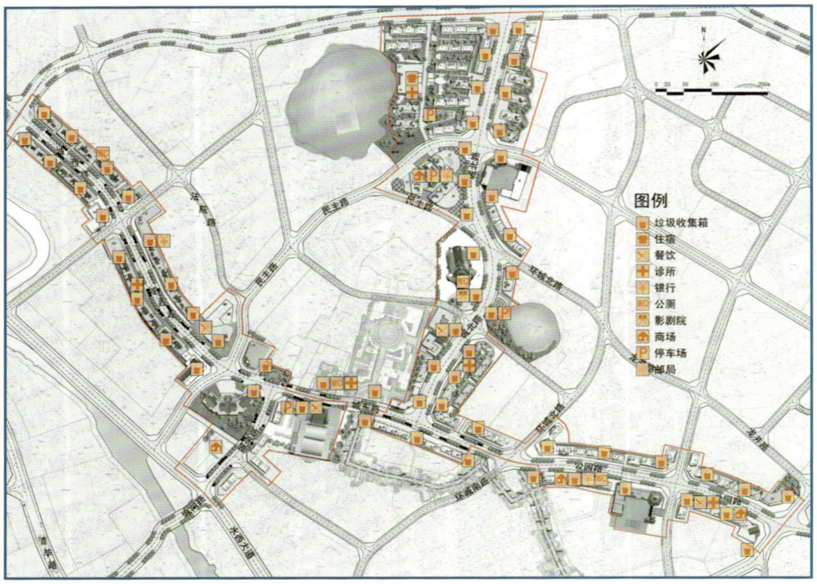

道路分析图

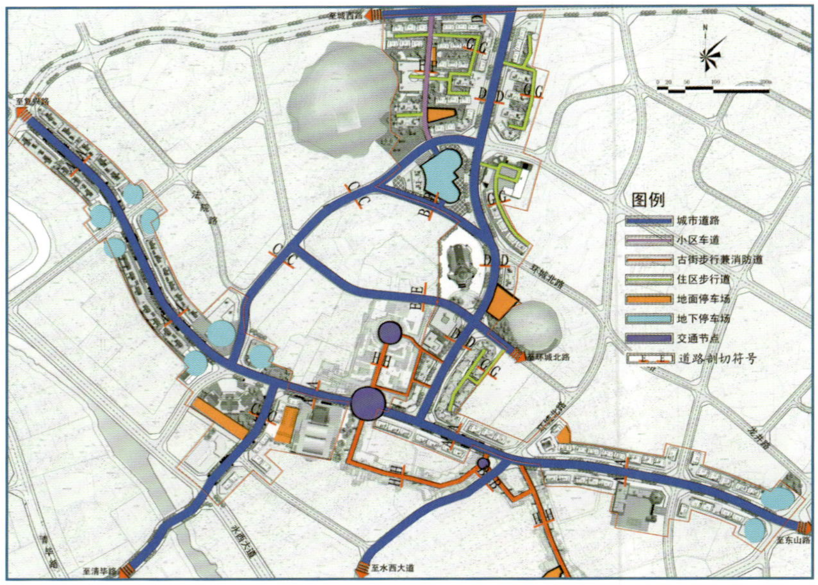

竖向设计图

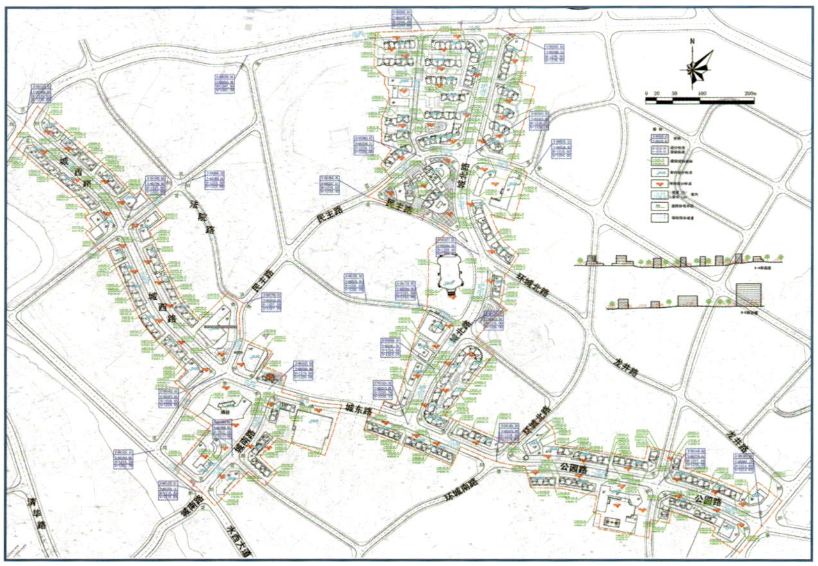

给水排水规划图

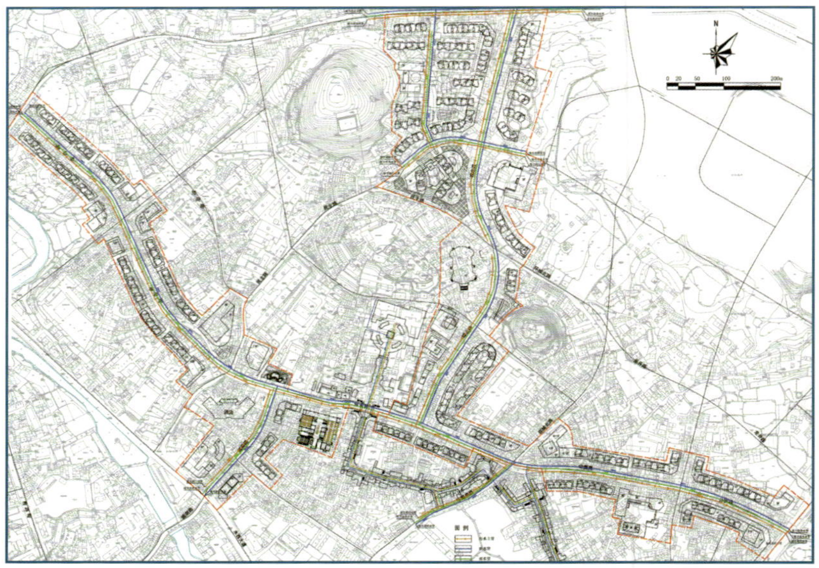

消防设施规划

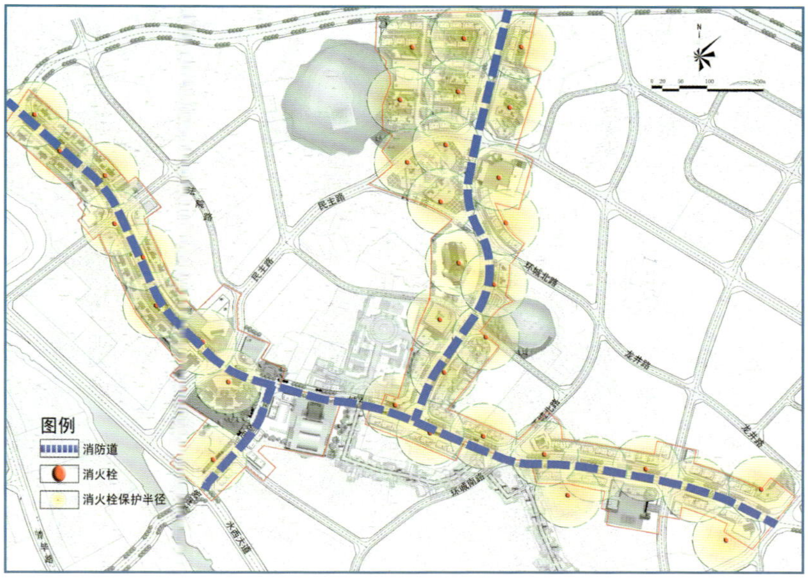

植物配置图

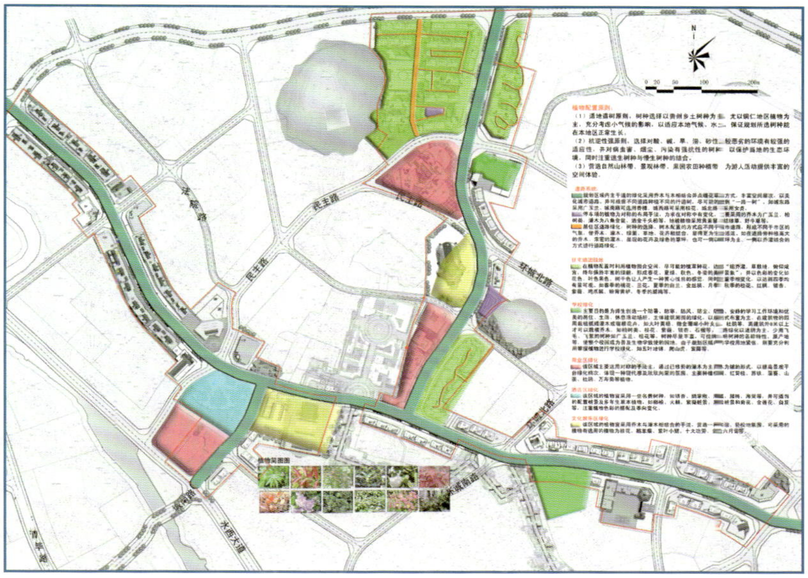

绿地规划图

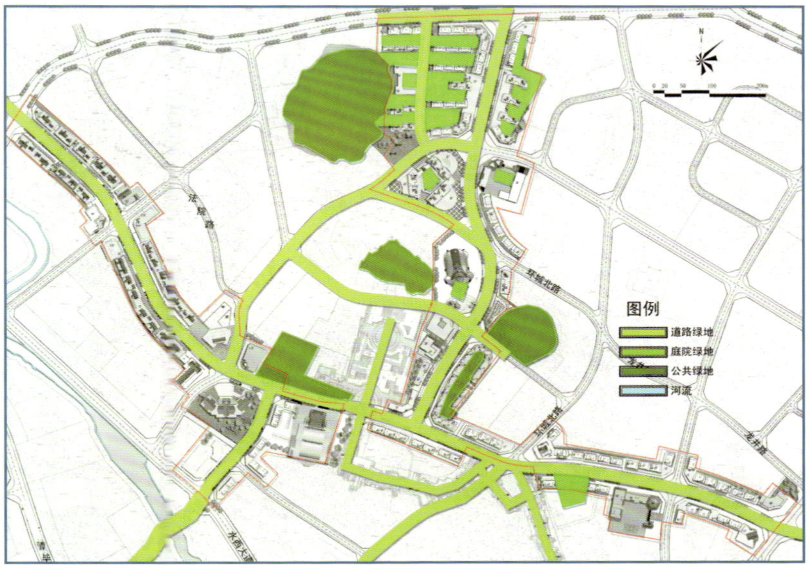

①

①街道透视图 ②街面建筑效果图 ③街面建筑效果图

①街道透视图 ②街面建筑效果图 ③街面建筑效果图

恭勤路
Gongqin Road

总平面图

公共服务设施规划图

街道改造透视效果图

街道改造透视效果图

改造后

改造后

改造前

①～②街道建筑改造效果图

改造前

①～②街道改造透视效果图

改造后

①～②街道改造透视效果图

①

②

③

①～③街道改造建成后实景图

黔西古城（风情街）规划设计

Plannig and Design of Qianxi City （the Ethnic Street）

一、现状 Ⅰ Present Situation

风情街位于老城区的中心位置，交通便捷，商业氛围浓厚。Ethnic Street is located in the center of old downtown with convenient traffic and thick commercial atmosphere.

风情街处于城东路、城北路和环城南路的交叉地带并与这三条道路交错，该片区交通发达，道路两侧建筑密度较大，多为商住楼及事业单位建筑，整条街的建筑质量较好，但风格不统一，缺乏整体性和协调性以及可识别性。

Ethnic Street, with developed traffic and dense constructions, is situated in the interface of Chengdong Road, Chengbei Road and South Huancheng Road and intertwined with these three roads. Most buildings along both sides of the street are used as offices of enterprises. With good quality but different styles of buildings, the street is lack of unity, wholeness and identification.

二、规划思路 Ⅱ Planning Idea

每个城市理应有一条具有代表性的街区来整体表现城市文化，集中展示当地特色文化的同时，为旅游者提供多样性的旅游服务。成都因为有了锦里而让游客流连忘返，阳朔因为有了西街而让漓江之游锦上添花，黔西将会因为有了风情街而使旅游者拥有了驻足于此的理由。

Every city should have a representative block to display entirely urban culture and, at the same time, provide visitors with diverse tourism services when fully shows local specialty. Jinli makes Chengdu attract visitors coming again and again; Beautiful West Street Traveling in Lijiang River, Yangshuo, is really the icing on the cake; Ethnic Street will surely keep tourists staying in Qianxi City.

利用本地块的景观条件，沿用现有道路，将之规划为步行街，形成高低错落，具有民族特色的黔西休闲娱乐亮点。

With existing landscape conditions and roads, the street is schemed as a pedestrian street in different height and with ethnic characteristics, so that it becomes the highlights of leisure and entertainment in Qianxi City.

规划利用步行街及周边外围环境共同构筑该片区旅游产业发展，用品位极高的风情古街去诠释一个令人心动的黔西。

The Pedestrian Street and its external surrounding environment will constitute to tourism development of this district in the planning as a high taste Yi Ethnic Street can reflect the attractive Qianxi City.

整个功能布局沿街布置，运用现代框架、现代结构加上水西彝族条纹装饰、斗拱装饰结合民情建筑形式，体现现代与民族的结合，形成黔西一道独特景观。

The whole functional layouts along the street are in modern frames and modern structures combining with strips decorations of Yi people of Shuixi, corbel bracket decorations and ethnic architectural shape to embody the integration of modern art and national culture and form a special view for Qianxi.

现状分析图

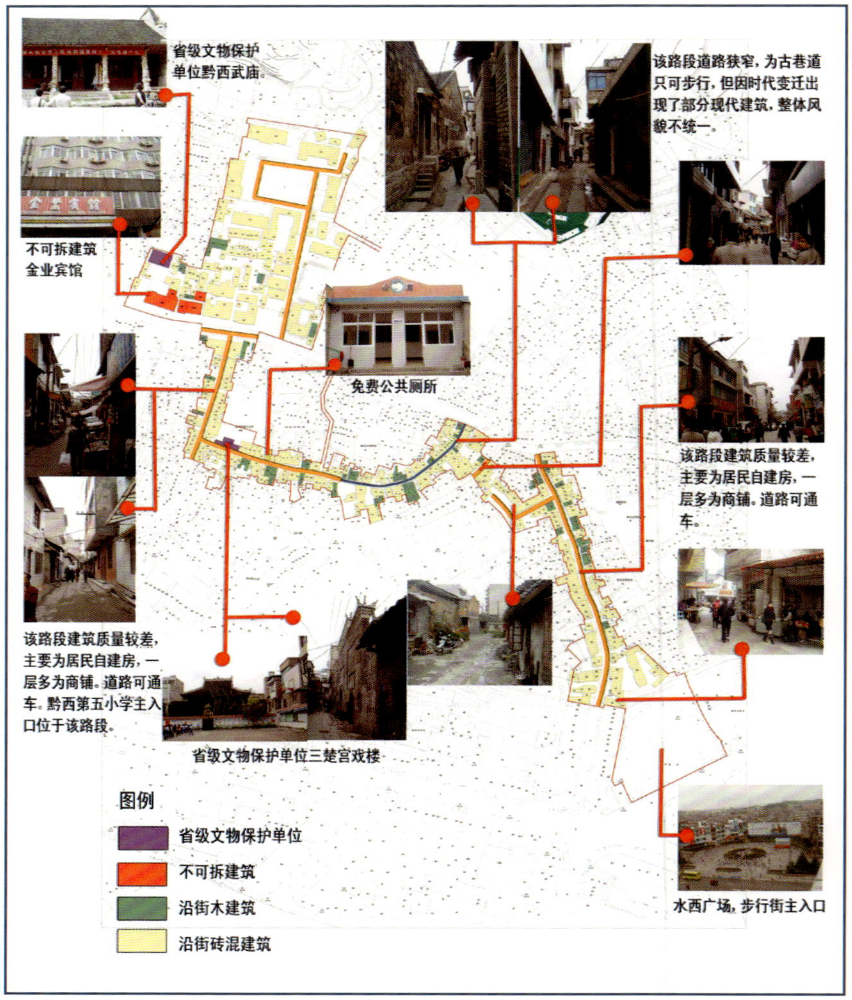

三、功能分区

将黔西风情步行街功能分为：入口景观区、工艺品展示区、土特产购物区、文化历史展示区、住宿餐饮娱乐区。（详见功能结构分析图）

Ⅲ Sectorization

The Ethnic Street can be divided into entrance landscape area, handcrafts displaying zone, local Products & shopping zone, culture & history displaying zone and board and lodging & entertainment area. (for details please refer to sectorization plan)

总平面图

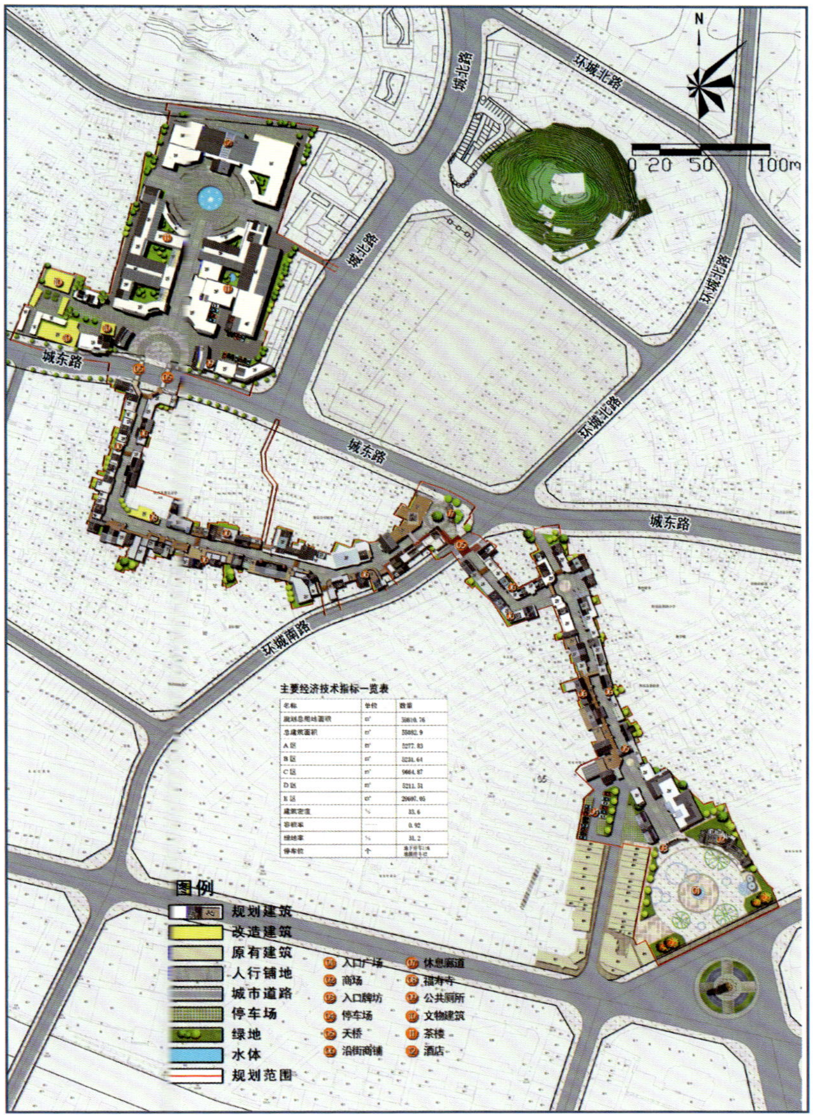

功能分区图

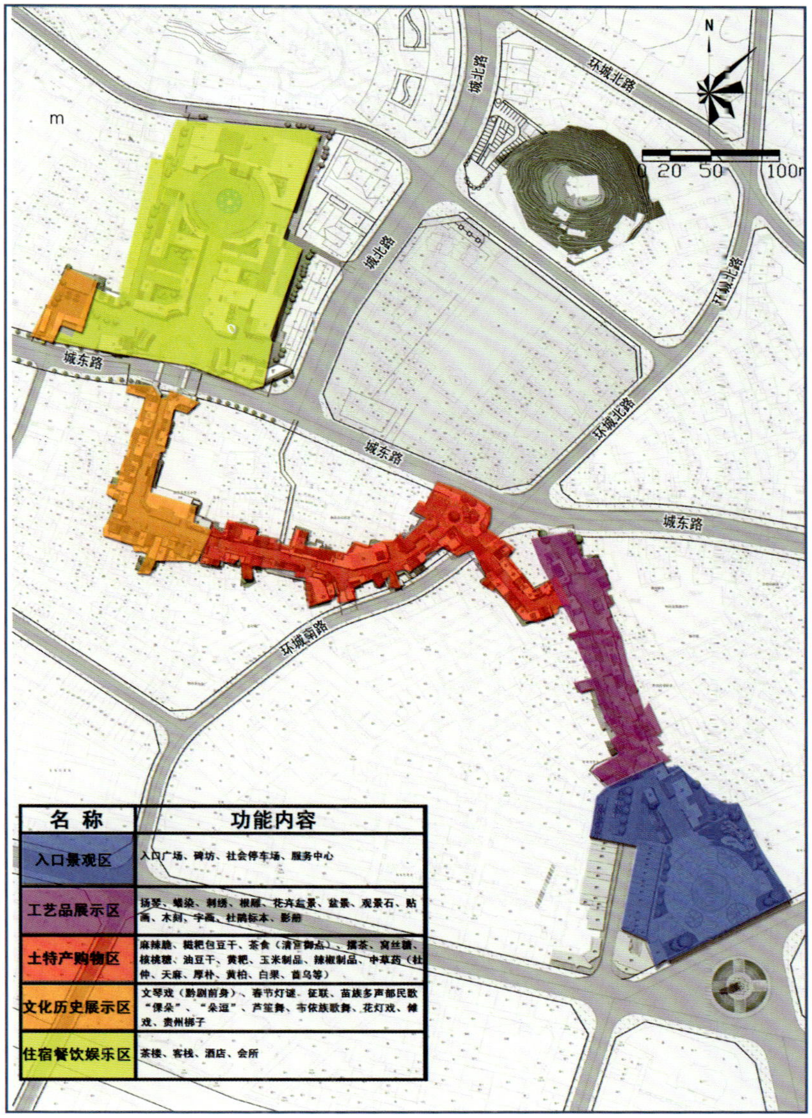

四、设计说明

1. 入口广场

风情街入口广场总体设计灵感来源于寓意着团团圆圆的圆形。广场整体形状为一个圆形，大圆里面又包含若干个小圆。其中，中间设置一个较小的圆形下沉广场，下沉广场周围是圆圆的台阶，中间为一个更小的圆形喷泉。三个以同一个点为圆心又互相包含的圆形分别用不同颜色、不同材质的石板或瓷砖作为地面铺装，既能够凸显出形状又体现了立体的感觉。

在下沉广场外面的圆环上，随意分布着以钢化玻璃为主的现代材质铺装而成的圆形荷叶、荡漾开的圆圆水波纹、花蕊造型的景观灯和荷花造型的雕塑，使得整个广场就像个圆形的荷塘。“荷塘”与对面路口的九狮闹莲雕塑相互呼应，体现了黔西九狮莲城的城市形象。

2. 入口牌坊

民族风情街主入口牌坊整体设计思路采用彝族元素为主。牌坊正上方以彝族服饰中常用到的纹饰图案作为装饰，支撑的柱子则用石材堆砌而成，两侧设立对称的小门，顶上以五彩球灯装饰，加上彝族特有的屋顶造型，与整条街区风格相呼应，相信会给游人以全新的震撼，让他将黔西铭记于心。

3. 风情街

风情街沿街建筑无处不在地彰显了彝族风格：采用了彝族建筑特有的马头墙，加上斗拱装饰，更凸显出民族特色；沿街建筑均以青瓦盖顶，外立面梁柱、墙面和栏杆以朱红色为主，体现彝族人民热情好客的性格特性；适当搭配文化石墙面，与青石板铺装的地面相呼应；廊道采用镂空设计，使建筑整体显得简约灵动；梁柱运用彝族服饰上的图案和彝族斗拱等进行装饰，华美大方的同时呈现黔西特色。

沿街设置独特造型的街灯点缀装饰整个街区，夜色下，五颜六色的彩色灯笼把整个街区装点得色彩缤纷，令人流连忘返。

IV Design Explanation

1. Entrance Square

The entrance square of the Ethnic Street is in a round shape, designed with the inspiration of reunion as ring has the meaning of reunion. The whole square is a big cycle including several small circles. A smaller subsided circle square in the big square surrounded with a circle step and surrounding a circle fountain. Three circles in different size with the same center can highlight the figure and the sense of three- dimension with different colors and paved with stone slabs or tiles in different material.

Some decorations of circle lotus leaves made from modern tempered grasses spread on the external circle of the subsided square, along with water waves, landscape lamps in stamen shape and lotus sculptures, the whole square is like a circle lotus pool. The "lotus pool" responds the sculpture of "Nine Lions Preserving a Lotus" embodying urban image of "Lotus City with Nine Lions" of Qianxi.

2. Entrance Arch

The entrance arch of Ethnic Street is designed mainly with Yi ethnic elements. The arch, with strips patterns decoration which is usually used by Yi people in their costumes, stone materials pillars, two symmetrical small doors on both sides, colorful ball lamps on the head and special roof of Yi People, perfectly interact with overall style of the street, which must bring visitors a new impact and leave a fresh impression to them.

3. Ethnic Street

Buildings along Ethnic Street show Yi style in details: Horse head wall, the unique architectural element of Yi people, is adopted along with corbel bracket decoration to display national characteristics; all the buildings along the street are with cyan tile roofs along with imperial red pillars, walls and railings to manifest the hospitality of Yi People; walls are embellished with cultural stones to reflect with the grounds paved with grey stone slabs; corridors are designed in hollow-out style to make the buildings simple and smart in the whole; beams and columns are ornamented with patterns deriving from Yi costumes and corbel bracket of Yi people to make buildings beautiful, and also demonstrate features of Qianxi.

Street lamps in special shapes are set to embellish the whole street, which, in the night, becomes a luminous peal with a sea of lights twinkling and leaves visitors deep impression.

公共服务设施

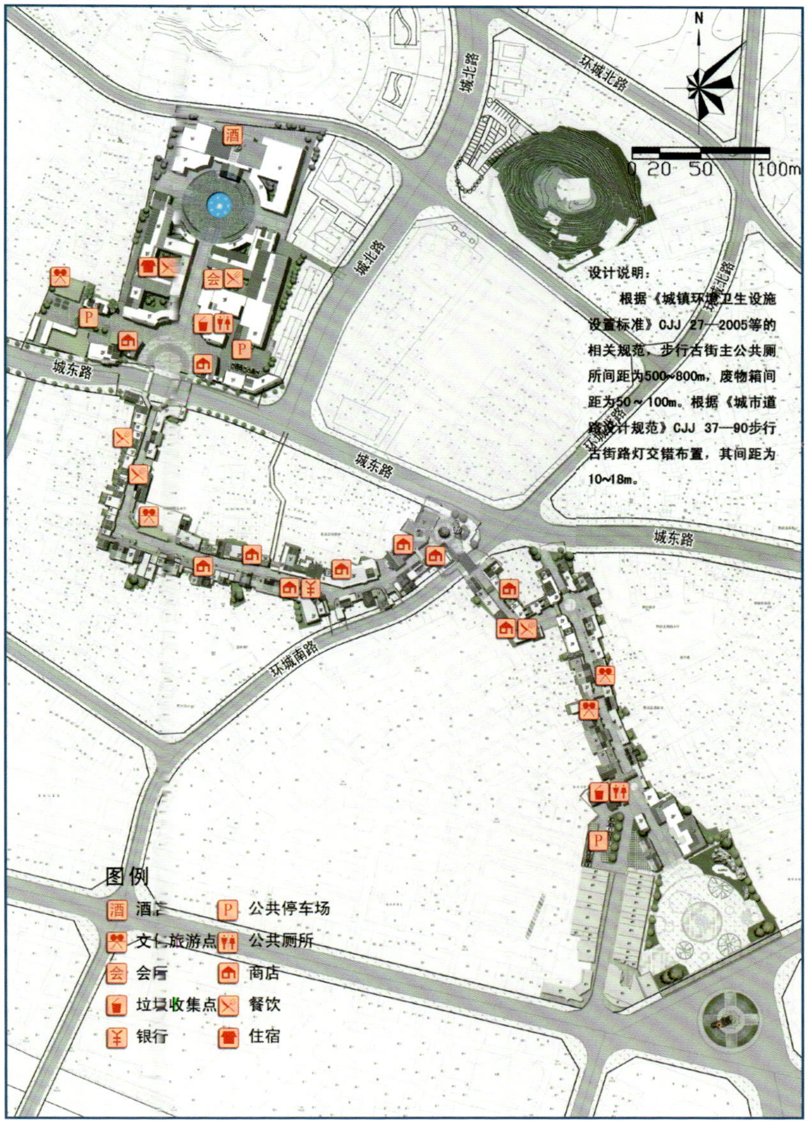

道路交通分析图

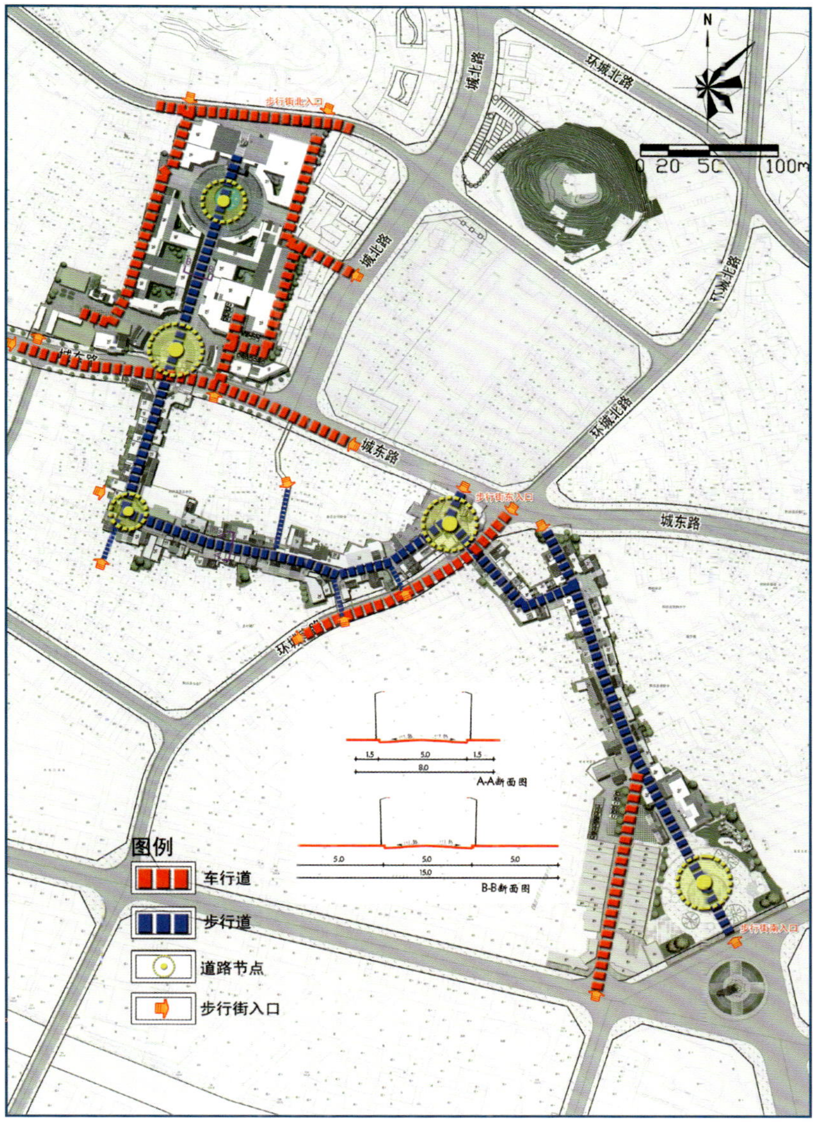

竖向规划设计图

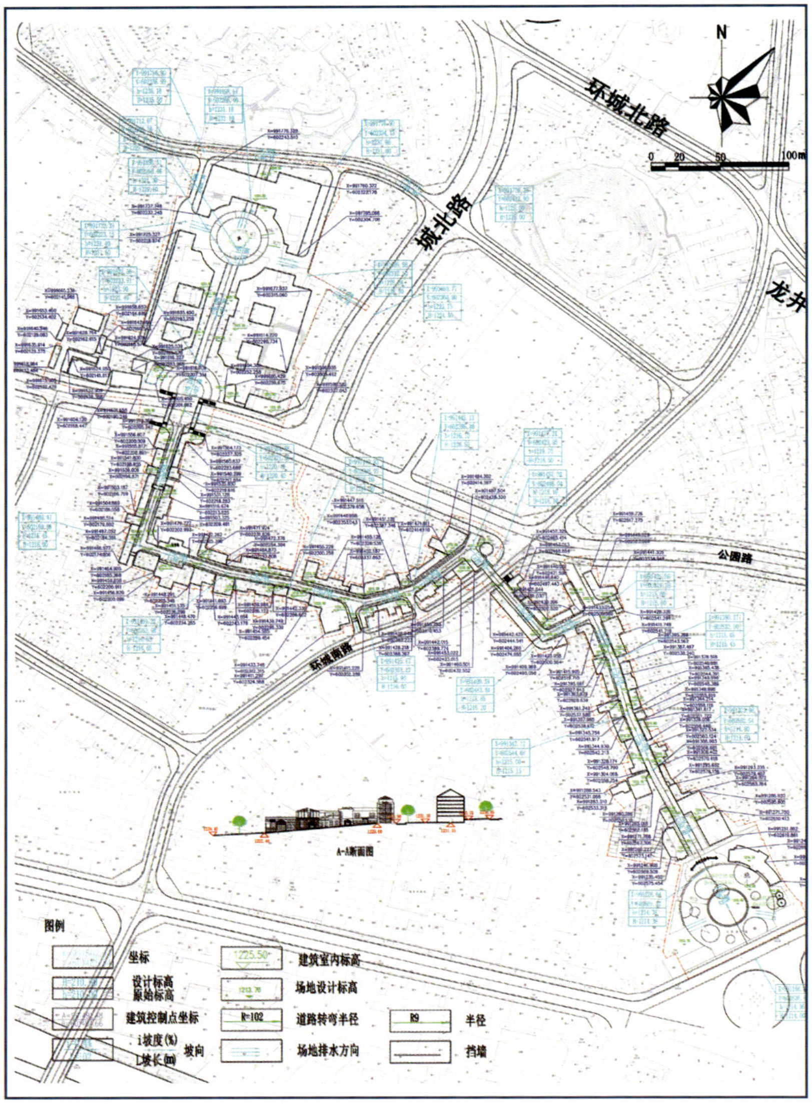

给水排水规划图

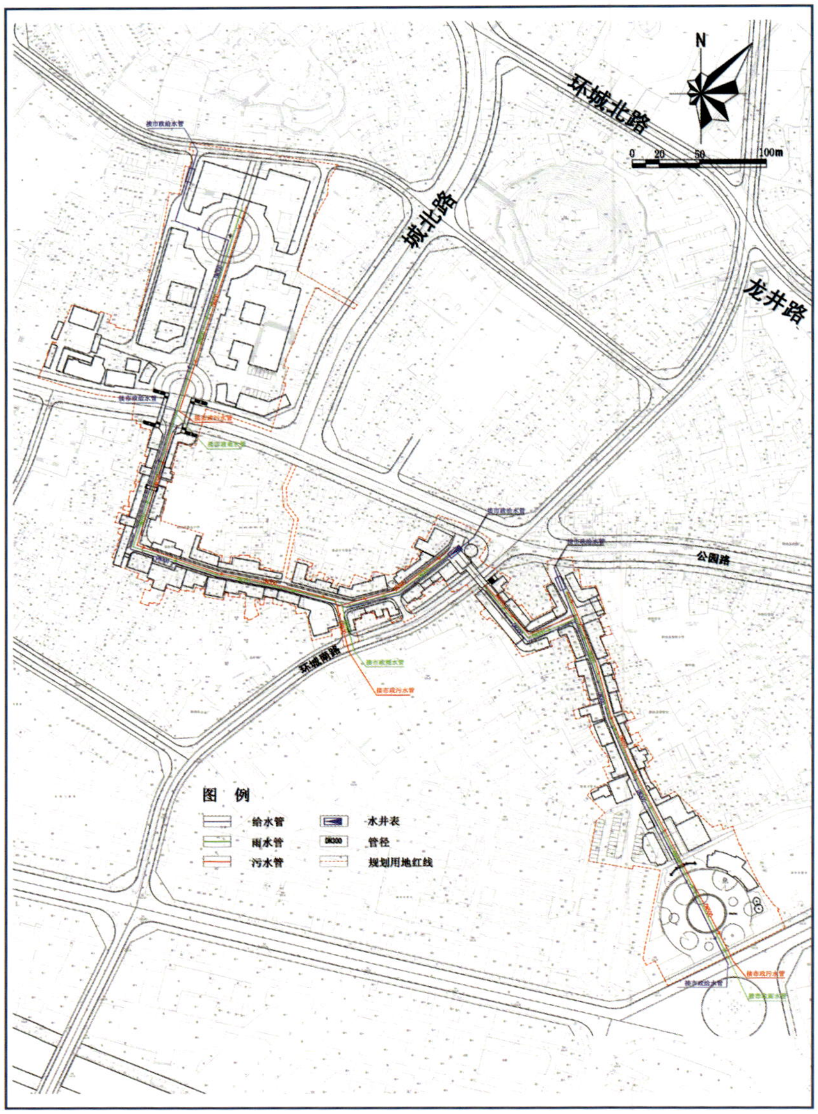

消防规划设计图

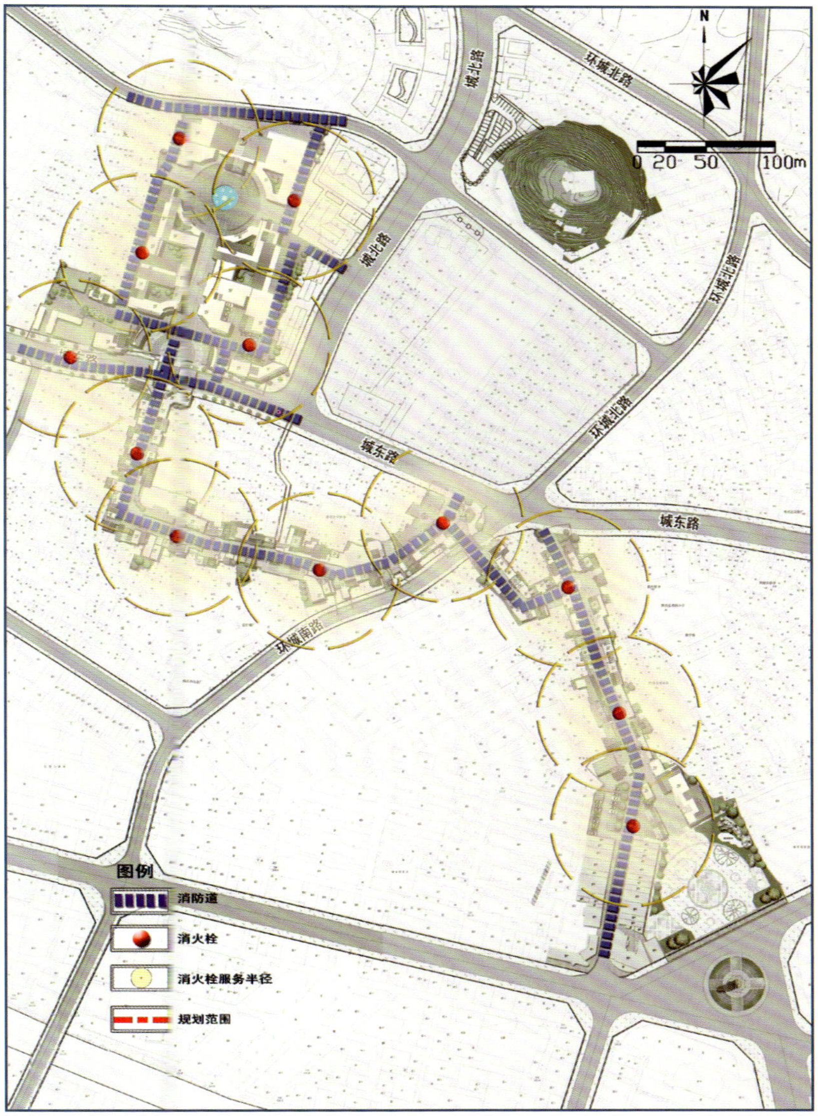

电力电信规划图

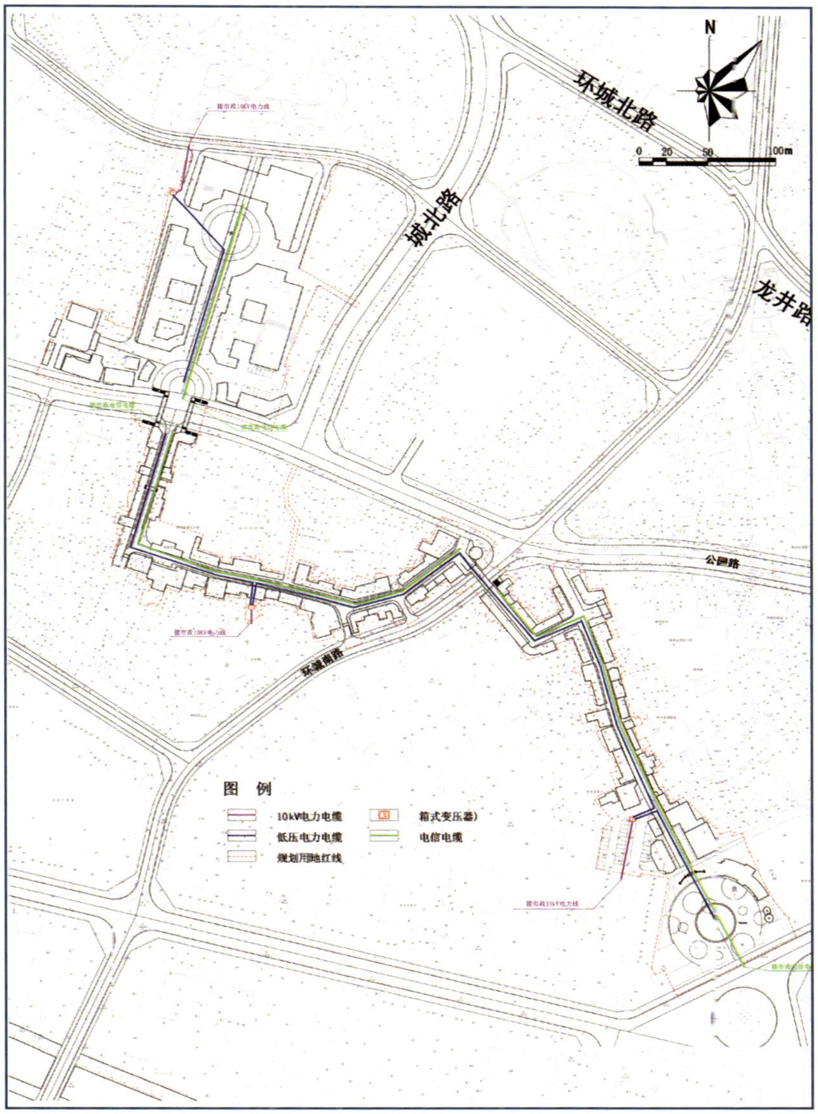

绿地规划图

N
0 20 50 100m
环城北路
城北路
城东路
环城南路
图例
步行街绿化
车行道绿地
庭院绿地
广场绿地

植物配置图

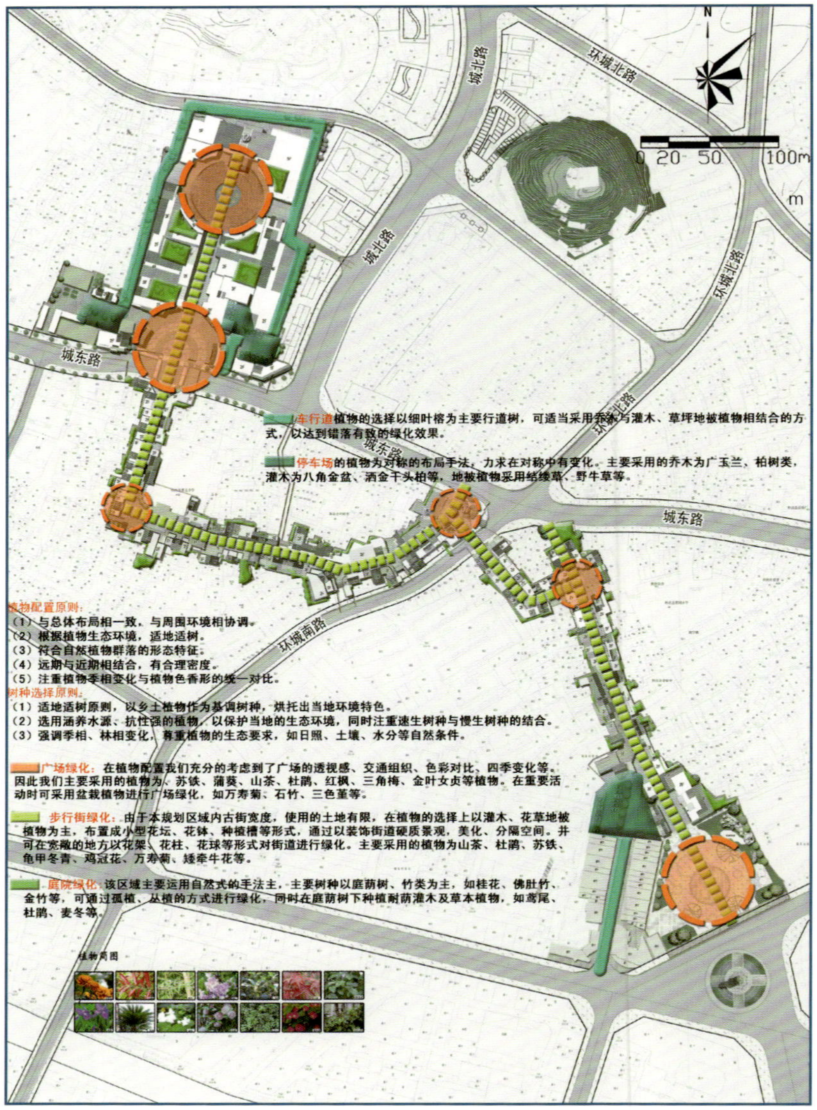

N
0 20 50 100m
环城北路
城北路
城东路
环城南路
车行道植物的选择以细叶榕为主要行道树，可适当采用乔木与灌木、草坪地被植物相结合的方式，以达到错落有致的绿化效果。
停车场的植物为对称的布局手法，力求在对称中有变化。主要采用的乔木为广玉兰、柏树类，灌木为八角金盆、洒金千头柏等，地被植物采用结缕草、野牛草等。
植物配置原则：
（1）与总体布局相一致，与周围环境相协调。
（2）根据植物生态环境，适地适树。
（3）符合自然植物群落的形态特征。
（4）远期与近期相结合，有合理密度。
（5）注重植物季相变化与植物色香形的统一对比。
树种选择原则：
（1）适地适树原则，以乡土植物作为基调树种，烘托出当地环境特色。
（2）选用涵养水源、抗性强的植物，以保护当地的生态环境，同时注重速生树种与慢生树种的结合。
（3）强调季相、林相变化，尊重植物的生态要求，如日照、土壤、水分等自然条件。
广场绿化：在植物配置我们充分的考虑到了广场的透视感、交通组织、色彩对比、四季变化等。因此我们主要采用的植物为、苏铁、蒲葵、山茶、杜鹃、红枫、三角梅、金叶女贞等植物。在重要活动时可采用盆栽植物进行广场绿化，如万寿菊、石竹、三色堇等。
步行街绿化：由于本规划区域内古街宽度，使用的土地有限，在植物的选择上以灌木、花草地被植物为主，布置成小型花坛、花钵、种植槽等形式，通过以装饰街道硬质景观，美化、分隔空间。并可在宽敞的地方以花架、花柱、花球等形式对街道进行绿化。主要采用的植物为山茶、杜鹃、苏铁、龟甲冬青、鸡冠花、万寿菊、矮牵牛花等。
庭院绿化：该区域主要运用自然式的手法主，主要树种以庭荫树、竹类为主，如桂花、佛肚竹、金竹等，可通过孤植、丛植的方式进行绿化，同时在庭荫树下种植耐荫灌木及草本植物，如鸢尾、杜鹃、麦冬等。

①

②

③

④

①风情街入口景观效果图 ②主体雕塑 ③风情街入口牌坊 ④服务中心

①

②

③

①～③风情街局部效果图

①

②

③

④

①~⑧风情街局部效果图

贵州土特产

④

⑤

⑥

①~⑥风情街局部效果图

贯城河景观设计
Landscape Design of Guancheng River

一、概况

穿城而过的贯城河起源于附廓水库，属于长江流域乌江水系，位于乌江一级支流野济河的上游。沿河两岸山清水秀，风景迷人，黔西人民对贯城河怀有深厚感情，亲切地称之为母亲河。

现有的贯城河水质及周边环境不尽如人意，但是县政府已经着力进行贯城河的治理，治理后的贯城河将成为穿越县城的精灵，成为黔西城市旅游发展的一大基石，同时也将是黔西提升城市品位、形成城市形象的一大法宝。

贯城河景观设计起于运煤大道上游 160m 处，止于贵毕公路皮甲河大桥污水处理厂处，总长约 7.6km，两岸景观用地宽度平均 15m，在部分重要地段，根据设计规范向周边用地扩展。

本区域设计主要以河岸景观设计为主，工程设计以佛山市市政设计研究院有限公司编制的《黔西贯城河综合治理工程》为依据。

I Overview

Guancheng River, with attractive view along its banks, is the upper stream of Yeji River, the first class branch of Wujiang River, and belongs to Wujiang River of Yangtze River Valley. It begins from Fukuo Reservoir and runs through Qianxi City where people have profound feelings to it and treat it as their "Mother River".

Both current water quality and surrounding environment of Guancheng River are not perfect. However, it has been recovered by the government and expected to be a fairy running through the city, the great fundament of developing tourism in Qianxi city and the precious asset of promoting urban taste and forming urban image for Qianxi.

The landscape design of Guancheng River is from 160m upper Yunmei Road to a waste water treatment factory of Pijiahe Bridge, Guibi Highway, the total length is 76 kilometers, and average width between the two banks used for landscape design is 15 meters and some important parts need to be extended in accordance with design norm.

This region is major in landscape design.

现状分析图

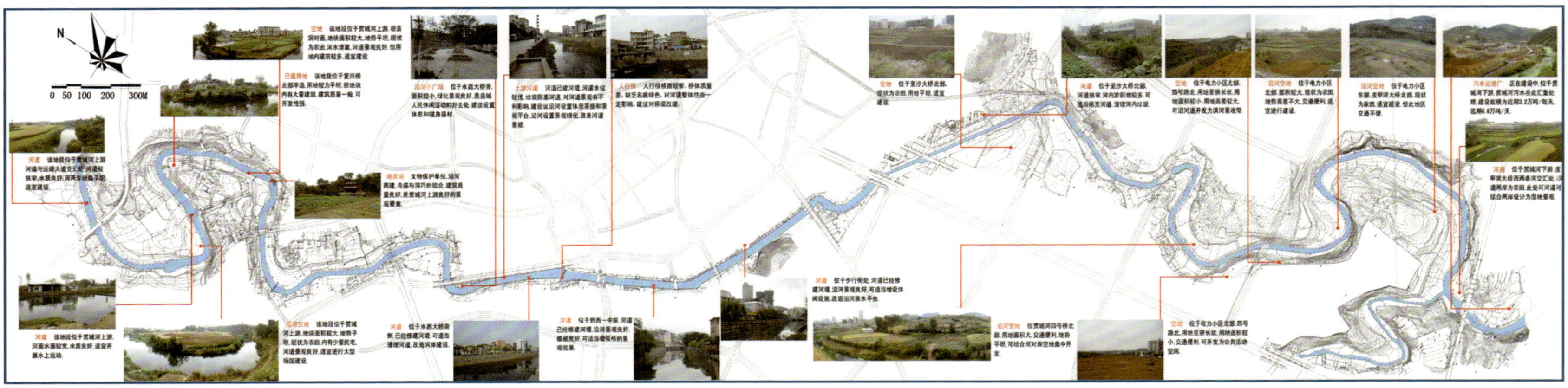

高程分析图

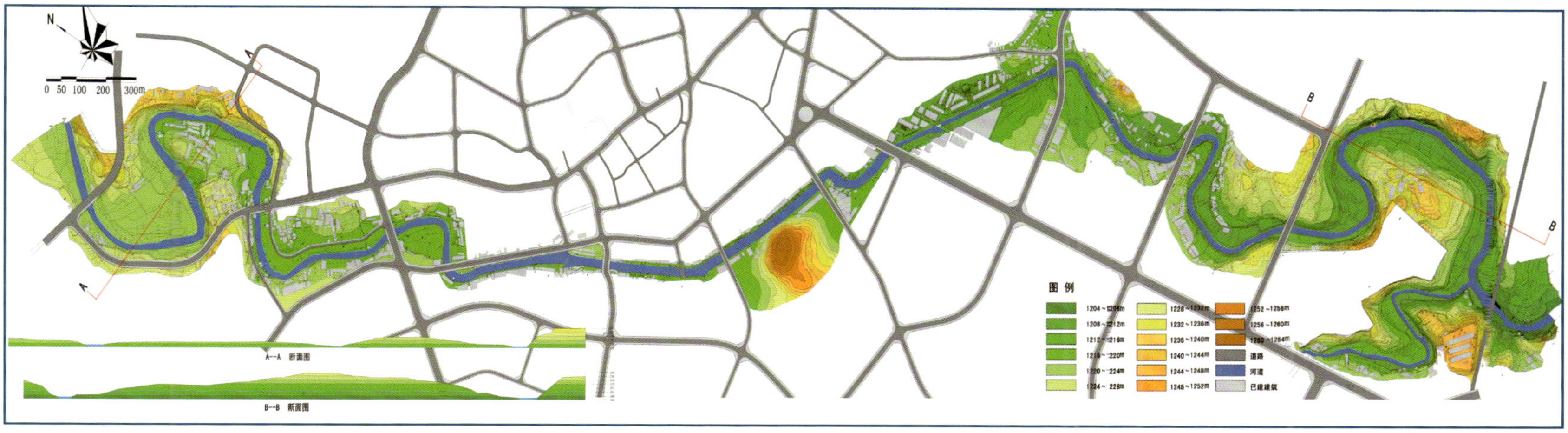

坡度分析图

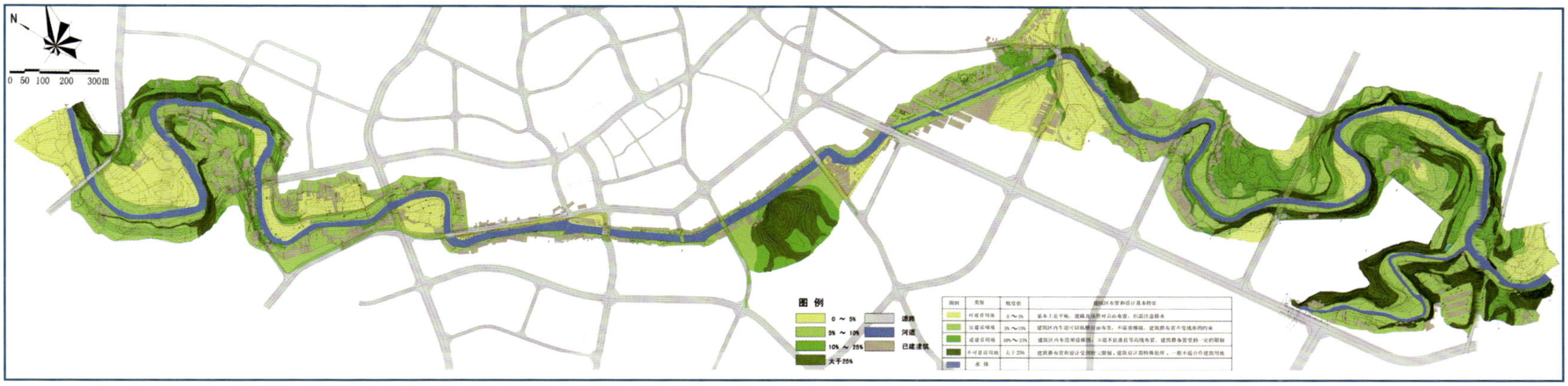

桥梁现状分析图

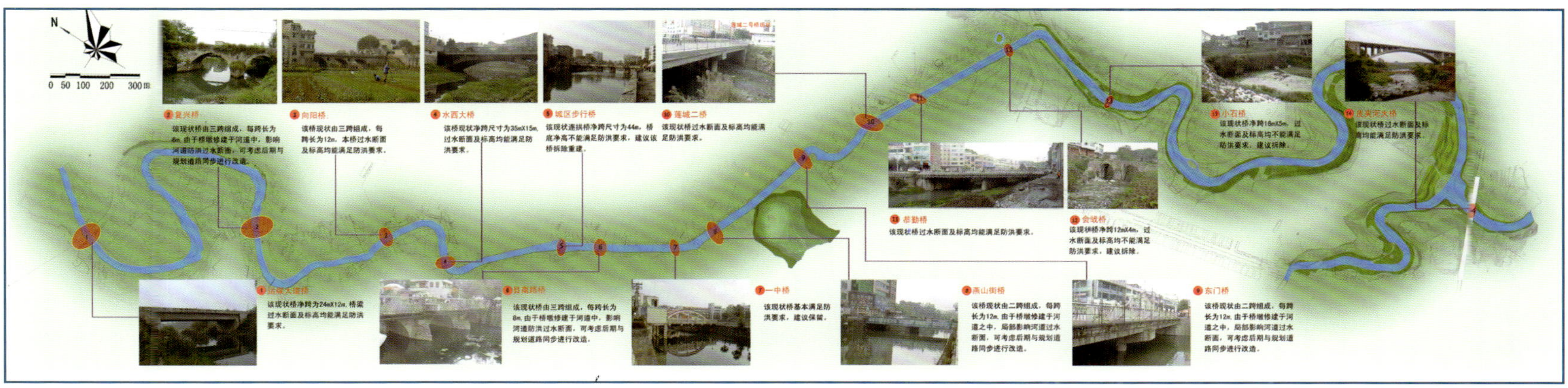

总平面图

二、规划思路

从许多滨河城市的经验来看，治理好江河，开发利用好江河资源，形成人与自然和谐相处的生态水环境和湿地系统，对提升城市品质，改善市民生活具有重要的现实意义。

黔西贯城河规划设计以生态改造为主，结合两岸现状，搞好商业设施选址布局，重点做好景观的规划设计，对人文景观的打造同样以彝族风格为主，但是表现手法更为浓重，在色彩、外观和造型上下工夫，突出美观、大方、便民的特点，与沿街建筑整体风格相协调的同时，突出重点，使整个黔西风格明确又有张有弛。努力将贯城河打造为集“吃住行游购娱”为一体的城市休闲带，黔西旅游的一道亮丽风景线。

整个贯城河设计了水面、绿化、建筑等多层岸线界面，打造极具观赏性的沿河轮廓，提升滨河区域整体形象。

重峦叠嶂的远山、岸边摇摆的柳条和清澈纯净的贯城河水融汇黔西独特的彝族风情、得天独厚的水西文化及历史文化为一体，形成自然景观与人文景观完美结合、地方特色浓郁的休闲度假之理想场所。

II Planning Idea

The developing process of many riverbank cities shows that recovering and making good use of river resources to form a harmonious ecologic water environment and wetland system of human and nature will surely promote urban quality and improve citizens' life.

The planning and design of Guancheng River is major in reconstruction of ecology. The location and layout of business facilities should be considered well based on the current situation of both banks, and landscape planning and design should be emphasized on. Humanistic landscapes should also be major in Yi ethnic style with more profound expression on colors, appearances and shapes to stress the properties of beauty, generosity and convenience, so that they can coordinate with the whole style of buildings along the streets and, at the same time, show sharply the important part, which make the entire style of Qianxi clear and with different degree. In all, Guancheng River will be created with great effort to be a city leisure belt including dinning, lodging, travelling, transporting, shopping and relaxing.

Multiple levels against the bank are designed such as water level, greenland level and building level, in order to create outlines along the river with great value of viewing, and promote entire image of river ban region.

The row upon row of verdant mountains in the distance, the winging willows branches by the river, and the clean and pure water of Guancheng River are integrated with special Yi flavor and precious Shuixi culture and historic culture to form the perfect combination of natural landscape and humanistic landscape and create an ideal place with thick local charm for leisure vacation.

三、功能布局

贯城河风光带建立在黔西县城内贯城河河段的基础上，由西向东贯穿县城全境。沿着两岸规划设计了休闲度假区、健身休闲区、滨河景观休闲区、娱乐文化休闲区和生态湿地休闲区等几个功能区，对黔西县城贯城河河段从形象到内涵上进行丰盈。

Ⅲ Functional layout

Guanchenghe landscape belt is constructed with the fundament of Guanchenghe River in Qianxi City, crossing the whole city from west to east. Specific areas include "Leisure Vacation Area", "Fitness & Leisure Area", "Riverbank Landscape Leisure Area", "Entertainment & Cultural Area" and "Ecologic Wetland leisure Area" on both banks, which enrich the Guanchenghe River from image to connotation in Qianxi.

功能分析图

名称	度假休闲区	健身休闲区	滨河景观休闲区	娱乐文化休闲区	生态湿地休闲区
理念	感受黔西之悠然自得	感受黔西之笑傲风月	感受黔西之闲情逸致	感受黔西之恬淡静谧	感受黔西之绿景怡情
内容	天然游泳池、会所、茶楼、玲珑洞、游泳馆、儿童戏水池、青少年宫	健身广场、棋牌室	三角健身广场、沿河景观步道、山顶景观休憩茶楼和景观塔	四星级酒店、水幕表演广场、休闲广场	民族食府、钓鱼岛、梯田景观

四、重点项目

1. 休闲度假区

理念：感受黔西之悠然自得

内容：天然游泳池、会所、茶楼、玲珑洞、游泳馆、儿童戏水池、青少年宫

沿江两岸绿柳成荫，岸边分布许多小巧而精致的休憩景观，这些景观顶盖青瓦、梁柱均为原木色、亭脊以彝族古朴粗犷的斗拱为装饰、形状各异，行人过客可在此逗留歇脚。

在贯城河岸边，还设置了天然游泳场和儿童戏水池，这个天然泳池与其他游泳池不一样的是，这里设计了人工的小海滩，人们不仅能体验到海边游泳的自由自在，还可以躺在沙滩上欣赏碧浪银滩的美景。在这里游泳的人们既不会像在海边游泳那样有呛水的危险，又不像在封闭无景的室内游泳池一样索然无味，而是既能够享受自然景色，又有着活力无限的体验。

游泳场边上还设置了彝族风情浓郁的服务站，小沙滩上还分布着几个顶上铺盖着茅草、风格古朴的小亭子，人们在游泳之余还可以在此休憩，享受悠然自得的惬意。

2. 健身休闲区

理念：感受黔西之笑傲风月

内容：健身广场

广场地面的整体形状为一片饱满的杜鹃花瓣，地面铺装使用方形瓷砖，青色鹅卵石镶嵌成杜鹃花瓣纹路，以绿篱围合而成的花蕊与彝族服饰上的螺旋纹饰不谋而合，其造型设计遵循了生态学和民族美学理论。

整个广场适当用绿树、花池、盆景绿化，还有小喷泉的点缀。在广场周边和贯城河边，还设置了颇具彝族风情的小亭子、凳子、垃圾桶和景观灯，这些小品设计都充分展示了人与自然的融合，突出彝族文化内涵。

IV Major projects

1. Leisure Vacation Area

Idea: feel leisure and relaxat on of Qianxi

Content: Natural swimming pools, clubs, tea houses, Linglong Cave, Swimming Gymnasium, Children's Playing Pool and Youth Palace

There are many smart and exquisite pavilions for rest on the banks along with frondent willows. These pavilions, with different shapes, are of grey tiles, wood color pillars, and its beams are decorated with the ancient and simple corbel bracket of Yi people. Visitor can stay in them for a rest.

On the banks of Guanchenghe River, there are natural swimming pool and children's playing pool which are different from others as artificial beach is designed for them, so that people can not just experience free swimming like in the sea, and also lie on the beach and enjoy beautiful views of blue water and silver beach.

Service centers with thick Yi flavor are set by the pools, as well as small pavilions covered with thatch and in ancient simple style, so that people can take a rest and enjoy themselves here after swimming.

2. Fitness & Leisure Area

Idea: feel nature and taste of Qianxi

Content: Fitness Square

The square ground is in the shape of a fresh azalea petal, and paved by square tiles with grey cobblestones outlining veins of azalea petal, while the stamen formed by green hedge response to helix decoration of Yi custom.

The whole square are embellished properly with green trees, flower beds, potted landscapes and small fountains. Also, around the square and by Guancheng River, there are the small pavilions, benches, litter bins and landscape lamps with Yi flavor, of which all the design fully demonstrate harmony of human and nature and stress on cultural connotation of Yi people.

在这里市民可以进行各种竞技活动比赛和体育健身活动，这些项目填补了城市的健身休闲和晚间室外娱乐活动的不足。通过互动参与，在交流情感、凝聚精神的同时，还展示和传承民族文化，是一处独特风景。

3. 滨河景观休闲区

理念：感受黔西之闲情逸致

内容：三角休闲广场、沿河景观步道、山顶景观休憩茶楼和景观塔

该区域充分利用地形设置茶楼、观景塔、景观墙、景观廊架等建筑小品以及其他休憩设施。在这些景观小品上充分使用彝族人民所崇拜的远古图腾——牛、羊、火和杜鹃花的花叶、花瓣、花蕊等元素作点缀，回忆彝族的远古历史，展示黔西灿烂的杜鹃花文化。

这些建筑小品均有适宜的尺度，其小巧、简洁、清新的建筑风格与山水相得益彰，使该区域成为人们远离都市喧嚣、享受清静、自然的滨水休闲空间。

4. 娱乐文化休闲区

理念：感受黔西之恬淡静谧

内容：四星级酒店、水幕表演广场、火把广场

在四星酒店前面是水幕表演广场，整个广场整体平面形似一把月琴，圆圆的琴身伸入河中，成为一个椭圆的水上看台。在广场临水一方则是如梦似幻的喷泉水幕电影。

在一片夜色中，音乐水景喷泉尽显其蓝色梦幻色彩，同时背景音乐回响在河水、广场上空；与此同时，水幕电影可以连续播放精彩的影像。通过各种高科技的舞美和特技效果给观众呈献出一台超越时空、超越感官体验的视听盛宴。整个广场以水为背景，是滨水生态景区特有的聚会、表演、游乐场所。

彝族人民把火作为具有神秘因素的超自然力的原始崇拜对象。在这个区域，设置一个彝族文化氛围浓郁的火把广场，火把广场中央，火红的雕塑形状为一簇熊熊燃烧的火焰，它代表了彝族人民的热情好客和黔西红红火火的发展势头。

广场周围同样设置了许多具有彝族特色的景观小品供市民和游客休憩。

5. 生态湿地休闲区

理念：感受黔西之绿景怡情

内容：民族食府、钓鱼岛、梯田景观

从选址开始，就注意借景、对景、观景的考虑，使之既可观景又可成景，以便和整体环境相协调。

Here, citizens can hold various athletic movements and do exercises for keeping fit, which complement leisure fitness and night outdoor recreational activities in the city, so that national culture can be displayed and passed down by people's participation.

3. Riverbank Landscape & Leisure Area

Idea: feel elegance and comfort of Qianxi

Content: Triangle Leisure Square, Landscape Path along River, Landscape Teahouse on the top of mountains, Landscape Tower.

Architectural accessories such as tea houses, viewing towers, landscape walls, and landscape corridors and other rest facilities are set by making good use of the terrain of this area. And on these landscape accessories, the ancient totems of Yi people like cow, goat, fire and the leaf, petal, stamen of azalea are used fully as decorations to represent distant history of Yi ethnic minority group and unfold brilliant azalea culture of Qianxi.

All these architectural accessories are in proper scale that their smart, simple and fresh architectural style match well with mountain and waters, which make this area be a riverbank leisure space for people to be far away from noisy and enjoy quietness and nature.

4. Entertainment & Culture Area

Idea: feel tranquillization and beauty of Qianxi

Content: Four-star Hotel, Water-curtain Performance Square, and Torch Square

In front of the Four-star Hotel, it is the Water-curtain Performance Square which is like a Yueqin (a kind of instrument in a full-moon shape) with its round body stretching toward the river and forming a elliptic ground stand in water. On the side near the river, there is a dreamlike water-curtain film.

In the night, There is a audio-visual banquet which is beyond time and sense presenting to visitors on the square and river pervading with dreamlike blue watermist and swirling over background music, as well as various images displayed on the water-curtain by all kinds of high technology. The whole square, with water as its background, is a special place for assembly, performance and entertainment in riverbank ecological scenic site.

Torch has been worshiped and treated as if it is of miserable factors and beyond nature by Yi people, which is the basis that we set a touch square with profound Yi culture in this region. The brilliant sculpture in the shape of a vehement flame reflects hospitality of Yi people and the perspectivity of Qianxi.

Many landscape accessories with Yi feature are set around the square for visitor to have a rest.

5. Ecologic Wet-land Leisure Area

Idea: feel green and scene of Qianxi

Content: Ethnic Restaurant, Fishing Island and Terrace Landscape

We've stressed the reflection, compare, and sight of views from the selection of location in order to form interactive landscapes and coordinate with the entire environment.

设置可以进入河中钓鱼岛的步行台阶，这些建筑低临水面，有一定的亲水感，与整体环境的建筑风格相协调，在其中布置一些能够体现彝族历史文化的壁画、置石、隔断等园林建筑小品，并尽可能和水有关系以便点题。

人们漫步于这亲水平台，在感受碧江拥怀的同时，还可以欣赏河边婆娑起舞的翠竹、随风飘摆的柳条、郁郁葱葱的水草，享受黔西绿景怡情的别样风景。

6. 会所、茶楼

会所和茶楼都以彝族建筑风格为主，由形状各异的小楼组合成群，建筑群体错落有致，色彩质朴协调，庄严大方。青砖灰瓦，朱梁黄墙，飞檐斗拱，雕梁画栋，气势恢宏，古色古香。

两组建筑均以现代的建筑结构为主体，大量采用了实木装饰和彝族特色图纹图案以体现古彝风格， 屋脊和尖峰刺天，正脊用砖瓦砌成空脊，主楼单檐歇山顶，门窗雕刻细致，斗拱古朴粗壮，似为明代构件装配而成，四周设廊。呈现在人们眼前的是少见的典雅古朴、精致美丽的建筑奇观，让人赞叹不已。

7. 民族食府

食府的造型与彝族民居风格一致，黄色墙面，青瓦屋顶，红漆廊柱、牛角斗拱，门窗简洁；梁柱、栏杆上适当配置古彝文化特点浓郁的精美木雕，上面雕刻的彝族元素符号、图案，古朴劲健，形态逼真，气韵生动。整个建筑都体现了彝族人民的审美情趣和建造艺术的独运匠心。

五、贯城河沿岸建筑改造

1.玲珑洞

玲珑洞位于黔西城西北郊， 原名神仙洞。洞畔杂树丛生，藤萝缠绕，洞前绿水荡回，波光荡漾，河形“太极图”，名“回龙水”，洞、水相映，秀丽美奇。

后来人们把它辟为佛教场所，现在仿古建筑群已初具规模，整天香烟缭绕，善男信女络绎不绝，特别是逢观音会期，更是人头攒动。已经成为黔西佛教的主要活动场所之一。

对于玲珑洞的改造主要是按原貌进行修缮为主，栏杆上适当加上花纹作为装饰。考虑到该区域人流量大以及贯城河改造后进入的道路被淹的情况，在河上架起可供进入的桥梁通道，入口设置彝族风格浓郁的牌坊，牌坊上以古朴粗犷的斗拱作装饰，与玲珑洞极其丰富的内涵相互映衬。

A path with steps towards the Fishing Island and closing to water in the river is built to coordinate with the overall architectural style, and some landscape accessories are needed to embellish this path such as wall pictures embodying historic culture of Yi minority group, stone settings and partitions, and other accessories as close with water as possible to be matched.

People can enjoy the verdant view of the dancing bamboos, swinging willows, gleaming green grass of Qianxi when they stroll along the water balcony and experience the hug of green river.

6. Clubs, Tea Houses

Clubs and tea houses, consist of small houses groups in different shapes, picturesque layout and harmonious colors, are mainly in Yi ethnic style. With grey bricks and tiles, emperor red beams, yellow walls, flying eaves and Dougong, and caved corridors, the buildings are of antiquity.

Both groups of buildings adopt modern architectural frame with decorations of actual wooden ornaments and pictures and patterns Yi ethnic to embody Yi Style. With a sharp concrete stick directing towards the sky or the hollow bricking ridge, delicate carved doors and windows and ancient and rough Dougong, the buildings are absolutely the rare architectural wonders, elegant, ancient and beautiful. It must earn people's compliment again and again.

7. Ethnic Restaurant

Ethnic Restaurant: The restaurant has the same style of sculpture with Yi houses. The whole building shows Yi people's aesthetic interesting and artistic uniqueness with the architectural elements i.e. yellow walls, grey tile roof, red-printed pillars of corridor, ox horn corbel bracket, simple and clear doors and windows; and exquisite wooden-carvings with thick cultural features of ancient Yi ethnic people on the beam-columns, etc.

V Reconstruction along the Banks of Guancheng River

1. Linglong Cave

Linglong Cave is located in the northwest of Qianxi with its original name "Fairy Cave". Covered with trees and vines, the front of the cave is surrounding with a river in the shape of a "Taiji Map".

Then it was regarded as a Buddhism place, and the constructions groups imitating the ancient ones have presented their primary shape and become one of the important locations for Buddhism events in Qianxi with a constant stream of worshippers and covered clouds of smoke. In particular, it is crowded with pilgrims during the ceremonies of the Bodhisattva's birthday.

Repairing the constructions according to their original style and decorating the railings with some patterns is the major solution when transforming Linglong Cave. As great numbers of people are moving around this area and the entering road will be submerged after the transformation of Guancheng River, a bridge will be constructed over the river with an archway at the entrance. The archway, decorated with the ancient and rough Dougong, is full of Yi charm and match well with the abundant connotation of Linglong Cave.

2. 桥

覆盖于贯城河的桥不少，桥上空间人流、车流密集，是居民在运动中观赏城市水滨景观的主要场所。

分别对各桥两侧栏杆加上原木镂刻的铜鼓和斗拱等彝族元素还有黔西特有的明清牌坊等元素以作修饰，使原来冷冰冰的钢筋混凝土桥梁与城市整体改造风格一致。

颇具特色的结构造型使两岸的景观集结成整体，把桥的形象与两岸城市形体环境、水道的自然景观特点有机结合。

2. Bridge

There are many bridges spaning Guangcheng River. With dense streams of people and vehicles, the bridges are the main places where residents enjoy urban riverbank landscape.

The reinforced concrete bridges can not be unified with the overall reconstruction style of the city unless added separately the guardrails on both sides with the decorations of Yi elements such as drums and Dougong, as well as arches of Ming and Qing dynasty which are only in Qianxi.

The special structural shapes links views of two banks and form a organic integration of the bridge, the urban landscape on two banks and the natural landscape of water.

公共服务设施规划

道路交通分析图

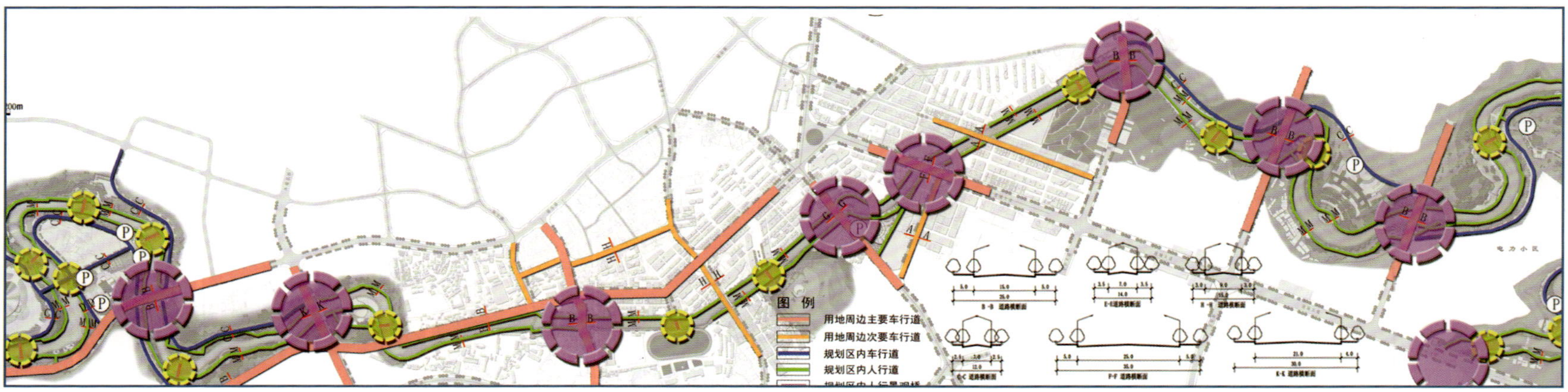

3. 其他建筑

贯城河岸边建筑改造与黔西县沿街建筑改造设计理念保持一致。建筑正面的墙面以暖黄色系为主，山墙面颜色偏灰色，顶上适当加上青瓦铺盖的马头墙，门头及窗户上用木质框架进行装饰，建筑整体风格大方典雅。

根据城市发展需要，如需利用道路两侧土地开发新建建筑，建筑风格建议同我方沿街建筑设计理念相符。

3. Other Buildings

Reconstruction of both banks of Guancheng River follows the same design concept of reconstruction of street architecure of Qianxi County. Buildings are mainly warm yellow on the front and grey the other side to the mountains, with some Horse Head Wall on the roof and some wooden frame decoration on its doors and windows, so that the overall architectural style is elegant and classic.

Any need of developing and constructing new buildings on both sides of the roads due to city development is kindly advised to be coordinated with our design concept.

绿地规划图

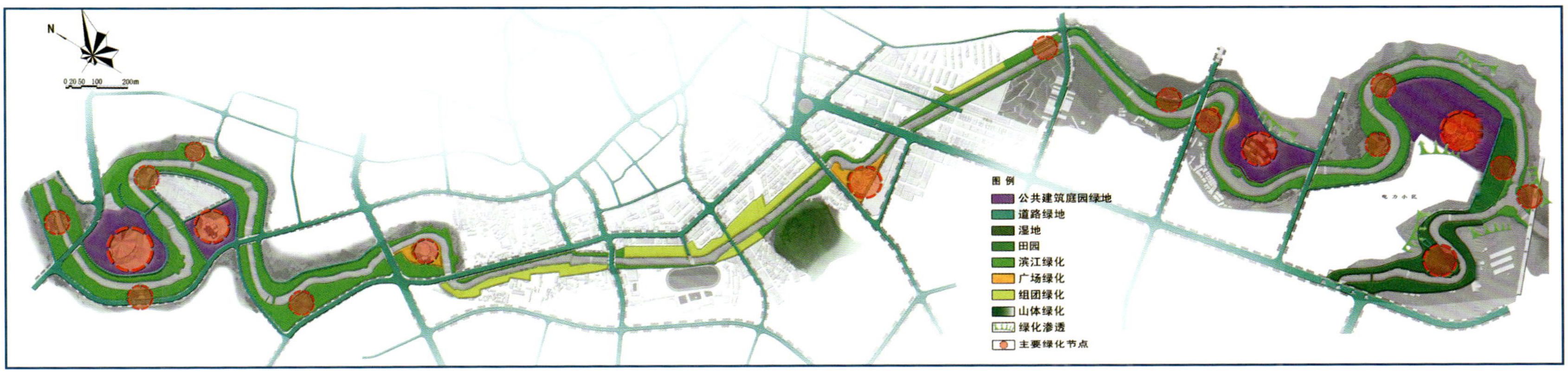

植物配置图

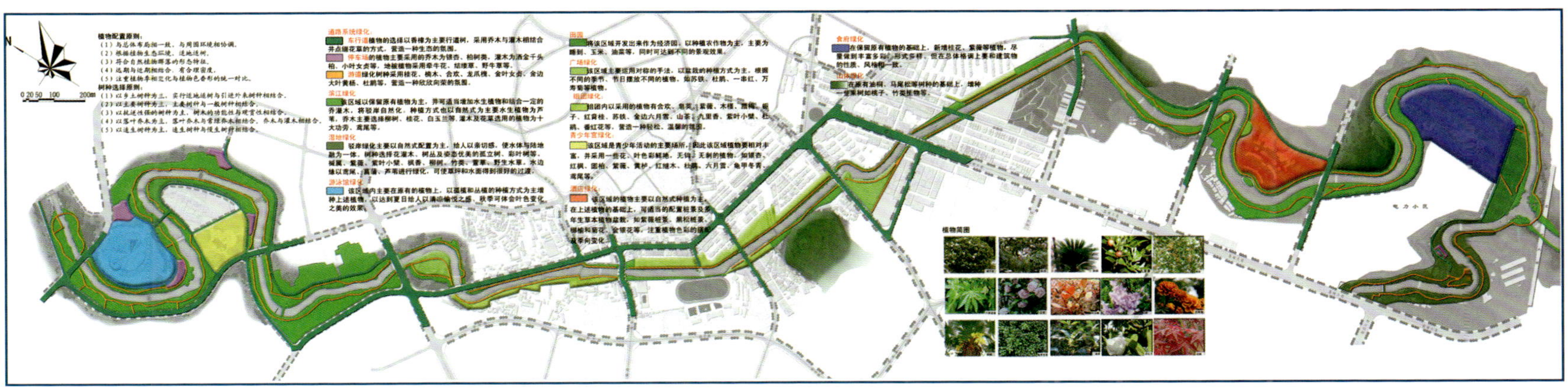

①

②

③

④

⑤

①会所鸟瞰图 ②休闲茶楼鸟瞰图 ③玲珑洞鸟瞰图 ④棋牌休闲区鸟瞰图 ⑤三角广场鸟瞰图

①

①水幕电影广场鸟瞰图 ②健身广场鸟瞰图 ③河道景观透视图

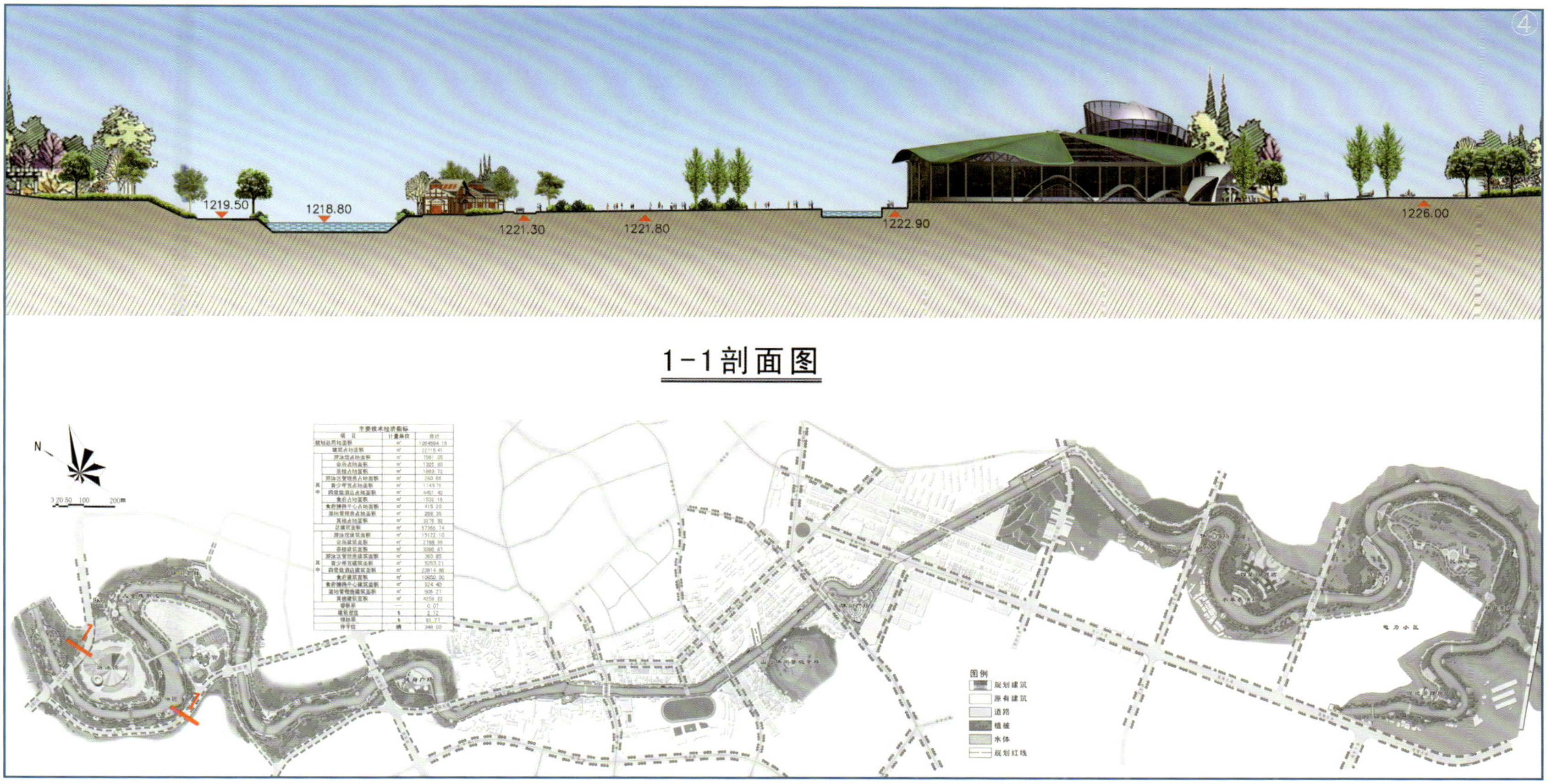

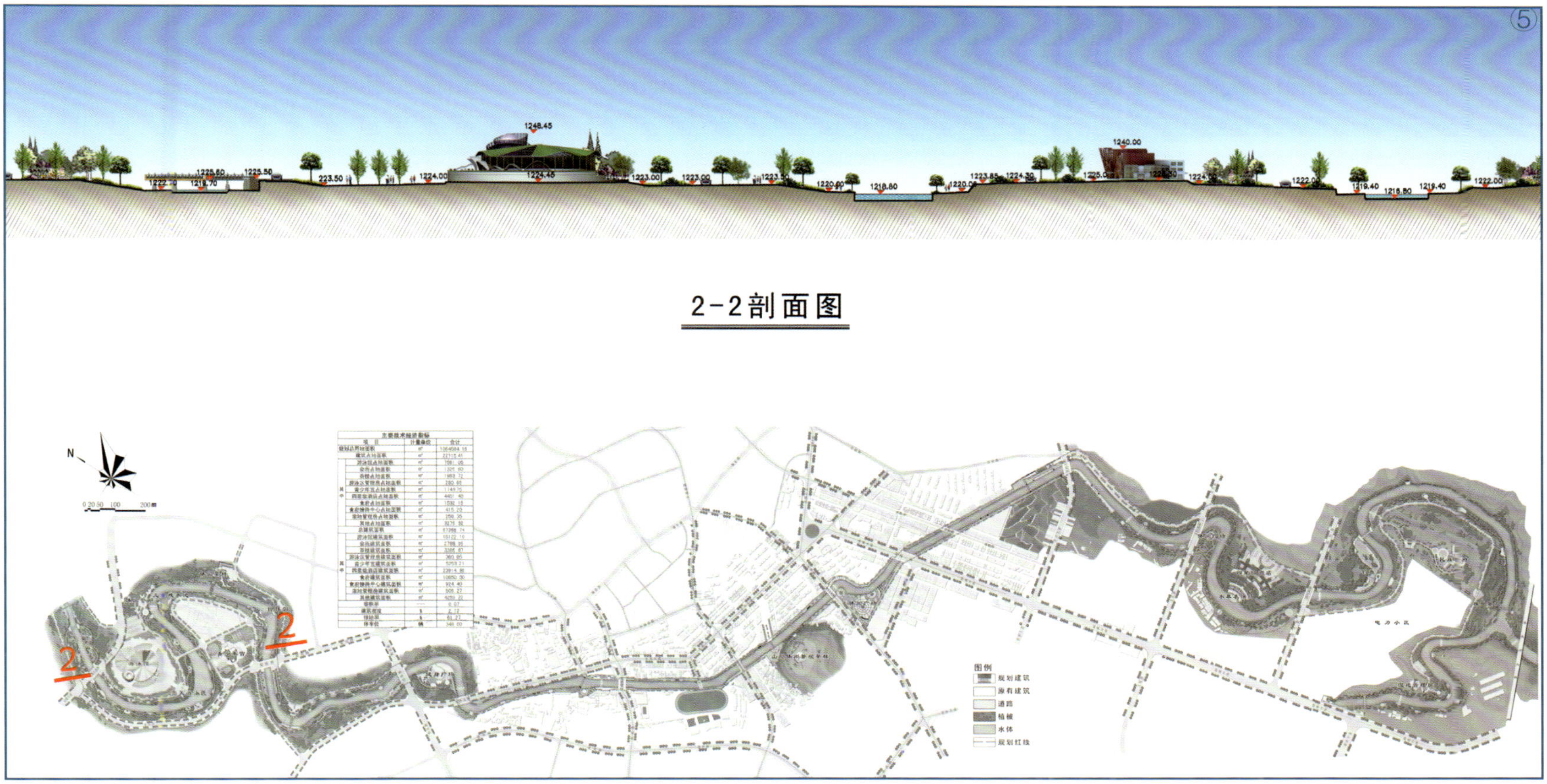

①河道景观透视图 ②河道景观透视图 ③河道景观透视图 ④剖面示意图 ⑤剖面示意图

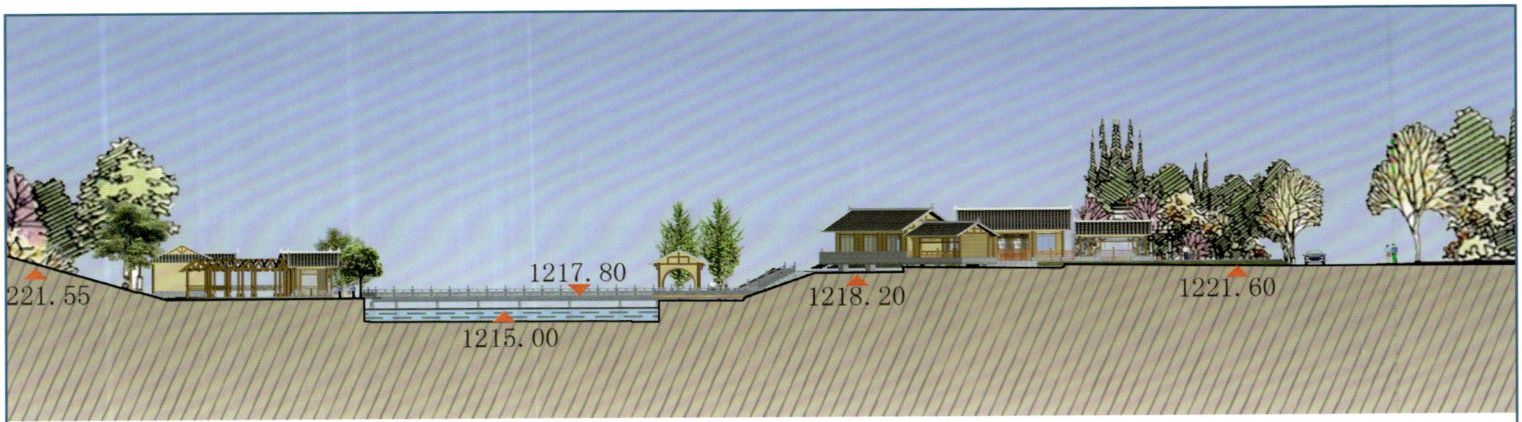
221.55
1217.80
1215.00
1218.20
1221.60

3-3剖面图

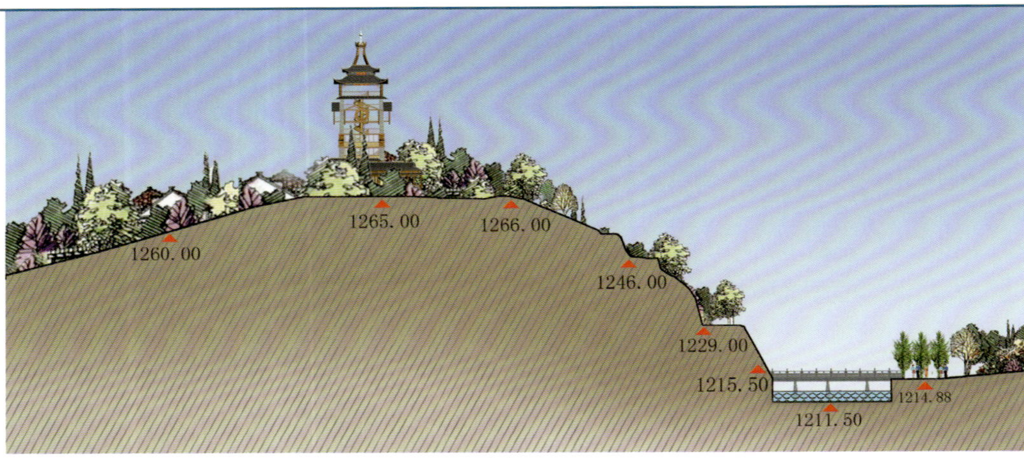
1260.00
1265.00
1266.00
1246.00
1229.00
1215.50
1214.88
1211.50

4-4剖面图

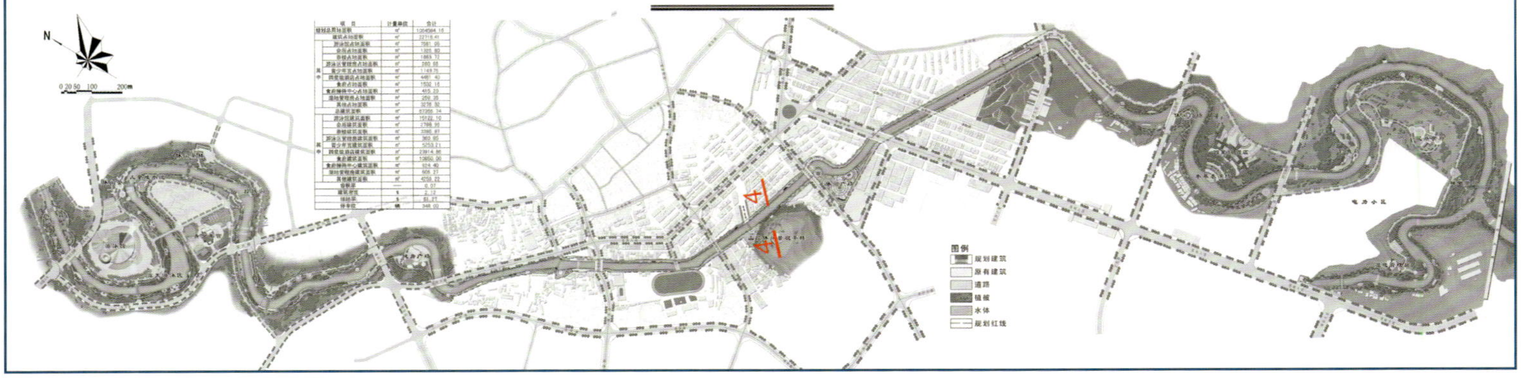

5-5剖面图

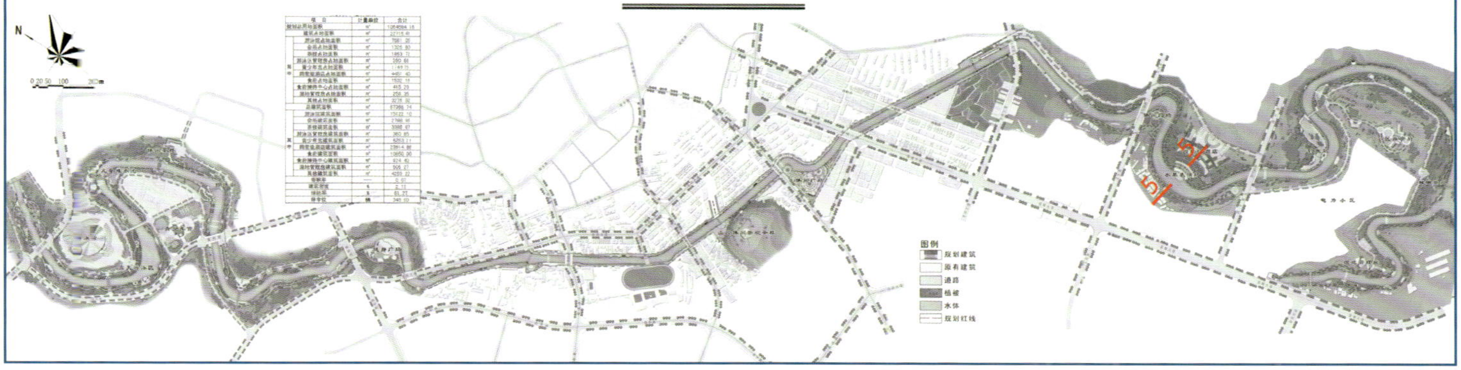

6-6剖面图

公共服务设施

Public Service Facilities

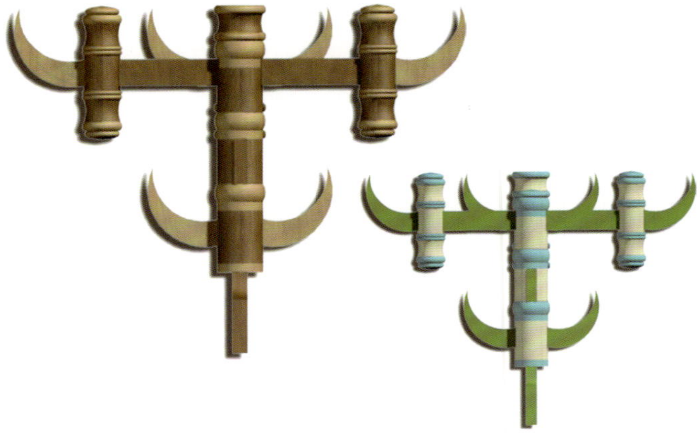

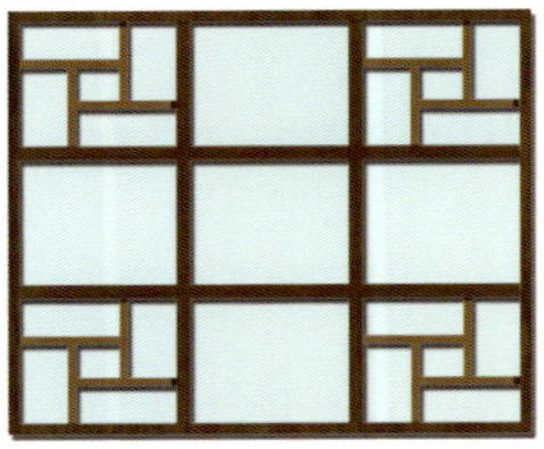

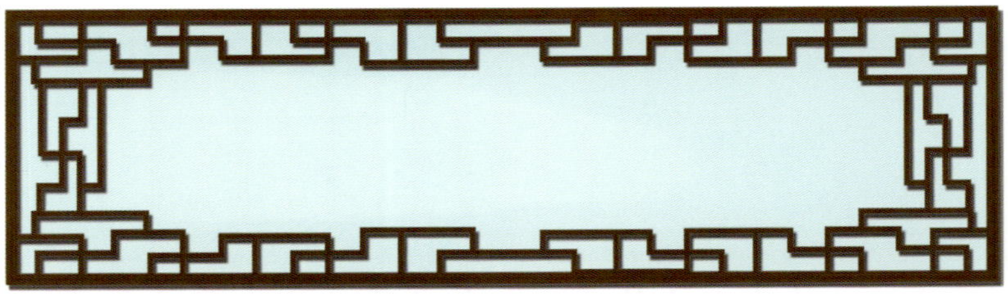

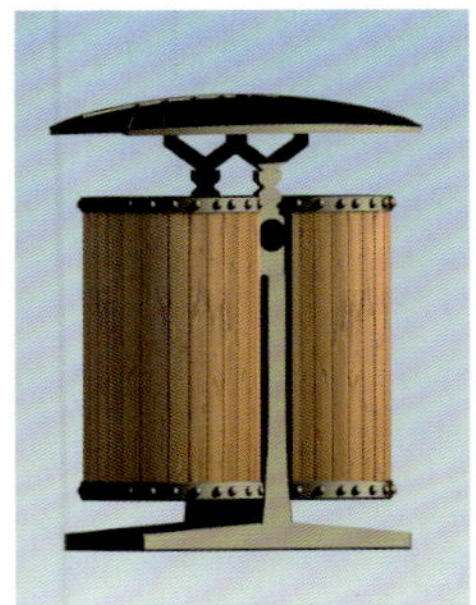

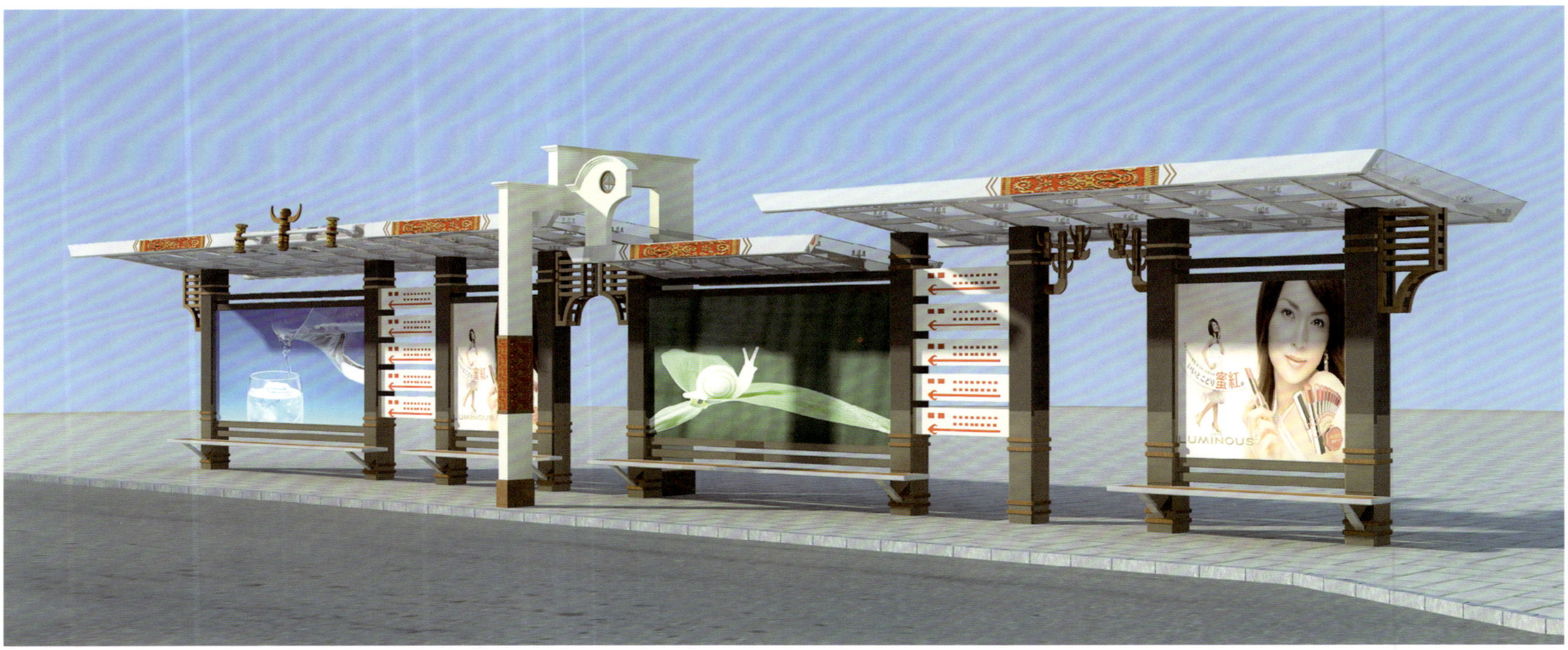

中国印象
青花瓷

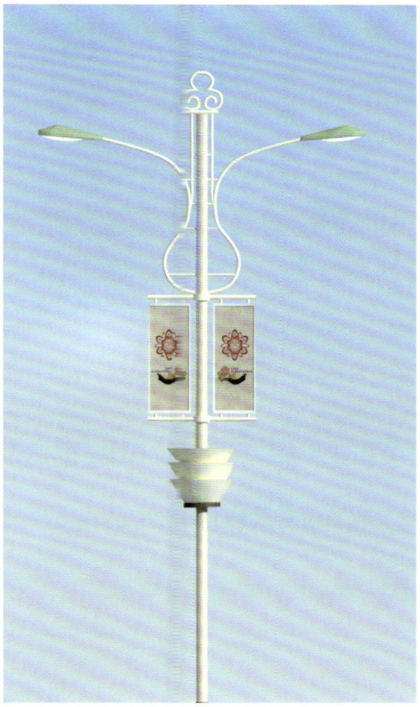

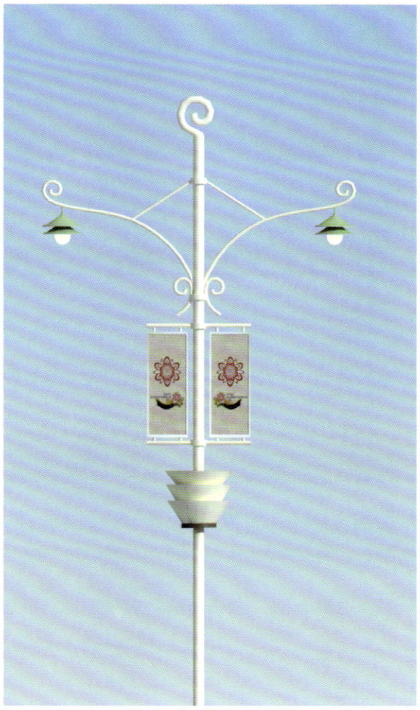

凯里篇 Kaili Profile

城市设计篇（第二版）

变城记

City Changing

前期策划
Prophase Planning

Preface

The best of our country is the common wealth of the world. Possessing the world class original national culture resources, Qiandongnan Miao and Dong Autonomous Prefecture should belong to the world. Developing strength but avoiding weakness, and getting the correct orientation, which are the key and esoterica to develop economy and strengthen a place. Qiandongnan Miao and Dong Autonomous Prefecture does not follow the trend of others, in stead, it aims to develop massively based on its unique and world class original national cultural tourism resource, which, undoubtedly, needs to be guided by extraordinary strategy and mastering ability.

With both advantages of advent of tourism age and prosperous tourism development in China, Qiandongnan Autonomous Prefecture has had the best opportunity of moving toward world and fully developing original national cultural tourism, which surely make Qiandongnan become the concentrated place of the world and must create incredible resplendence.

序言

民族的就是世界的。黔东南苗族侗族自治州拥有世界级的原生态民族文化资源，黔东南苗族侗族自治州理应是世界的。扬长避短，找准自己的发展方向，这是发展经济、振兴一方的肯綮和秘籍。黔东南苗族侗族自治州不追风赶势，而是咬定自己独有的、世界级的原生态民族文化旅游资源不放松，立志在唯我独有的世界级原生态民族文化旅游资源上大做文章，无疑具有超常的战略发展眼光、大势掌控能力和把握能力。要做活黔东南苗族侗族自治州，要做强黔东南苗族侗族自治州，尤其需要这种超常的眼光和能力，尤其需要一种超常的胆略和魄力。

随着全民旅游时代的到来，面对中国已经成为世界旅游大国的大好时机，黔东南苗族侗族自治州无疑已经迎来了走向世界、全力发展原生态民族文化旅游产业的绝佳机遇，黔东南苗族侗族自治州无疑会因此成为世界聚焦的地方，并借此走向出人意料的辉煌。

发展定位

贵州原生态民族文化旅游集散中心和展示中心。

目标愿景

1. 将黔东南苗族侗族自治州打造成中国乃至世界最具原生态意蕴的民族文化旅游胜地。

2. 将凯里市打造成贵州少数民族文化都城，使之成为贵州乃至中华民族文化旅游城市示范地和省外游客进入贵州旅游的首选城市。

3. 将黔东南苗族侗族自治州打造成最佳整合地域资源、独具慧眼确立发展方向，通过发展原生态民族文化旅游产业推动地域经济发展、提升城市品位、铸就一方辉煌的典范，使之不仅在贵州，而且在全国都具有一定的影响力和示范意义。

初步构想

1. 凸现、彰显凯里市民族文化风韵：凯里市作为贵州苗族、侗族首府理应具有浓郁的民族文化气息，整个城市毋庸置疑地应该具有强烈的民族文化感染力，而要做到这一点，就必须对整个城市进行统一的、极具民族意蕴的包装和渲染，通过画龙点睛的修饰和点缀，即通过规范得体的、恰到好处的“穿衣戴帽”等方式，强化和突出凯里市民族文化符号和意蕴（建筑风格的民族化、街道名称的民族化、城市雕塑的民族化等），赋予凯里市原本就应该拥有的极具视觉冲击力的民族文化风韵。

2. 扩张、丰盈凯里市民族文化体量：作为贵州少数民族首府，仅凭“穿衣戴帽”式的城市改造还远远不能显现少数民族都城应有的气派和品质，凯里市要想成为名副其实的贵州少数民族首府，打造少数民族都城，就必须拓展城市规模，充实城市民族文化内涵，即通过城市窗口（西出口）的打造，城市民族广场、城市民族民俗博物馆的打造，城市高档迎宾接待区的打造，城市民族工艺作坊街、民族商品购物街的打造，城市民族民居社区和民族豪宅区的打造，城市现代民族游乐园的打造，使凯里市的城市规模得以延展和扩张，使凯里市的民族文化内涵更为丰满和充盈。

3. 整合黔东南州旅游资源，提升黔东南州景区品位：谁拥有资源，谁就拥有话语权。黔东南苗族侗族自治州作为贵州少数民族集聚地，民族文化资源富足充裕不言而喻，自然生态资源亦同样多姿多彩，令人惊羡，黔东南苗族侗族自治州毋庸置疑地拥有实施旅游强州战略的话语权，毋庸置疑地拥有发展旅游产业、构筑旅游强势的得天独厚的先决条件。但仅有此还远远不够。条件终归只能是条件，要想把条件变为现实，真正使黔东南苗族侗族自治州的旅游资源转化为经济财富，就必须对州内旅游资源进行充分的整合，即丰厚的民族文化资源与多姿多彩、神奇瑰丽的自然生态资源的整合，以及各具特色、交相辉映的各景区相互关联、形成一体的整合，既要把民族文化资源和自然生态资源完美地融合在一起，更要对州内各景区进行统一规划和统一开发，金线串珍珠般地把州内景区串联起来，使之形成有机的旅游产业链，从而真正形成黔东南苗族侗族自治州的旅游强势。

Orientation of Development

To be the tourism distribution and displaying center of original national culture of Guizhou.

Objective

1. To create Qiandongnan Miao and Dong Autonomous Prefecture to be a national cultural resort with the most original ecological charming of China and even the world.

2. To create Kaili City to be the cultural capital of Guizhou ethnic groups, and make it to be the national cultural tourism demonstration base of Guizhou Province and even China, and the favorite choice of visitors from foreign provinces.

3. To create Qiandongnan Miao and Dong Autonomous Prefecture to be the example of the best integration of regional resources, unique orientation of development and promoting regional economy, enhancing urban taste and achieving glory in a region, so that it can be of certain impact and guidance not only in Guizhou, but also the whole country.

Initial Idea

1. To highlight national cultural flavor of Kaili City. As the capital of Miao and Dong ethnics of Guizhou, Kaili City should be with thick national cultural atmosphere and strong national cultural impact, which asks for a unified and special decoration and embellishment to the whole city. Specifically, Kaili City needs to be ornamented through the way of properly "clothing" to strengthen and stress on national cultural marks and charming including national architectural style, national street names and national urban sculptures and so on, and to endow Kaili City with national cultural flavor with great vision impact that it should have owned.

2. To expand and enrich national culture of Kaili City. As the capital of ethnic groups of Guizhou Province, Kaili City has not achieved the air and quality as it should have had if only reconstructed by "clothing", but also need to expand city scale and enrich nation cultural connotation by construct the city window (west entrance), the national square and folk museum, the superior grad reception area, the folk street with workshops, the shopping street with national commodities, the national residential area and national great houses area and the national modern amusement park.

3. To integrate tourism resources and promote regional taste of Qiangdongnan Prefecture. The one who possesses resources have speech privilege. Being the home of ethnic minority people of Guizhou, Qiandongnan Miao and Dong Autonomous Prefecture has rich national cultural resources. Along with its diverse and plentiful natural ecological resources, it owns the speech privilege of implementing the strategy of tourism strengthening this region and the premise of forming tourism strength. However, it is not enough to turn the tourism resources into economic income unless integrating the rich national culture with diverse natural resources and uniting the different scenic spots.

策划版块

1. 核心圈：三条主要街道、民族民俗博物馆、仰阿莎广场、民族高档迎宾接待区。

2. 延展圈：西出口、迎宾大道、民族工艺制品作坊街、民族商品购物街、民族民居社区、民族豪宅区、现代民族游乐园、镰刀湾景区、三棵树景区。

3. 辐射圈：巴拉河景区、西江景观、雷山景区、下司景区、三宝景区、隆里景区、肇兴景区、剑河景区、镇远景区、岜沙景区。

形象描述

市内主要街道：经典的、独具风韵的民族建筑符号，无不充盈着浓郁的民族都市风情，与凯里这座少数民族都城应有的品质配型。

民族民俗博物馆：经典的民族文化符号与现代的建构理念演绎出精妙绝伦的民族民俗文化展示殿堂，民族现代化，现代民族化将在这里得到最好的诠释和演绎。

仰阿莎广场：美妙的民族传说雕塑着人们心中的美好祈愿，宽阔的广场最宜各民族人民祭祀庆典、喜迎宾客和释放激情似火的豪情。

民族高档迎宾接待区：民族的既是世界的，民族的也必然会成为高品位的。高品位的、极具民族风韵的接待中心最适宜恭迎远道而来的贵客和嘉宾。

西出口：这里是人们认识凯里、感受凯里的窗口，这里更是人们感悟黔东南苗族侗族自治州浓郁的原生态民族文化风韵的窗口，它所具有强烈的视觉冲击力和难以抗拒的诱惑力，无不像磁力强大的磁场一样牢牢地吸引住人们，没有人不会不在它面前驻足，没有人不会不在它面前怦然心动。

迎宾大道：匠心独具的城市家具的陈设与布置无疑是愉悦前来观光旅游的宾客的一道亮丽风景线。

民族工艺制品作坊街：留住一门手艺，或许就是留住一段历史；铸就一件饰品，或许就是再现一个民族的生存状态，打造州内各民族民间工艺品作坊一条街，无疑是在修复久远的历史，还原古朴率真的原生态状态。

民族商品购物街：琳琅满目的商品是展示各民族多姿多彩的生活，更是在彰显绚烂夺目的原生态民族文化和浓郁诱人的原生态民族风情。

民族民居社区：走进这里，您就走进了一个现代版的民族村落。您尽可以在这里漫步徜徉，细细玩味现代文明与原生态民族文化的完美融合。

Planning Part

1. Core Part includes the three major streets, National Folk Museum, Yang'asa Square, superior grade national reception area.

2. Expanding Area includes the West Exit, Welcome Avenue, National crafts and Workshops Street, National Residential Area, National Great Houses Area, National Modern Amusement Park, Sickle Bay Scenic Spot, and Sankeshu Scenic Spot.

3. Radiation Area includes Balahe Scenic Spot, Xijiang View, Leishan Scenic Spot, Xiasi Scenic Spot, Snabao Scenic Spot, Longli Scenic Spot, Zhaoxing Scenic Spot, Jianhe Scenic Spot, Zhenyuan Scenic Spot and Yisha Scenic Spot.

Image Description

Major streets in the Chhity: Classic and special ethnic architectural signs enrich and thicken the ethnic flavor of the city, and match with the properties that Kaili, the capital of ethnic groups, should have.

National Folk Museum: A splendid exhibition palace for national folk culture is constructed with classic national cultural signs applied by modern architectural concept. The concept of modernizing nationalism and nationalizing modernism is explained and displayed thoroughly here.

Yang'asa Square: Beautiful legends are expressed through sculptures to record people's best wishes. And the wide square is the best place for people of different nations to hold sacrifices and rituals, welcome guest and release passion.

National Great Reception Area: The best of our country is the common wealth of the world, will be of high taste. The reception center with high taste and national charm is proper to welcome the guests from distance.

West Exit: This is the window with which people know and feel Kaili City. What's more, it is the window where people experience the thick original national cultural charm of Qiandongnan Miao and Dong Autonomous Prefecture. Nobody can resist its strong vision impact and attraction and refuse to stay and enjoy.

Welcome Avenue: The innovative layout of urban furniture must form brilliant landscape fro the visitors.

Traditional Crafts Street: To keep a craft may keep a piece of history; to make an ornament may represent the life situation of a certain nation. Thus, to construct a street containing folk crafts of all nations in this prefecture is to let the ancient history and simple original situation reappear.

Shopping Street with National Commodities: Various commodities display colorful lives of all nations and emanate brilliant original national culture and attractive folk charm.

National Residential Area: You must enter into a modern ethnic village when you visit this area, then you can stroll along the path and taste thoroughly the perfect integration of modern civilization and original ethnic culture.

民族豪宅区：民族建筑绝不是遮风避雨的陋室，更不是破败、简陋的代名词，当它一旦构筑成华丽的殿堂和豪宅，定然会给人带来意想不到的视觉冲击力和心灵震撼力。

现代民族游乐园：点石成金，化腐朽为神奇，将废弃的变为大有作为的，利用废弃的厂区为凯里人民构筑一座充满奇思妙想、充满乐趣的公园，这无疑是一个资源再生、颇具示范意义的创举。

镰刀湾景区：将民族的打造成现代的，将朴拙的、原生态的打造成时尚的，融山水风情于一身，集民族现代于一体，体验现代版的原生态民族情调与风韵，镰刀湾一定会让人眼前一亮。

三棵树景区：这里是进入巴拉河沿线旅游景区的门户，但这里毋庸置疑的更应该是一道自成一体、风韵独具的人文景观，它无疑同样会让人眼前一亮，给人以意外的惊喜。

巴拉河景区：对于本土民族来说，这里或许是他们生存的依托；而对于游客而言，它或许就是人们洗去铅华、还原本真、陶冶性情、摆脱烦恼、寻求精神慰藉的绝佳去处。

西江景区：历史的苦难或许就是今天的辉煌，面对如此壮观的原生态民族村落，没有人不会不为它所震撼，没有人不会不对它产生由衷心的敬畏。

雷山景区：作为拥有“中国民间艺术之乡”和“中国苗族第一寨”美誉，以及被联合国教科文组织誉为人类保存最完好的一块生态和文化净土，无疑是人们返璞归真、回归自然本真的理想王国。

下司景区：这里是进入黔东南苗族侗族自治州的门户之地，这里原本就具有厚重的原生态民族文化底蕴，因而，这里理应成为游人感悟中国现代民间画乡和龙舟大赛的精彩的去处。

肇兴景区：“诗意般的居住”从海德格尔的嘴里说出来是一句美妙的诗句，但“诗意般的居住”在这里却是实实在在的原生态生存状态，没有人不会不为这种诗意般的生存状态所迷恋。

隆里景区：这里曾是朱元璋屯军的城堡，这里至今还保留着当年汉人兵士的生活遗风，在少数民族集聚的地方竟然还保存着如此完整的汉文化孤岛，不能不说是一道奇异的景观。

岜沙景区：走进中国最后一个带枪部落，领略镰刀剃头的精彩，感悟这里的苗人对自然生态高于生命的敬畏和崇拜，以及他们对生与死诗意般的领悟和体验，不能不给人带来太多太多的思考。

镇远景区：各种文化和信仰的汇集，使其成了久远的、过去的漂泊者们的精神寓所，更凝固成了今日古镇厚重的文化底蕴；而泛舟潕阳河畔，观沿岸山光水色，品古镇闲淡的生活方式，不能不说别有一番情趣。

剑河景区：全新的、充满民族风韵的移民新城，重塑的不仅仅是一座城池，更是在重塑一份自信，它无疑会像它所拥有的温泉一样永远地温暖和滋润着这一方土地和人们，吸引着远道而来的游客。

Ethnic Great Houses Area: Ethnic constructions are definitely neither the shelter from the wind and rain, nor the synonym for simpleness and ruin. Once construct them to be magnificent palace and great houses, they must bring incredible vision and mind impact.

Modern Ethnic Amusement Park: This park is constructed on a discarded factory with the idea of "turn stone into gold", which is an amazing work of reusing sources and setting example.

Sickle Bay Scenic Spot: It must be shining people's eyes integrating with water and mountains, ethnic and modern style.

Sankeshu Scenic Spot: This is the entrance of other scenic spots along Balahe River. Nevertheless, it is a single and special cultural landscape which brings visitors unexpected joy.

Balahe Scenic Spot: For local people, it may be their life reliance, while for visitors, it is the best place for them to be free from annoyance and seek spiritual comfort.

Xijiang Scenic Spot: The bitter of history may be the glory of today that today's Xijiang is such a spectacular origin-ecologic ethnic village.

Leishan Scenic Spot: Enjoying the fame of "Chinese Folk Artistic Village" and "The Best Miao Village", and being known as the Perfect Ecologic and Culture Pure Land by the United Nations Educational, Scientific and Cultural Organization, it is surely a kingdom of ideal for people returning to nature and truth.

Xiasi Scenic Spot: This is the entrance of Qiandongnan Miao and Dong Autonomous Prefecture. With massive original cultural connotation, it should be the ideal place where visitors experience modern folk painting village and dragon-boat race.

Zhaoxing Scenic Spot: "Poetic dwelling", a beautiful idiom described by Martin Heidegger, and it is the real description of the origin-ecology living status.

Longli Scenic Spot: It has been the castle where Zhu Yuanzhang, the first emperor of Ming Dynasty, collected his army whose life style has been rained even by now. It's really rare that such entire Han culture remaining in the ethnic minority people's place.

Yisha Scenic Spot: The last tribe where people can take guns with them of China. Here, you can feel the worship and respect of Miao people to the nature and their cognition and understanding of life and dead, which makes you think more and more.

Zhenyuan Scenic Spot: Collecting various culture and faith, it has become spiritual settles of the ancient floaters and coagulated the massive cultural connotation of this ancient city.

Jianhe Scenic Spot: A new city reconstructed for migrations with full ethnic charm. What reconstructed is not just a new city but its self-confidence which will surely warm and moisten its people and land like the spring it owns, and attract visitors from distance.

设计元素（苗族建筑）
Design Element

设计元素（民族建筑）

Design Element

设计元素（文化）
Design Element

城市总说
Outline

第一章 区域背景概况

第一节 自然地理环境背景

一、区位及交通通信条件

1. 区位

凯里市位于贵州省东部，是黔东南苗族侗族自治州首府驻地。地处清水江上游、苗岭东北麓，在东经 107°41′～108°12′，北纬 26°24′～26°48′之间。东西长 51.7km，南北宽 44.3km，行政区域面积 1305.9km²。与台江、雷山、丹寨、麻江、福泉、黄平等市县接壤。

2. 交通通信

凯里市作为黔东南州交通运输和通信的枢纽，境内湘黔铁路株六复线过境 53.5km。320 国道、麻（江）凯（里）高速公路、凯（里）铜（仁）高速公路贯穿全境；村村通公路，并与周边省道、县道相连通，构成市内交通网络。境内清水江、重安江内通航 72km。杭——福——贵——成光纤电缆横贯境内，实现村村通电话。城乡电网覆盖率达 100%，变电容量超过 130 万千伏安，10kV 及以上输电线路总长超过 2200km。

二、自然地理条件

1. 地质地貌

本地区地处云贵高原东侧的梯级壮大斜坡地带，横跨黔中丘原和黔东中低山丘陵两个地貌区。境内山地面积占 66.8%，丘陵面积占 29.5%，山间平坝面积占 3.7%。地势呈西北、西南部和东南部较高，中部和东北部较低的走势。境内平均海拔 868m。

2. 气候、水文

本地区属中亚热带湿润季风性气候，年均气温 15℃，年均降水量 1240.4mm，年日照时间 1289h，无霜期 287d。

区域内主要河流有清水江、重安江、巴拉河等，其中长 10km 以上，流域面积 20km² 以上的河流 13 条。境内河流均属长江流域沅江水系。

3. 生物资源

境内几乎拥有亚热带到温带的所有生物，其中粮食作物有 9 大类 130 多种，蔬菜有 11 大类 100 多种，果类有 6 大类 70 多种，野生牧草有 100 多种，珍稀古树有 10 多种，野生动物有数百种，药用动植物有 400 多种。

Chapter Ⅰ Regional Background Profile

Section Ⅰ Environmental Background of Natural Geography

Ⅰ Condition of Location, Traffic and Communication

1. Location

Kaili City is located in the eastern part of Guizhou province and the capital of Miao's and Dong's autonomous prefecture in eastern Guizhou. It is in the upriver of Qingshui River and the northeast of Miao Mountain. The city lies between 107° 41'~108° 12'east longitude and 26° 24'~26° 48'northern latitude. The city is about 51.7 kilometers long from east to west, 44.3 kilometers wide in south-north direction, and 1 305.9 square kilometers of total administration area. It borders on the cities and counties like Taijiang, Leishan, Danzhai, Majiang, Fuquan, Huangping and so on.

2. Traffic and Communication

As the hinge of traffic and communication in southeast Guizhou, there is the Zhuliu Line of Xiangqin Railway which has the total length of 53.5km in Kaili. The 320 national highway, Makai speedway and Kaitong speedway run through all the city areas. In addition, the roads exist in every villages and connect the border's province road & county road which constitute the city traffic network. It's open to navigation traffic which has the total length of 72km within the Qingshui River and Chongan River. The Hang-Fu-Gui fiber optic cable in the domain helps achieve the availability of telephone for every village. The coverage of urban and rural power grids is 100%, the power transformation capacity exceeds 1.3 million kva, and the total length of power transmission line (above 10kv) exceeds 2200km.

Ⅱ Natural Geography Conditions

1. Geography

The region lies in the rundle great incline zone of eastern Yungui Highland, bestriding the Qianzhong Hilly Plain and Qiandong low and middle hilly land. In the territory, the area of mountain is 66.8%, the area of hilly land is 29.5% and the area of Pingba is 3.7%. The trend of the hypsography changes with higher altitude in northwest, southwest and southeast, and lower altitude in middle area and northeast. The average altitude is 868m in the field.

2. Climate, Hydrology

This area belongs to the humid subtropical monsoon climate with the annual mean temperature of 15℃, the annual precipitation of 1 240.4mm, the annual sunlight time of 1 289 hours and frost-free days of 287 days.

In the zone, the main rivers include Qingshui River, Chongan River and Bala River and so on. Among these, there are 13 rivers which has the drainage area of above 20km2. All the rivers belongs to the Yangtze Valley or Ruan River water system.

3. Biology Resource

Within the field, it owns nearly all kinds of creatures from subtropical to temperate zone. Among them, there are more than 130 species in 9 categories for crops, over 100 species in 11 categories for vegetables, over 70 species in 6 categories for fruits, over 100 species for wild pastures, over 10 species for rarely old trees, hundreds of wild animals and over 400 species of medical animals and plants.

第二节 区域社会经济及历史文化概况

一、社会经济概况

1. 区位

凯里市现辖 2 个乡、8 个镇、5 个街道办事处、31 个居民委员会、209 个村民委员会。区内有汉、苗、侗、布依、壮、回、水、彝等多个民族。2004 年末全市总人口 45.78 万人，其中少数民族人口 34.68 万人。

2006 年，全市实现国民生产总值 38.2 亿元人民币，人均国民生产总值达 8312 元，比上年增长 10.65%，实现财政总收入 3.5 亿元，比上年增长 16.54%，其中地方财政收入 1.81 亿元。城镇居民人均可支配收入达 7707 元，比上年增长 9.1%，人均消费支出 5731 元；农村人均纯收入 2177 元，比上年增长 9%，人均消费支出 1371 元。

二、历史沿革

凯里自古以来即为少数民族聚居区域。“凯里”系苗族语音译，意为木佬人的田。春秋时期，属南蛮牂牁国；战国属夜郎且兰国；秦汉属且兰县；隋属宾化县；元始设治，有凯里安抚司，恭焦溪蛮夷军民长官司，臻洞长官司等，至今已有 700 多年。明洪武十四年（公元 1381 年）置清平堡；二十二年（公元 1389 年）升为长官司，属平越卫。二十三年六月初四（公元 1390 年 7 月 16 日）置清平卫于司北，改隶清平卫。弘治七年五月二十七日（公元 1494 年 6 月 23 日）置都匀府，同时清平长官司废，置清平县，隶都匀府。嘉靖八年二月二十八日（公元 1529 年 4 月 6 日）凯里安抚司由四川布政司改隶贵州布政使司，属清平卫。万历二十九年（公元 1601 年）改隶平越军民府。万历三十五年(公元 1607 年)，降凯里安抚司为凯里长官司，康熙七年七月安二（公元 1668 年 8 月 9 日），清平县入麻哈州。康熙十年，恢复清平县，凯里长官司隶清平县，雍正十年（公元 1732 年）开辟苗疆。进一步改土归流，兵设凯里卫，隶属丹江厅，领十三屯堡军。雍正十二年，添设清平县丞一员分驻凯里，称凯里县丞。乾隆三年十一月（公元 1739 年 1 月），将镇远县臻洞长官司革除，以该司所属臻洞司，赏郎等江 21 寨就近改隶清平县凯里县丞。民国 3 年（公元 1914 年），因清平县与山东省东昌府清平县重名，故以境内有香炉山改称炉山县，凯里县丞改称凯里分县。

1949 年 11 月，炉山县解放。1951 年 1 月，成立凯里苗族自治区人民政府。1956 年 4 月，国务院批准建立黔东南苗族侗族自治州，7 月宣告凯里作为州府所在地。1983 年 8 月，国务院批准，撤销凯里县，设立县级市。

1984 年 1 月，凯里市正式成立。

Section Ⅱ Conditions of Local Society, Economy, History and Culture

Ⅰ Conditions of Society and Economy

1. Location

There are two villages, 8 towns, 5 street offices, 31 resident committees, and 209 village councils. Within the area, it includes many nationalities like Han, Miao, Dong, Buyi, Zhuang, Hui, Shui, Yi and others. At the end of 2004, the total population reaches 457 800 which include 346 800 of minorities.

In 2006, the GNP of the city is RMB 3.82 billion and the Per Capita GNP is RMB 8 312 which increases 10.65 year on year. The gross financial income reaches RMB 350 million which increases 16.54% over the last year, including the local financial income RMB 181 million. The per capita disposable income is RMB 7 707 which increases 9.1% over the last year for urban residents, while the per capita consumption is RMB 5 731. The per capita net income is RMB 2 177 which increases 9% over the last year for rural residents, while the per capita consumption is RMB 1371.

Ⅱ History Evolution

From very old time, Kaili is the assembly and living area for minorities. Kaili comes from the Miao language meaning the farmland of Mulao people. In Spring and Autumn Period, it belonged to Zangke of Nanman; In Warring States Period, it belonged to Qielan of Yelang; In Qin & Han dynasty, it belonged to Qielan County; In Sui dynasty, it belonged to Binhua County; At the beginning of Yuan 700 years ago, there was Zhan County including Kaili Anfu Official, Gongjiaoxi Manyi Junmin Chancellor, Zhendong Chancellor and so on. In Minghongwu 14 (A.D. 1381), there was Qingpingbao; In 22 (A.D. 1389), it was upgraded to Zhangguansi belonged to Pingyueyi. In lunar calendar of 4 June 23 (A.D. 16 July 1390), Qingpingyi was set up in Sibei and belonged to Qingpiingwei. In Hongzhi 27 May 7 (A.D. 23 June 1494), it established Duyun Mansion and eliminated Qingping Chief. Qingping County was also established and belonged to Duyun Mansion at that time. In Jiajing 28 February 8 (A.D. 6 April 1529), Kaili Anfu House was changed to belong to Qingpingwei, Guizhou Buzheng House from Sichuan Buzheng House. In Wanli 29 (A.D. 1601), it was changed to belong Pingyue Junmin Mansion. In Wanli 35 (A.D. 1607), Kaili Anfu House was changed to Kaili Chancellor House. In Kangxi July 7 (A.D. 9 August 1668), Qingping County belonged to Maha state. In Kangxi 10, Qingping County renewed and Kaili Chancellor House belonged to it. In Yongzheng 10 (A.D.1732), Miaojiang was pioneered. The policy of Gaitu Guiliu was underway and the Arming House who led Thirteen Tunbao armies were stayed in Kaili Yi belonged to Danjiang Hall. In Yongzheng 12, one Qingping County official was dispatched to Kaili called Kaili Xiancheng. In Qianlong November 3 (A.D. January 1739), the Zhendong Chancellor House of Zhenyuan County was dismissed, and put the 21 villages near the Langdeng River into the management of Kaili Xancheng of Qingping County. In ROC 3 (A.D. 1914), since Qingping County was same with the name of Qingping country, Dongchang Fu in Shandong Province, it's changed to Lushan County and the Kaili Xiancheng was changed to Kaili Fenxian.

In November 1949, Lushan County was released. In January 1951, the People's Government of Kaili Miao Autonomous Prefecture was founded. In April 1956, the Miao & Dong Autonomous Prefecture of Southeast Guizhou was approved by State Council and Kaili was announced as the capital in July. In August 1984, Kaili County was withdrawn and county-level city was underway by State Council.

In January 1984, Kaili City was formally founded.

三、民族文化

凯里是一个“以苗族为主体,多民族聚居”的多民族花园,被誉为“苗岭明珠”。同时,凯里是“芦笙的故乡”、“歌舞的海洋”,素有“百节之乡”的美称,以其得天独厚的自然景观、古朴浓郁的民族风情和丰富独特的民族文化,被确立为贵州东线民族风情旅游中心。

1. 苗族节日

凯里规模较大的苗族节日有过苗年、吃新节、闹冲节、爬坡节、牯藏节等。在这些节日中,大都开展跳芦笙、跳铜鼓、斗牛、赛马、唱歌、斗雀等集会活动。

2. 民族村寨

(1) 南花苗寨

位于凯里市三棵树镇东南部,距离凯里市中心 13km,与炉(榕)省河道隔河相望。迄今,南花村已分别被中央、省、州授予“全国先进党组织”,“先进文明村寨”、“农业部生态富农示范点”等称号。

(2) 舟溪民族村寨

位于凯里城南 20km,昔日可行舟而得名,是苗族、仫佬族等民族聚居区,其芦笙制作、织锦、刺绣,久负盛名,是黔东南最大的芦笙堂之一。

(3) 青曼民族村寨

位于凯里城南 23km,长岭岗东南麓,是黔东南州民族风情旅游景点之一。其纺纱、织布、织锦、刺绣工艺精湛,制作方法古老,为国内鲜见。

(4) 麻塘革家寨

位于凯里城北 21公里,虎(庄)洛(榜)公路东侧,𫟼家聚居,𫟼家拥有内容极为丰富的传说故事、古歌、情歌、童话;拥有别具一格的音乐、舞蹈、服饰、银饰、婚姻习俗。此外,麻塘𫟼家妇女擅长蜡染,其图案巧妙、线条古朴,颇具欣赏价值。

Ⅲ Ethnic Culture

Kaili is a multinational garden with the majority of Miao and many other nationalities called "Pearl of Miao Mountain". At the same time, it's the "hometown of Lusheng", the "ocean of singing and dancing", and the "place of hundred festivals". Depending on its advantaged natural sight, old and plaint ethnic expression and the abundant and unique ethnic culture, Kaili is established as the center of ethnic style tour in east Guizhou.

1. Miao Festival

The large-scale Miao festival includes Miao Year, Chixin Festival, Naochong Festival, Slope Climbing Festival, Guzang Festival and so on. Among these festivals, most of them develop the assembly activities such as jumping Lusheng, jumping tymbal, bullfighting, horse racing, singing, sparrow fighting and so on.

2. Ethnic Stockaded Village

(1) Nanhua Miao village

It is located in the southeast part of Three Trees Town 13km away from the city center, and look over the Lusheng river way separated by a river. So far, Nanhua village has been awarded the name of "National Advanced Party Organization", "Advanced Civilization Village", "Ecology Farmer Richness Demonstration Spot of National Agriculture Ministry" by the central government, province and state.

(2) Zhouxi ethnic village

It's located 20km away from south of Kaili city and famous as boating available before. Miao, Mulao and other nationalities live here. Its Lusheng is noted in terms of production, brocading, and embroidering. It's one of the largest Lusheng Houses in the southeast of Guizhou.

(3) Qingman ethnic village

It lies 23km away from south Kaili and southeast mountain foot of Changling Hillock. It's one of the ethnic style tour spot in southeast Guizhou. It's good at spinning, cloth weaving, brocade and needlework with the ancient production method which is rare in the country.

(4) Matang Gejia village

It lies 21km away from north Kaili and the east side of Huluo Road. Gejia live here and own an abundant stories, old songs, love songs and fairy tale. They also own the unique music, dancing, garment, silver decoration, marriage custom. In addition, the women here do well in wax dyeing with artful pictures and plain lines which owns appreciative values.

第二章 规划设计总则

第一节 指导思想

一、社会、经济、环境效益综合发展的理念

二、区域整体发展理念

三、城市特色发展理念

第二节 规划设计原则

一、彰显文化,有机更新的原则

二、以人为本,构建人性化城市空间的原则

三、生态优先,建设生态化城市的原则

四、兼顾经济,持续发展的原则

Chapter Ⅱ General Principles of Layout Design

Section Ⅰ Guidance Thought

Ⅰ Integration development idea of society, economy and environmental benefit

Ⅱ Development idea of regional integration

Ⅲ Development idea of city features

Section Ⅱ Principles of Layout Design

Ⅰ Principles of cultural outstanding and organic renewal

Ⅱ Principles of people-oriented and construction of humane city space

Ⅲ Principles of ecology first and construction of ecological city

Ⅳ Principles of economy consideration and sustainable develoment

第三节 规划设计依据

一、国家、省、市、县相关法律法规

二、项目委托书

第四节 空间结构布局

本次规划所涉及的范围在空间上呈现“三点一线”的整体格局，即“以西出口为起点，途经迎宾大道，到中点仰阿莎广场后，再经友庄路、宁波路，或由博东路、或由博西路进入终点万博广场”的凯里城市主干道整体景观规划。

第三章 规划思路

当前，旅游业的竞争已不只是单个景区、景点的竞争，而是旅游目的地之间的竞争。而由“旅游城市”到“城市旅游”的定位上的转变，正是将标的城市由“旅游集散地”转变为“旅游目的地”。

凯里市作为黔东南苗族侗族自治州的首府驻地和黔东南地区主要的旅游集散地，拥有良好的旅游接待服务环境和充足的过境游客基数，具有发展城市旅游的基本前提和绝对优势。

同时，对城市旅游进行开发，需要对城市形象和城市意象进行塑造。所谓城市形象，是指城市所具有的各种外形特征或特点。城市意象，是指观察者（包括当地居民和旅游者）对城市的各种外形特征或特点所产生的心理图像或印象，是一种“公众印象”，是城市居民中多数人拥有的对某个城市的共同心理图像。

在城市旅游中城市意象的塑造上，不仅要考虑居民的需求与感受，更要着重关注旅游者活动的特点及其体验，不仅要从城市景观方面展现其特色与魅力，更要从活动、设施、氛围、文化等方面进行打造，使城市本身成为一个让游客流连忘返的旅游目的地，一个能不断满足居民和游客不同的休闲游憩需求的城市。因此在本次规划中主要从以下两方面做到统一：

一、将相对独立的项目从内容到形式上都作出了统一的规划，不仅对所涉及的规划范围作出了更为完善更为系统的布局，形成科学可行的规划思路，还在较大范围内统一了城市风格，更有利于彰显城市的人文内涵和精神品质。

二、苗族、侗族建筑风格和文化符号的使用，始终贯穿整个规划。其所代表的民族精神是整个规划的灵魂所在和根本指导，同时也是凯里市的文脉所在，凯里市的城市性格象征，凯里市的城市发展方向。

Section Ⅲ Proof of Layout Design

Ⅰ Relevant laws and regulations of county, city, province and country

Ⅱ Project letter of attorney

Section Ⅳ Layout of Space Structure

The overall layout of this Kaili main road project is reflected as "3 points stand in one line" in space, namely, starting from the western exit and going through the Yingbin Avenue to the middle point of Yangasha Square, then going via Youzhuang Road and Ningbo Road or Bodong Road or Boxi Road to enter the terminal of Wanbo Square.

Chapter Ⅲ Layout Thought

Nowadays, the competition of tourism is not only depending on the sole scenic spot and resort competition, but the competition among the tourism destinations. Hence, the positioning change from tour city to city tour is just to transfer the city from tour hub to tour destinations.

As the capital of Miao & Dong autonomous prefecture and the main tour hub in southeast Guizhou area, Kaili owns the good tour reception service environment and enough cross-border tourist radix. It makes Kaili possess the basic precondition and absolute advantage to develop city tour business.

Meanwhile, it needs to build the city image and artistic conception of city to develop the city tour business. City image means the various physical characteristics of the city. Artistic conception of city means the picture or impression caused by the various physical characteristics of the city from the observers; it's a kind of public impression and the common mental pictures of a particular city from the majority of city residents.

In the process of building the artistic conception of city in city tour, we not only need to consider the needs and feelings of the residents, but also pay more attention to the characteristic, needs and experience of the tour activities. In addition, we not only need to show the attraction and feature of the city from the sight, but also need to improve it in terms of activities, devices, atmosphere, culture and so on. By doing so, the city can become a place where makes the tourists enjoy themselves, and also a place where can satisfy the different needs of entertainment and fun from residents and tourists. Hence, the layout should achieve the unification of following two aspects:

First, we should make the unified layout of those relatively independent projects from content to form. This can not only help make a more comprehensive and systematic distribution of the related layout scope to form the scientific and available layout thought, but also unify the city style in a greater scope to better show the humanistic connotation and spiritual quality.

Second, we should make sure that the use of Miao and Dong architecture style and cultural signs run though the whole project all the time. This is because its representing ethnic spirit is the soul and essential guidance of the whole project. Besides, it's the cultural context of Kaili, the city character symbol and the direction of city development.

迎宾大道沿街建筑改造

Rebuilding Constructions along Yingbin Avenue

一、凯里市迎宾大道规划设计说明

1. 规划设计思路

凯里市迎宾大道作为进入凯里市区的必经之路，对展示凯里城市形象，突出凯里人文风貌有着重要的现实意义。因此，我们在规划中，充分考虑了凯里的区域特色和地方文脉，打破常规的道路景观设计理念，将凯里市迎宾大道打造成为进入凯里市区的第一个旅游景区（点），不仅为城市铸造一条景观公路，也是在为城市打造一个景区（点）。

在设计上，运用黔东南少数民族的吊脚楼、风雨桥、鼓楼等建筑符号形式，对迎宾大道两侧的建筑进行相应的外立面包装，从外形上赋予其民族文化内涵；同时在迎宾大道的隔离绿化带中穿插一些颇具黔东南民族特色的雕塑小品，不仅是对凯里市城市景观的丰富、文化内涵的提升，也是在向亲临凯里市的各方宾客形象的推介黔东南。

I Layout Design Explanation of Yingbin Avenue in Kaili City

1. Layout Design Thought

As the only entry road to Kaili city area, Yingbin Avenue is significant to show the city image and outstand the humanistic sence for Kaili. Hence, we try to make the Yingbin Avenue as the first scenic spot of Kaili city area by breaking the regular design idea of road sight under the full consideration of Kaili's regional feature and local context. It's not only an establishment of a scenic road, but also a scenic spot for the city.

In the aspect of design by applying the northeast Guizhou minority's architecture sign forms such as Hanging House, Fengyu Bridge and drum-tower, it makes the corresponding outer elevation package of the buildings on the two-side of Yingbin Avenue to be endowed with national culture meaning from shape. At the same time, we cut in some ethnic feature sculptures of southeast Guizhou in the Yingbin Avenue's isolation green belt. It can enrich the city resorts and increase the cultural meaning; it can also introduce the southeast of Guizhou to the coming guests from everywhere.

区位分析图

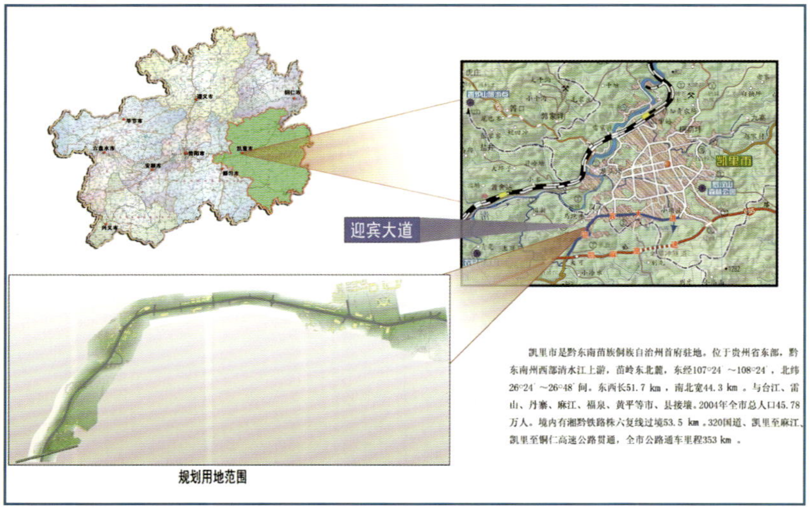

道路现状分析图

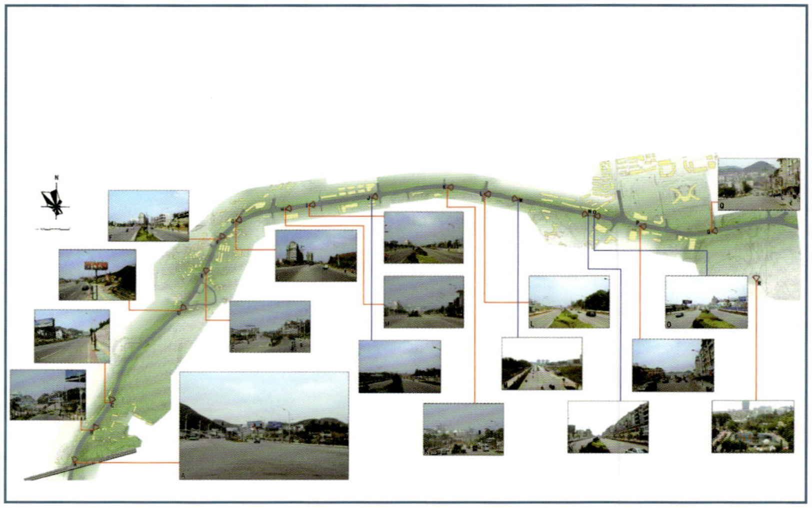

2. 细部介绍

(1) 公共服务设施

包括迎宾大道两侧的公交车站台、路灯、垃圾桶、休闲长椅，以及人行天桥。其中，公交车站台和人行天桥充分运用侗族风雨桥的建筑符号进行造型；路灯、垃圾桶、休闲长椅则以凸显地方民族文化符号和传说，同时彰显城市时尚文化为主要设计创意思路。

(2) 街口牌坊

给游客进入和走出迎宾大道以提示，同时通过与众不同的牌坊，将汉文化的建筑符号与苗侗文化的民族符号相融合，兼容并蓄，使之显得既民族又时尚，不仅彰显了多民族融合的城市性格，还提升了城市时尚的民族化品位。

2. Details Introduction

(1) Public service establishment

It includes bus station, road light, rubbish bin, leisure bench and overpass. Among these, the bus station and overpass fully use the architecture sign of Dong's Fengyu Bridge; road light, rubbish bin and leisure bench are under the main design thought of outstanding local ethnic cultural signs and tales while showing the fashion culture of the city.

(2) Street tablet

It is used to give reminder to tourists who enter or leave the Yingbin Avenue. Through the different tablets, they can combine the architecture signs of Han culture and the ethnic signs of Miao & Dong culture to make it ethnic and fashionable. It not only shows the city nature of multinational combination, but also boosts the ethnic taste of city fashion.

总平面图

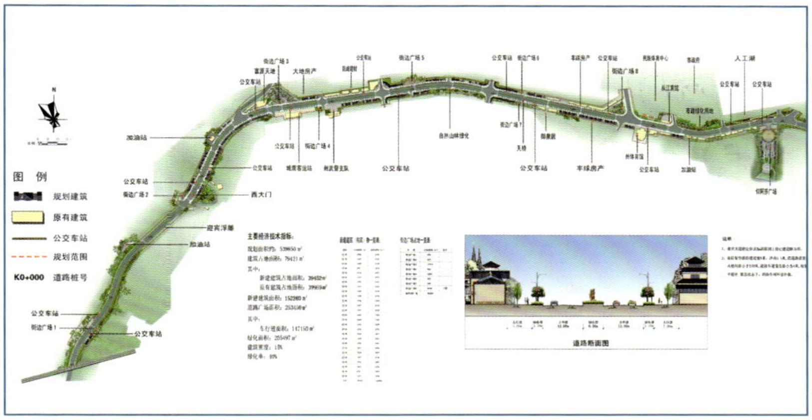

功能分析图

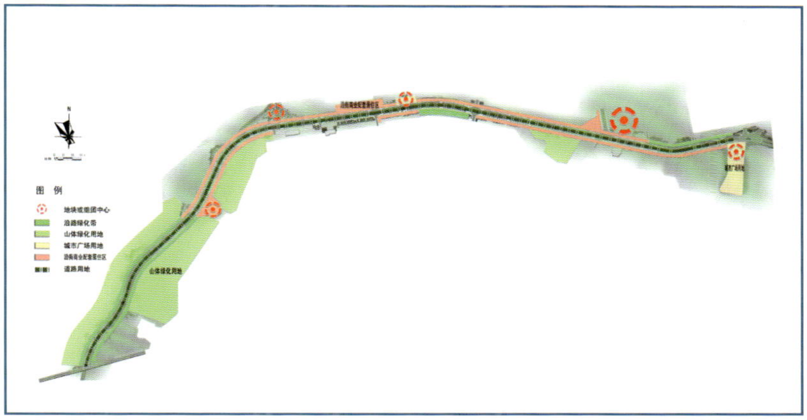

公共设施规划图

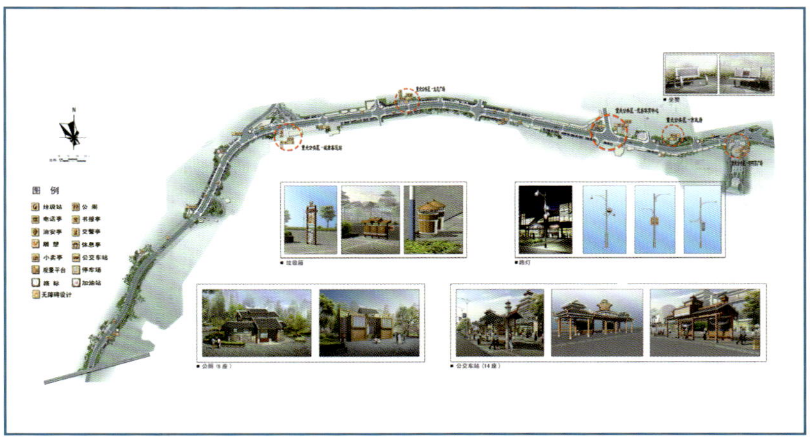

道路交通规划图

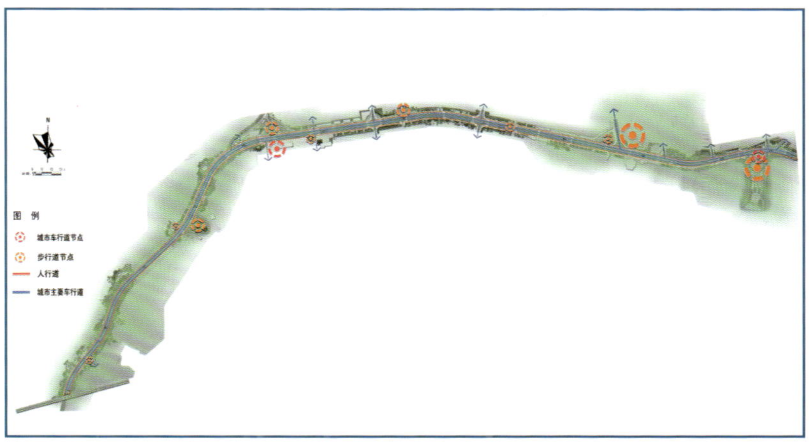

(3) 雕塑小品

为彰显和宣传地方民族文化的点睛之笔，主要包括两种不同类型：一是体现当地民俗节庆场面的，二是反映当地民族历史文化传说的。

特别值得介绍的有施洞独木龙舟雕塑，它紧靠迎宾大道上的一处充满民族特色的人行天桥，与周边的建筑和背景的桥浑然一体，体现了“人从桥上过，舟在水中游”的意境，为炎炎夏日带来了一丝清凉。

而岜沙枪手的主题雕塑，则以一种抽象的表现手法将岜沙汉子打猎归来，正逢宾客迎门，集体鸣枪迎客的场景展现在我们的面前，使我们能迸发出无尽的想象空间，从而更加突出雕塑所蕴涵的那种充满野性的原生之美。

“踏春奔牛”双面透雕群，则灵活运用中国传统艺术表现形式之一——剪纸的表现手法，将苗族农耕文化中的耕牛的形象和苗族历史文化传说故事有机地结合在了一起，以一种喜庆的色彩向人们述说着凯里的民族历史文化故事。

如此雕塑，在迎宾大道上可谓信手拈来，实为旅游者之幸、城市之福。

二、凯里市迎宾大道主要经济技术指标

序号	用地类别	面积（万平方米）	比例（%）
1	总用地面积	53.99	100%
2	建筑占地面积	6.90*	12.78%
3	总建筑面积（新建）	12.62	——
4	道路广场面积	25.35	46.95%
5	景观绿地面积	21.60	40.01%

注：* 建筑占地面积，其中原有 4.09 万平方米，新建 2.81 万平方米

容积率：——

建筑密度：12.8%

绿地率：38.0%

(3) Sculpture works

They are made to outstand and promote the local ethnic culture. There are two main types: to reflect the scenes of custom and festival; to reflect the historical culture story of local people.

A special sculpture needs to introduce is called Shidong Dumu Dragon Boat. It leans on an ethnically featured overpass on Yingbin Avenue and seems to blend into the surrounding building and background. It can reflect the scene of "people walking on the bridge and boat flowing in the river" which brings coolness to the hot summer.

Another sculpture with the Basha Gunmen theme describes that the Basha men who just come back after hunting are welcoming the guests with nonobjective method. The picture of collective gunshot welcome makes us endlessly imagine so that it outstands more about the wildly natural beauty from the sculpture.

The clustered double-side sculpture called "Tachun Benniu" applies paper-cut technique which is the traditional art form. It integrates the image of furrow bull in Miao farming culture and the legend story of Miao history and culture, to tell people the ethnic history and culture of Kaili to people under happy mood.

II Main Economic and Technological Index for Yingbin Avenue in Kaili

Serial No.	Land type	Area (sm'0000)	Percentage (%)
1	Total land area	53.99	100%
2	Construction area	6.90*	12.78%
3	Total construction area (new)	12.62	——
4	Road square area	25.35	46.95%
5	Sight & greenbelt area	21.60	40.01%

Note: *Former construction area is 40,900sm, new construction area is 28,100sm

Plot ratio: ——

Construction density: 12.8%

Greenbelt ratio: 38.0%

绿地系统分析图

景观视线分析图

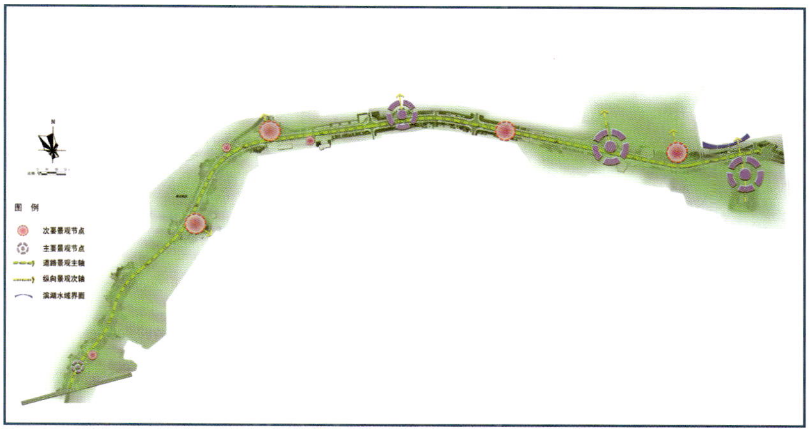

植物配置图

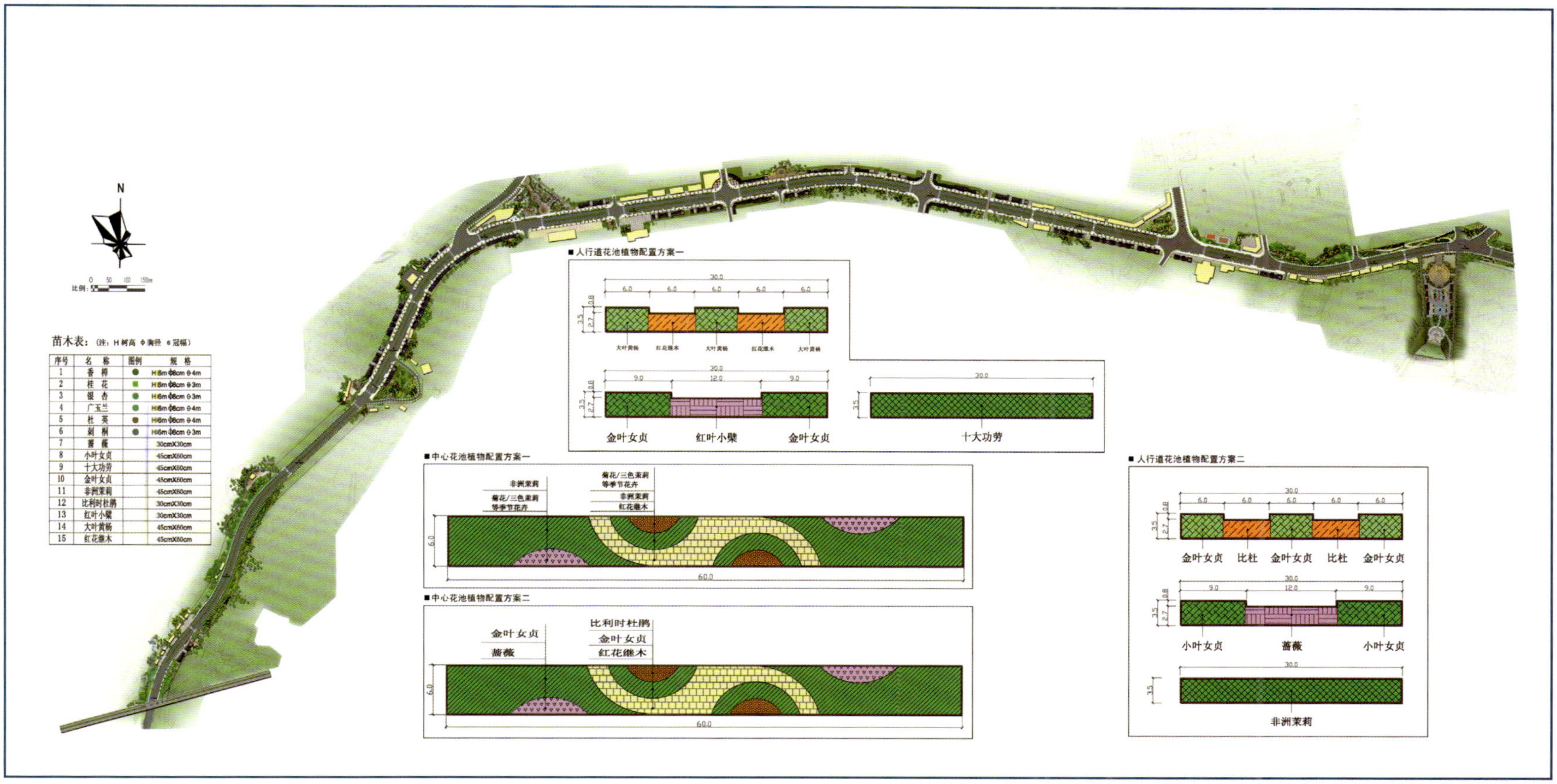

苗木表：（注：H 树高 Φ 胸径 θ 冠幅）

序号	名 称	图例	规 格
1	香 樟	●	H6m Φ8cm θ4m
2	桂 花	●	H6m Φ8cm θ3m
3	银 杏	●	H6m Φ8cm θ3m
4	广玉兰	●	H6m Φ8cm θ4m
5	杜 英	●	H6m Φ8cm θ4m
6	刺 桐	●	H6m Φ8cm θ3m
7	蔷 薇		30cmX30cm
8	小叶女贞		45cmX60cm
9	十大功劳		45cmX60cm
10	金叶女贞		45cmX60cm
11	非洲茉莉		45cmX60cm
12	比利时杜鹃		30cmX30cm
13	红叶小檗		30cmX30cm
14	大叶黄杨		45cmX60cm
15	红花继木		45cmX60cm

景观布置图

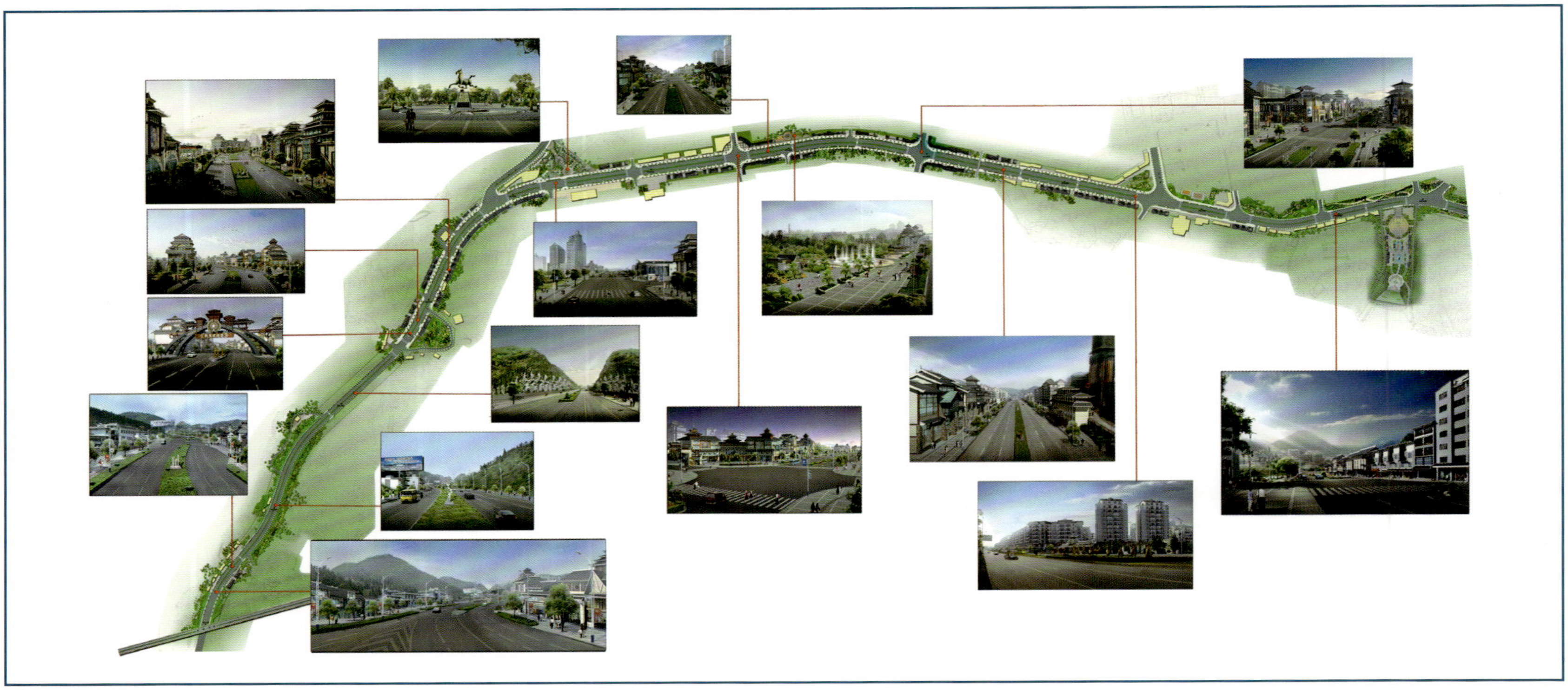

牌坊布置图

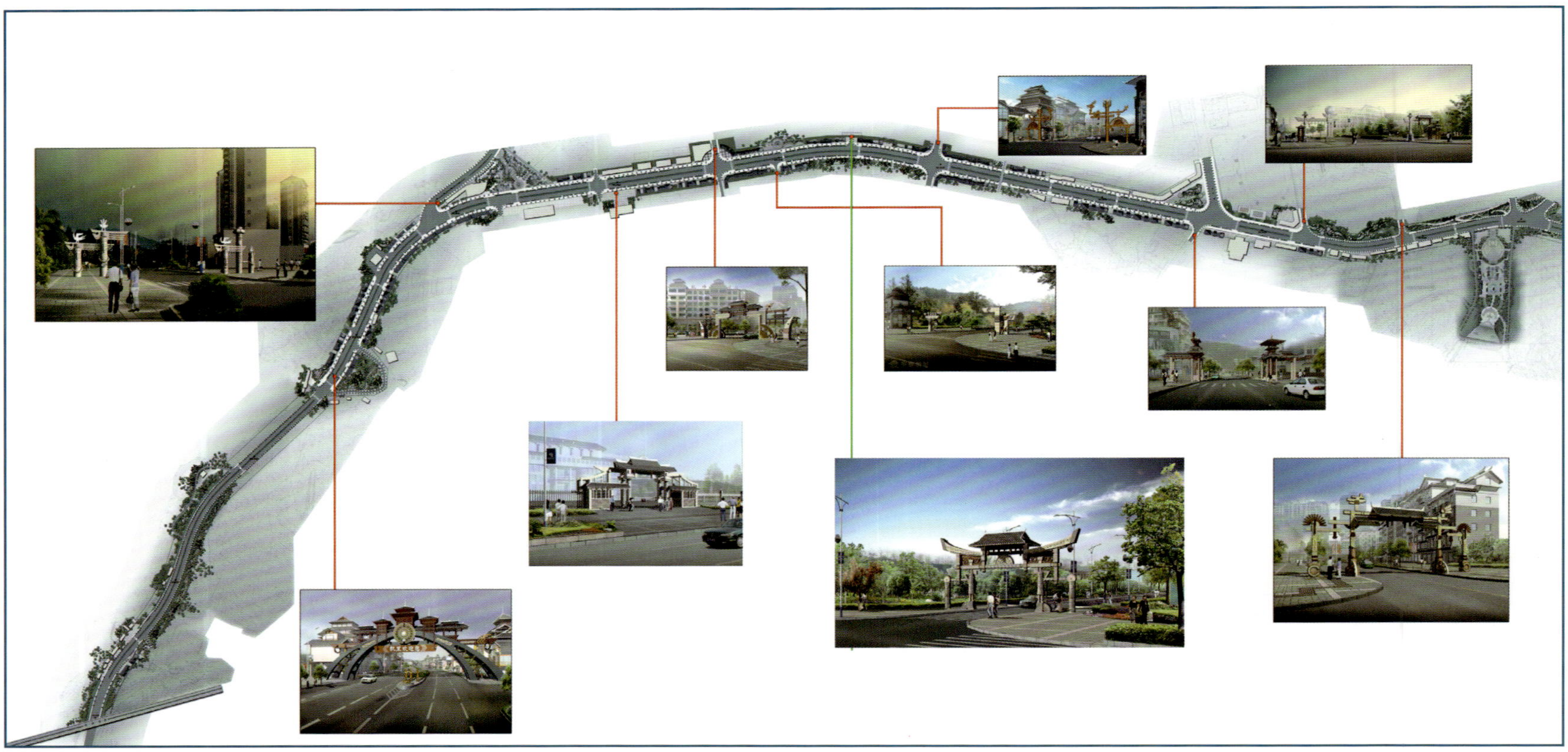

雕塑小品布置图

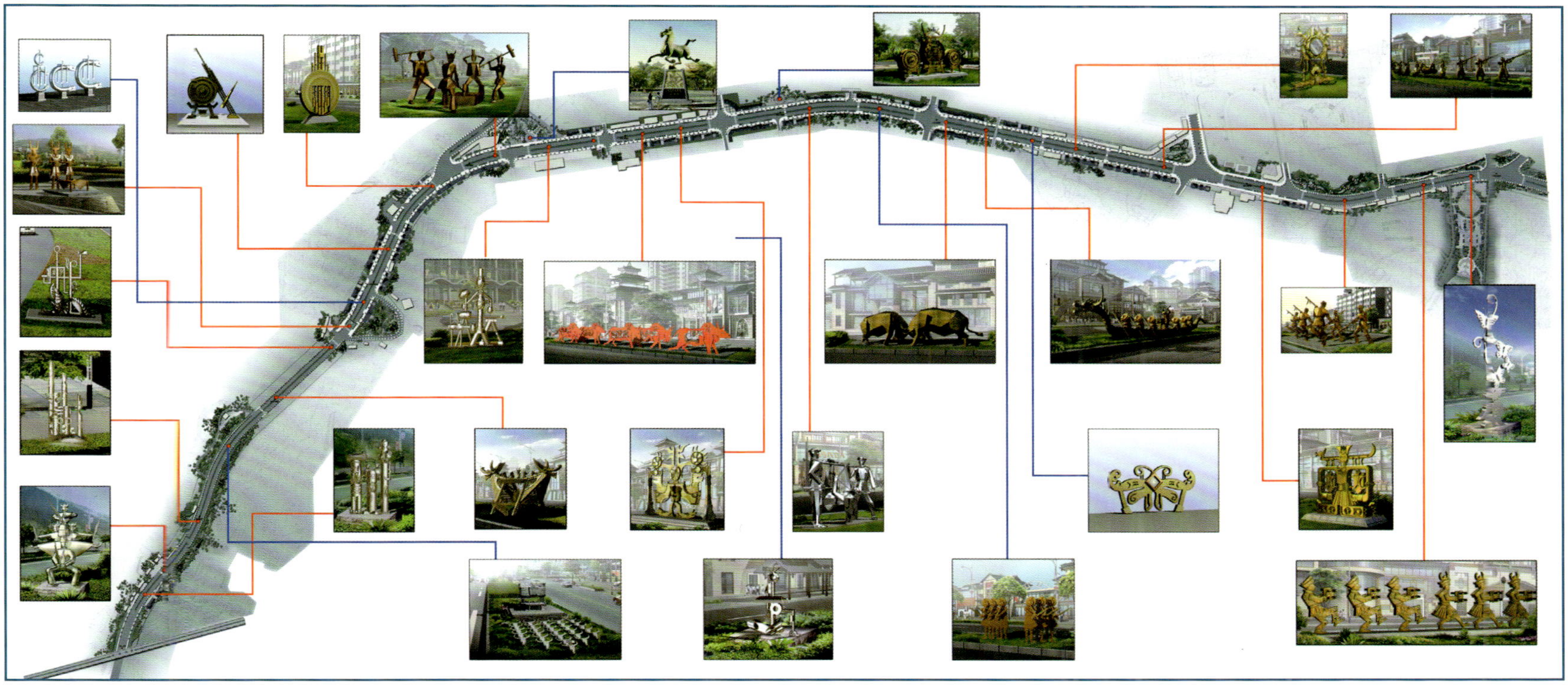

雕塑小品尺寸及材料一览表

材料：不锈钢 尺寸：300cm×180cm	材料：不锈钢 尺寸：200cm×120cm	材料：不锈钢 尺寸：300cm×180cm	材料：铸铁 尺寸：200cm×150cm	材料：铜 尺寸：200cm×100cm	材料：混凝土 尺寸：150cm×90cm	材料：不锈钢 尺寸：150cm×90cm	材料：玻璃钢 尺寸：160cm×200cm
材料：铜 尺寸：200cm×200cm	材料：铜 尺寸：150cm×150cm	材料：铜 尺寸：200cm×180cm	材料：玻璃钢 尺寸：200cm×100cm	材料： 尺寸：150cm×1000cm	材料：不锈钢 尺寸：200cm×150cm	材料：铜 尺寸：200cm×80cm（个人）	材料：铜 尺寸：150cm×100cm
材料：不锈钢 尺寸：150cm×120cm	材料：不锈钢 尺寸：200cm×120cm	材料：铜 尺寸：180cm×120cm	材料：玻璃钢 尺寸：150cm×120cm	材料：铜 尺寸：180cm×120cm	材料：铜 尺寸：180cm×120cm	材料：不锈钢 尺寸：120cm×200cm	
材料：铜 尺寸：200cm×600cm	材料：不锈钢 尺寸：180cm×1000cm		材料：玻璃钢 尺寸：200cm×80cm（个人）		材料：铜 尺寸：180cm×150cm	材料：铜 尺寸：1200cm×550cm	材料：不锈钢 尺寸：350cm×90cm

沿街建筑立面对照图

改造前

改造后

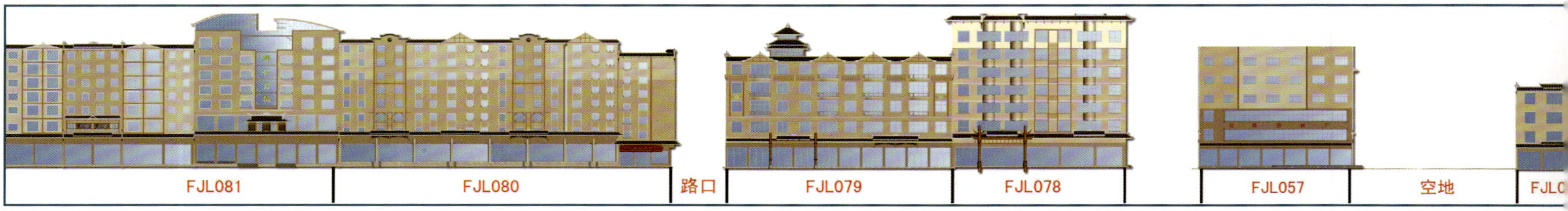

平面图

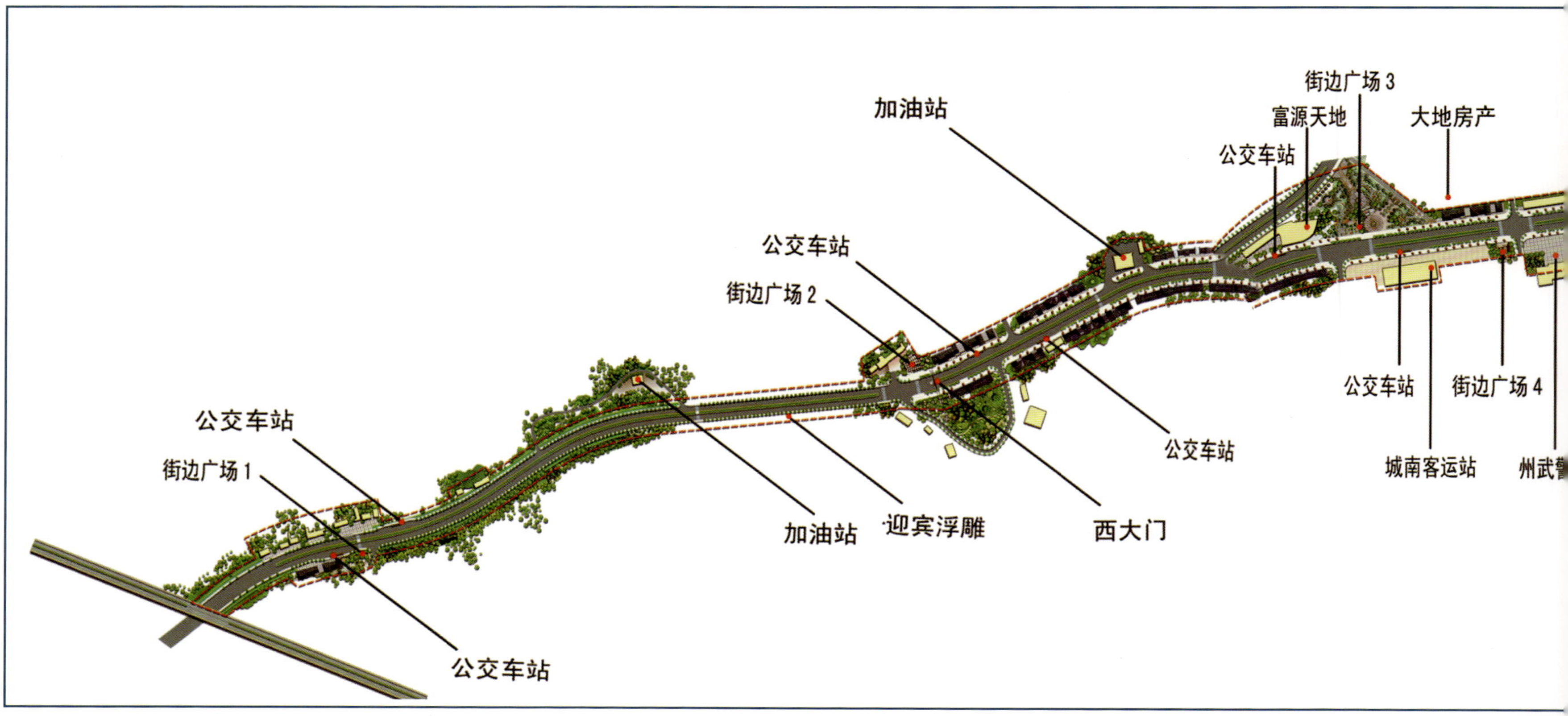

FJL054
FJL053
路口
FJL052
FJL045
FJL044
路口
FJL043
FJL042
路口
FJL041

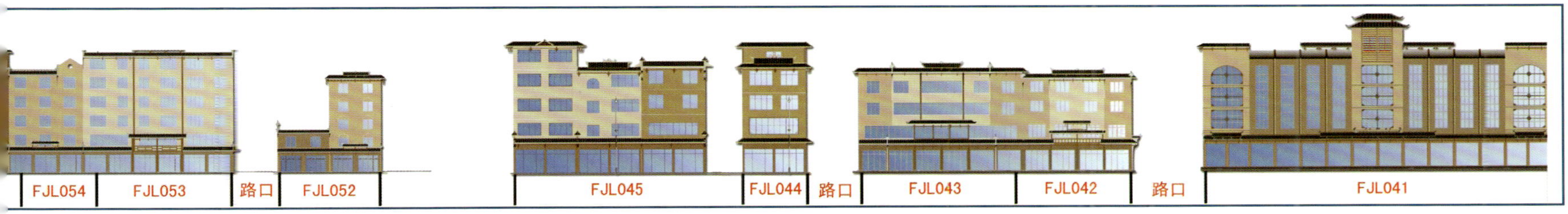
FJL054
FJL053
路口
FJL052
FJL045
FJL044
路口
FJL043
FJL042
路口
FJL041

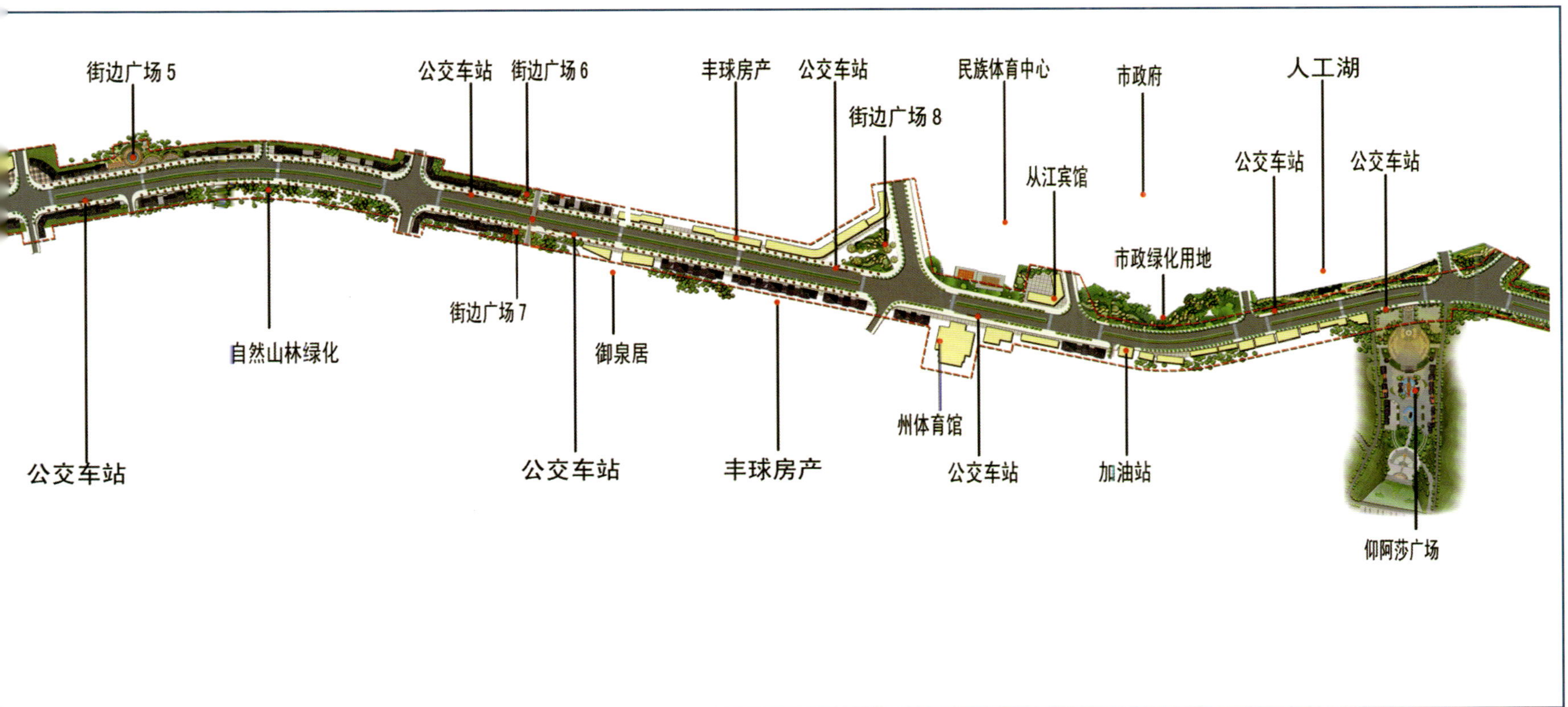
街边广场 5
公交车站
街边广场 6
丰球房产
公交车站
街边广场 8
民族体育中心
市政府
人工湖
从江宾馆
公交车站
公交车站
市政绿化用地
自然山林绿化
街边广场 7
御泉居
州体育馆
公交车站
公交车站
丰球房产
公交车站
加油站
仰阿莎广场

入城大门效果图

①

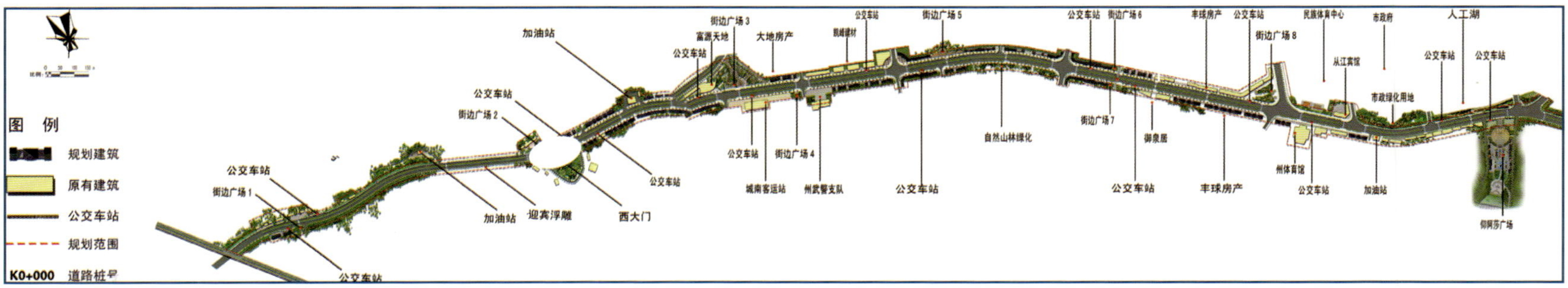
图 例
规划建筑
原有建筑
公交车站
规划范围
K0+000 道路桩号
公交车站
街边广场 1
公交车站
加油站
迎宾浮雕
街边广场 2
公交车站
西大门
公交车站
加油站
公交车站
富源天地
街边广场 3
大地房产
公交车站
城南客运站
街边广场 4
州武警支队
公交车站
街边广场 5
公交车站
自然山林绿化
公交车站
街边广场 6
街边广场 7
公交车站
御泉居
丰球房产
丰球房产
公交车站
街边广场 8
州体育馆
公交车站
民族体育中心
从江宾馆
市政府
市政绿化用地
加油站
公交车站
人工湖
公交车站
仰阿莎广场

改造前

改造前

①街道透视图 ②局部效果图

①

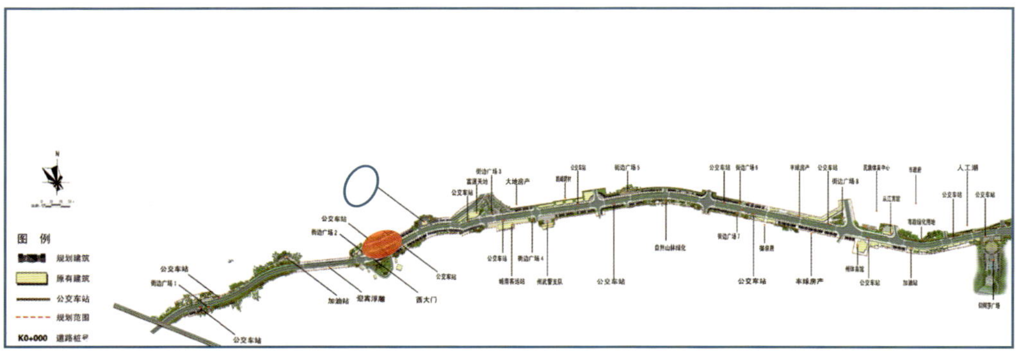

改造前

改造前

①街道透视图 ②街面建筑效果图

街道景观效果图

友庄路、宁波路沿街建筑改造

Rebuilding Constructions along Youzhuang Road and Ningbo Road

一、凯里市宁波路、友庄路两侧建筑外立面改造规划设计说明

为了丰盈凯里市民族特色建筑的体量，烘托“贵州少数民族都城”的城市形象，同时延伸迎宾大道所表现的民族特色。故而对凯里市友庄路、宁波路的楼房外立面进行改造，既统一了整体风格，又突出了民族特色。

在设计上，充分运用当地少数民族的特色建筑符号，以控制节约成本为原则，以最少的投入获取最佳的回报，从而提升城市形象、增添城市景观、重塑城市特色、提高城市软实力与影响力。

二、凯里市宁波路、友庄路两侧建筑外立面改造相关数据说明

1. 凯里市宁波路、友庄路两侧建筑外立面改造投资估算

由于对友庄路、宁波路的楼房外立面改造因其在建筑材料上使用有很大的周转余地，因此对其投资的估算上也有很大的弹性，可根据资金情况灵活调整。因此，本规划所给的“友庄路、宁波路的楼房外立面改造”的估算值仅是一个基值，主要起到的是参考和指导作用。

通过测算，友庄路、宁波路的楼房外立面改造所需的资金大约为 49559.60 万元人民币。

2. 凯里市宁波路、友庄路两侧建筑外立面改造主要经济技术指标

由于凯里市入城主干道两侧建筑外立面改造，主要是对友庄路、宁波路临街楼房外立面进行改造，并未涉及常规的经济技术指标，在此就不再赘述。

Ⅰ Rebuilding Layout Design Explanation for Outer Elevation of Two-side Architecture of Ningbo Road and Youzhuang Road in Kaili City

In order to enrich the scale of the ethnic-feature architecture, set off the city image of the minority capital of Guizhou and extend the ethnic-feature of Yingbin Avenue, the rebuilding program for outer elevation of two-side architecture of Ningbo Road and Youzhuang Road in Kaili City is made. It not only unifies the overall style but also stands out the ethnic features.

In the terms of design, it fully uses the special architecture signs of local minorities under the principles of obtaining the optimal return at lowest costs. Hence, it can boost the city image, increase the scenic spots, rebuild the city feature and enhance both the soft power and influence of the city.

Ⅱ Relevant Data Explanation for the Rebuilding of Outer Elevation of Two-side Architecture of Ningbo Road and Youzhuang Road in Kaili City

1. Investment Estimates for the Rebuilding of Outer Elevation of Two-side Architecture of Ningbo Road and Youzhuang Road in Kaili City

Due to the large turnover space of the building materials for the rebuilding of outer elevation of two-side architecture of Ningbo Road and Youzhuang Road, the investment estimate is also be flexible which can be adjusted according to the capital condition. Hence, the estimate is just a basic value to have the effect on reference and guidance.

Through the estimate, the project needs the roughly RMB 495 596 000.

2. Main Economic and Technological Index for the Rebuilding of Outer Elevation of Two-side Architecture of Ningbo Road and Youzhuang Road in Kaili City

Since the rebuilding of outer elevation of the two-side buildings of city-entry road mainly focuses on nearby street buildings in Ningbo Road and Youzhuang Road with no references to the regular economic and technological index, there is no detail about it.

①街道改造透视图 ②街面建筑改造效果图 ③街面建筑改造效果图

改造前

改造前

①～④街面建筑改造效果图

①

②

①街道建筑改造效果图 ②街道建筑改造后实景图

③街道建筑改造效果图 ④街道建筑改造后实景图

①～④街道改造后实景图

③

④

①

②

①～②街道改造后实景图

③万博广场鸟瞰图 ④万博广场透视图

丘北篇 Qiubei Profile

城市设计篇（第二版）

变城记

City Changing

前期策划
Prophase Planning

Qiubei County, located in the southeast of Yunnan province, is famous for the unique carst rural scenery, the special tourism product of "Water-sprinkling Festival every day" and the developing marvelous "Thousands of Mu of Lotus". Qiubei County is the settlement of seven different ethnicities including Han, Zhuang, Miao, Yi, Yao, Bai and Hui, the ethnic minorities is 62.52% of the total population.

The completion of the Yungui Express, Putan Route, Qiuguang Route and Qiushi Route will bring an unprecedentedly developing opportunity to Qiubei County.

Qiubei county is full of resources which, however, have been exploited at the primary stage. The urban construction is lack of specialty and necessary connection with tourism development, which will not keep the pace with the whole development of the future of Qiubei County.

The government have to face up and think about the major issue of how to maintain the regional favorable opportunity and transform the disadvantages to advantage so as to make the local economy and society develop rapidly and the local people earn much income.

This planning is on the basis of the intercoordination between urban construction and tourism development, so as to create Qiubei city to be a famous tourism city with complete tourism service facilities and urban infrastructure, good urban environment and high-level tourist industry.

丘北县位于云南省东南部，境内的“普者黑”喀斯特山水田园风光在区域内独树一帜，“天天泼水节”的特色旅游产品远近闻名；开发中的“万亩荷园”婀娜壮观。丘北县境内居住着汉、壮、苗、彝、瑶、白、回等七种民族，少数民族人口占总人口的62.52%。

随着云桂快速铁路以及普炭线、丘广线、丘师线的建成通车，将使丘北县获取前所未有的发展机遇。

丘北县境内资源丰富，然而对资源的开发利用却仅仅停留在初始阶段，城镇建设缺乏特色，旅游开发与城镇建设各说各话，缺乏必然的联系，这与丘北县未来的大区域环境显然是不相匹配的。

如何把握区域发展的良好机遇，将优势转换成胜势，使当地经济、社会能够借此机遇实现跨越式发展，为当地民众谋得更多的实惠，则是当地政府当前必须面对和急需思考的一个重大问题。

本次策划是处在云南省丘北县正在积极申请建立普者黑市的背景下，因此本次策划站在普者黑市城市建设与旅游业发展相互协调的战略高度来进行考虑，结合“十二五”期末丘北撤县设普者黑市的思路进行形象定位，力图将丘北县打造成为旅游服务设施及城市基础设施配套齐全、城市环境基础性良好，旅游产业发展水平高的山水旅游名城和宜居生态城市。

SWOT 分析

一 优势

1. 自然资源具备规模

核心景区普者黑规模较大，囊括范围广，所涉及的景点丰富。

在建中的万亩荷园构思深远，规模空前。

2. 自然资源具有一定的品位及知名度，开发具备一定基础

核心景区普者黑为文山州唯一的 4A 级旅游景区，在文山州及周边地区具有一定的知名度；景区开发数十年来，形成了一定基础，具备初级旅游功能，形成初级旅游产品。

3. 自然资源形成组团式发展

自然资源连成一片，整合力度较强，加之周边自然生成及人工营造的环境优异，促成丘北成为一个真正意义上的山水园林城市。

4. 主要资源距离城市较近

丘北县主要旅游资源为普者黑景区，而这一景区距离城区仅仅数公里，为景区与城区的联动发展提供了良好的条件。

5. 自然景观具备一定的区域特色

云南省是旅游大省，但其自然景观主要体现为高原喀斯特景观，在云贵高原上呈现出一片江南水乡、田园风光实属难得，此类景观具备一定的区域特色。

6. 城镇建设具备一定规模

城区在经过一定强度的开发后具备了一定规模，基础设施建设较为完善，城市框架基本形成。

二 劣势

1. 开发不当对资源造成破坏

普者黑是丘北县的核心资源，但是由于多年的开发不当，致使景区承受了多重打击。首先是乱搭乱建对景区整体环境的破坏；其次是景区从事旅游开发的无序状态对景区生态环境以及旅游品质带来了巨大的负面影响。

2. 人文资源挖掘不够，开发空缺

人文资源较为匮乏，而且仅有的人为资源都无法体现在旅游者的行程中，无法贯穿于城市的建设中，无法对当地居民及旅游者形成印象，事实上文化才是旅游的灵魂。

3. 景区内的建设开发与资源品位不相符

普者黑景区是 4A 级国家级风景名胜区，但是景区内的开发建设却处于一种无序状态，缺乏必要的规划指导，致使整个景区风格不协调，脏乱差现象比比皆是，缺乏与资源品位相符的开发建设。

4. 对景区内原住民的疏导不够

社区居民是旅游开发过程中面临的一个难点问题，景区要有序开发，景区内居民的生活要得到改善，二者似乎不能共存，顾得了一方就难以为继另一方的利益，目前丘北县正

I Advantage

1. Large scale of nature resources

Puzhehei, the core resort is grand and with rich scenic spots.

Wanmu Lotus Pond, being constructed, is unprecedented large and with deep concept.

2. Nature Resources are of certain elegance, fame and conditions to develop

As the core scenery area, Puzhehei is the only 4A-grade tourism resort of Wenshan Prefecture, so that it is fair famous in Wenshan Prefecture and the surrounding region; With some decades of developing, the resort is of primary tourism functions and products.

3. Nature resource developing together

The connected nature resources and the surrounding natural and artificial beautiful atmosphere contribute to a real garden city with mountains and water of Qiubei.

4. Main nature resources are close to the city

Puzhehei Tourism Resort, as the main tourism resource, is several kilometers away from the city, which is a good chance for the resort and city to mutually develop.

5.Natural landscape is of certain regional features

Yunnan is an important tourism province, famous for the carst landscape of Yungui Plateau. It is seldom to see the charms of watery town and pastoral scene on Yungui Plateau. Such landscape is of regional features.

6. Construction of towns is of certain scale

This city region is of certain scale after developing to a certain extent with a no more than perfect infrastructure and urban frame.

II Weakness

1. Damage to the resources due to improper exploiture

Being the core resource of Qiubei County, Puzhehei Resort suffered a lot due to many years of improper exploiture: firstly, it is the damage to the whole environment of the resort by disordered constructions; secondly, it is the situation of incorrect tourism development that cause huge negative impact to the ecological environment and tourism products of the resort.

2. Cultural resource shortage

This region is fair short of cultural resource, moreover, the few human resources can be used in the tourism, nor in the city development and leave the tourist a deep impression. Actually, culture is the soul of tourism.

3. Unmatched construction between scenery development and resource type

Puzhehei, a 4A-grade National Scenery was developed, however, without necessary guide of plan, thus the whole scenic spot is not in a coordinate style, but in a mess and without any architectures matched with its resources.

4. Improper way to evacuate origin residents

The residents living in the scenic spot is a tough problem for developing tourism. It is hardly impossible to consider both the perfect development of the scenery spot and the amelioration of these residents.

面临的便是这一问题，事实上，这一问题的核心是在于对景区内原住民缺乏正确的疏导。

5. 旅游开发与城市建设脱节

相关部门尚没有认识到城镇建设与旅游发展之间的联系，事实上二者之间是相辅相成的，是可以相互促进的。然而目前丘北县的城镇建设与旅游开发却无法有机地结合在一起，更有甚者，就连旅游资源都分属城建、林业、水利、宗教、文化等多部门管理，各自为政，各行其是，尚未形成合力。这样做的后果便是重复建设、不必要的建设大规模的产生，造成时间、金钱以及城市形象的多重损失。

6. 城镇建设千城一面，缺乏自身特色

近年来丘北县发展势头迅猛，城市建设速度惊人，但是仍然走进了城镇建设的一大误区——只注重城镇建设的功能，从而出现了千城一面的城镇建设现状，缺乏自身独有的风格，难以形成视觉上的冲击，难以使旅游者留下深刻的印象。

三 机遇

1. 区位优势越发明显

云桂快速铁路以及一系列一级公路的建成贯通，将使丘北县处于通边达海的交通枢纽地位。

2. 城镇建设与旅游发展步伐不断加快

近年来随着旅游业的三大效益越来越显现出来，优先发展旅游业已经成为各个地区的共识。而与此同时城镇建设也步入了一个新的台阶，在此基础上，丘北县也提出了“旅游活市，城镇富市”的发展方针。但是事实上，城建和旅游二者之间是相辅相成的，中央经济工作会议上提出的“通过旅游推进城镇建设”的战略方针正好佐证了这一思路，也为城建和旅游的共同发展提供了基础。

四 挑战

规划理念能否落实到位。在考察的过程中我们发现丘北县规划先行的理念在城市发展的过程中得到了重视，但是在实际操作过程中却无法体现定位，因此在进行旅游城市规划设计的过程中我们落实规划内容才是重中之重。

In fact, it is essential to evacuate the original residents in a correct way.

5. Tourism development divorces from city construction

The relevant governmental departments haven’t realized the connection between city construction and tourism development which, actually can move forward mutually. Nevertheless, the current city construction and tourism development of Qiubei County cannot connect with each other organically, moreover,

6. City construction lack of specialty

Qiubei County has developed rapidly in recent years, but in a wrong way which only pay attention to the functions, while ignore its own style.

Ⅲ Opportunity

1. location advantage will be More and more obvious

With the completion of Yungui Express and a series of first grade highways, Qiubei County will be the place of traffic hub.

2. Town construction and tourism development will speed up

Tourism industry has created more and more in three important aspects, thus it enjoys the priority to develop in every prefecture. Meanwhile, construction of towns has entered a new stage. Considering these circumstance, Qiubei government made the developing policy of “Tourism Activates City, While Towns Construction Enrich City”. In fact, towns construction and tourism mutually develop, which was proved and on the basis of “to improve towns construction through tourism” proposes on the Central Economic Work Conference.

Ⅳ Challenge

Planning concept could be implemented or not. We found that the prior planning concept of Qiubei County had drawn attention in the process of city development, but had not been implemented in actual practice. Therefore, it is very important for us to fulfill the planning in the planning and design for a tourism city.

战略定位

以自然环境为依托，以城镇发展为载体，以文化资源为内涵的山水园林旅游城市。

目标愿景

通过旅游城市的打造形成文山州乃至云南省特色旅游城市。

通过旅游城市的打造，带动、促进其他产业的发展，从而实现丘北县城社会、经济、文化的全面发展。

将丘北县自然资源的价值体现于景区中，同时也要将人文资源的价值体现在城市建设中，形成景区发展与城镇建设的双赢。

发展思路

（一）更新景区开发的观念

打破景区开发小而全的观念，杜绝景区具备“吃住行游购娱”全部功能.保持景区的参观功能,以确保景区的生态性，将其余功能放置在城中，以此形成旅游业对经济的拉动。

内容：将核心景区周边居民迁出景区，加强景区的植被覆盖率，保持景区的生态性，保护丘北县核心资源的完整性。

在此需要强调的是对于从景区迁出的居民，政府不是放任不管，而是要保证景区居民在迁出景区后同样能从旅游业的发展中受益。例如由政府出面投融资，对于城镇发展旅游配套项目过程中所需的劳动力或者是出让、出租的商铺、作坊优先考虑社区居民，保证社区居民能够获取最大利益。

（二）确定城市风貌，从视觉上塑造特色城镇

通过本土文化的挖掘，进行现代艺术手法的提炼、加工，确立、构建自身特色，形成自身品牌，提升城镇品质和人居环境，增强游客的可识别性，提升当地居民的自豪感。同时还可借助城镇建设的推进，促进旅游业的发展。

（三）配套城市公共场馆

随着城市的发展，人民生活水平的提高，精神上的享受日益迫切，需要城市具备相关的公共场所满足居民的需求，同时也可以通过城市公共场馆的建设形成独具特色的标志性建筑，提升城市的品位。

（四）打造民俗风情商业街区

民俗风情商业街是城市发展不可或缺的，阳朔因为有了西街而让人流连忘返，丽江因为有了众多古街而让人乐不思蜀，它们都在彰显着当地的文化，因此我们旨在通过风情街区的打造彰显城市文化，形成丘北县城真正意义上的商业中心、服务中心、文化中心。

Strategic Orientation

To construct Qiubei to be a garden city for tourism on the background of natural environment, town construction and cultural resources.

Objection

To be a special tourism city in Wenshan Prefecture even in Yunnan Province.

To improve the other industries and realize the overall development of the society, economy and culture in Qiubei City.

To display the value of natural resources and human resources of Qiubei County in the scenic spots and the urban construction in order to obtain the win-win situation of the development of scenic spots and city construction.

Developing Idea

1. Update the concept for developing the scenery

The concept of developing the scenery spot with comprehensive services and in small scale should be broken, thus tourists can only enjoy view in the scenic spot without other facilities such as restaurants, hotels, shops and clubs, so as to keep the scenic spot in ecology, while the other facilities should be in the city to realize the objection that tourism accelerate economy.

Content: To move out the residents around the core scenic spot, to increase vegetation cover rate, kee of the scenic spot, to protect the core resources of Qiubei Couty

We want to point that government should care about the residents moving out from the scenic spot and make sure that they can also get benefit from tourism. For example, the government could make investment in the projects relevant to tourism development to create more jobs for these people, or give or rent these people shops or workshops so that they can make profits as many as they should be.

2. Fix city image to shape specialty for the city

Excavating local culture to shape specialty for the city with modern art measures, form brand, enhance city quality and living environment, increase city identification for tourist and promote local people’ s sense of pride.

3.Venues for major public city activities

With the development of cities and the improvement of living quality, people ask for spiritual enjoyment eagerly. Therefore, city should be of relevant public facilities to meet the residents’ need, and some of the constructions for the public can be outstanding enough to be the symbol of the city and to enhance city taste.

4. Construct the Folk Business Streets

Folk Business Street is needed in a city, such as the West Street of Yangshu and the ancient streets of Lijiang, both of them display local culture. Thus we aim to highlight the city culture and form the real business, service and culture center by creating the Folk Business Street.

功能布局

（一）自然生态旅游区

发展方向：为游客提供真正意义上的生态旅游产品，不掺杂任何人为的破坏，仅仅为游客提供游览服务。

内容：简单生态的接待设施。

（二）老城改造区

发展方向：对城市现有沿街建筑进行外立面的改造，以丘北县当地文化为核心，经过抽象与创意，最大限度地彰显当地文化的内涵，形成独具丘北县特色的沿街建筑。

内容：沿街建筑外立面改造、街景绿化、指示系统等。

（三）游客服务中心

发展方向：为游客提供全方位的旅游配套服务，与普者黑景区联动发展，形成游于景区，娱于城市的发展局面。

内容：包括游客咨询中心、旅游车站、酒店、娱乐休闲场所等。

（四）民俗风情商业街区

发展方向：将民俗风情街区打造为以文化休闲为主要发展方向的商业街区，在显现商业功能的同时更注重文化性及休闲性，形成丘北文化的集中展示地。同时风情商业街区还承担着接收普者黑景区迁出居民的重担，将普者黑迁出居民安置于此，以实现景区发展、城镇建设以及社区居民的共同利益。

内容：民俗演艺、休闲娱乐、购物、客栈等。

市场运作思路

（一）利用规划方案，变“原料”为“产品”直至成为“商品”

通过规划方案明确的发展定位，超前的战略构想，整体的开发思路，把“生地”变成“熟地”，将“原料”变为“成品”，再通过特殊的营销手段，将成品转换成为“商品”。以规划方案对土地进行挂牌，为地方经济带来更多利益的同时，还能有效地避免重复建设和对资源的粗放型利用。

（二）利用规划方案，作为招商引资的亮点

依托丘北县丰富多彩的旅游资源，政府巨细兼顾的政策支持，利用规划方案高远、大胆、创新的战略构想，吸引国内外大财团、大投资商作为战略投资伙伴，为丘北县旅游品质提升档次，旅游强市的打造埋下伏笔。

（三）利用成功的招商，带动地方经济发展和城区建设

通过旅游资源的整体开发与打造，利用项目投资的乘数效应，走上一条以旅游拉动城市其他投资，以税收增强城区建设的可持续发展之路。

Functional Layout

1. Natural Area for Tourism

Orientation: to provide tourists with the exact ecological tourism product just for traveling and sightseeing without any damage by human being.

Content: sample ecological reception facilities.

2. Transformation of the old city

Orientation: to transform the current street architecture in facade

Content: transformation of street architectures in facade, greening street view, and direction systems and so on.

3. Tourist Service Center

Development orientation: to provide tourists with overall tourism relevant services and develop with Puzhehei Resort so as to form the pattern of travelling in the resort and entertaining in the city.

Content: Tourism Inquiry Center, Tourism Bus Station, Hotel, Leisure Club and so on.

4. Folk Business Street

Development orientation: The folk street will be created to be a business block based on culture and leisure to display Qiubei culture convergently. Meanwhile, the business block is the settlement of the residents moving out from Puzhehei Scenic Spot, so as to realize overall benefits of the scenic spot, city construction and the residents.

Content: folk performance, leisure entertainment, shopping, inns, and so on.

Market Operation Ideas

1. To transform “original material” into “product”

With clear developing orientation, advanced strategic concept and entire exploitive ideas, the planning proposal is the guidance to transform “uncultivated land” into “cultivated land”, “original material” into “product”, and finally into “commodity”. The land will absorb investment with the planning proposal, which will not only bring more benefit for local economy, but also avoid effectively building redundant project and waste of resources.

2. Highlight the planning proposal to attract investment

Qiubei County will be promoted in tourism quality and strength depending on the diversity of tourism resources, governmental policy of giving consideration to both big and small and the advanced and innovative strategic ideas of the planning proposal to attract investment of big corporations and big investors in China and abroad.

3. Make use of successful investment to accelerate local economy and city construction

Qiubei County should move forward in a sustainable way where tourism boosts other investment of the city and revenue strengthens regional construction through the multiplier effect of exploiture of tourism resources.

设计元素
Design Element

（彝族建筑）

设计元素

Design Element

（非物质文化）

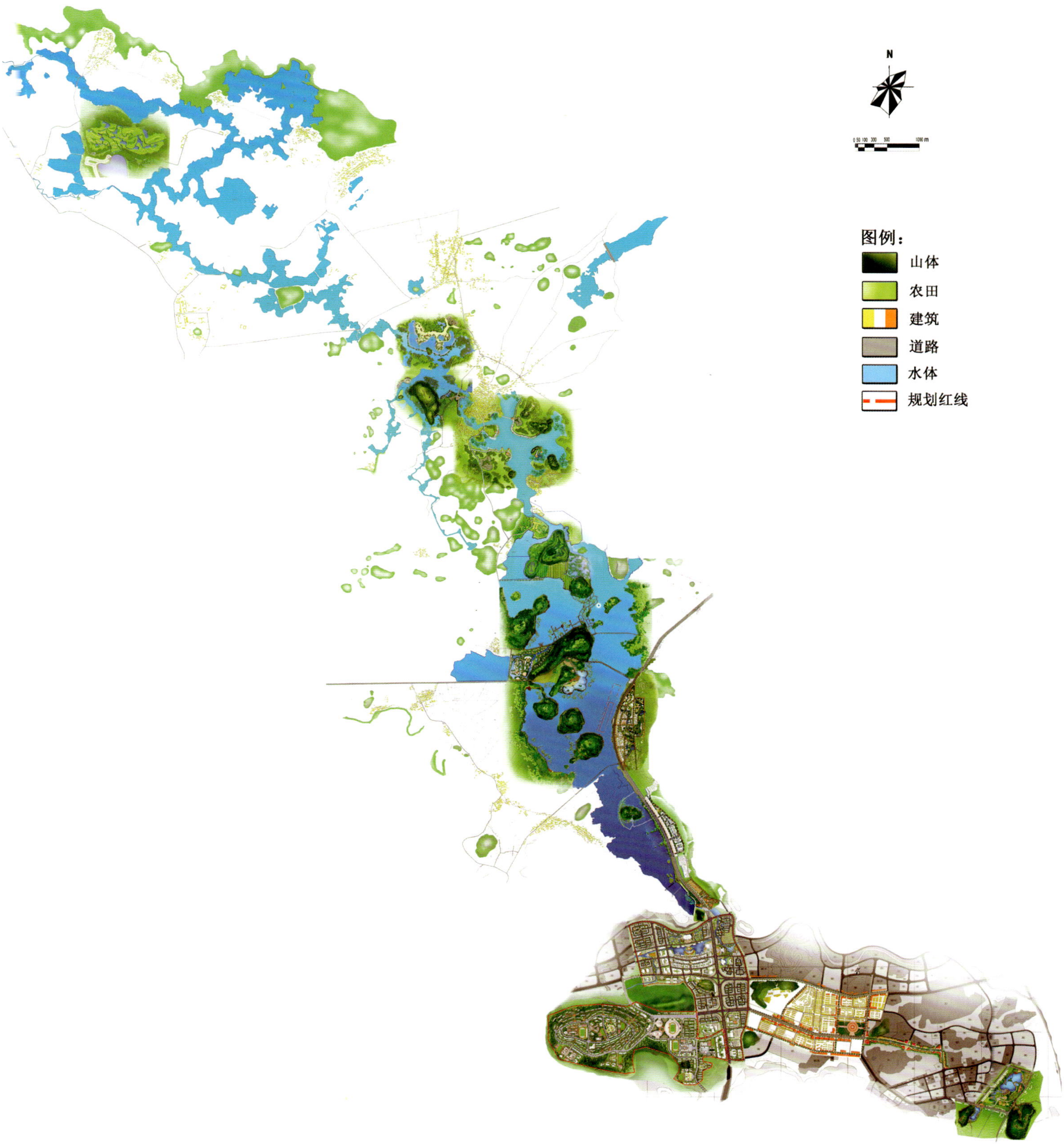

丘北县旅游城市规划总平面图

丘北县全域鸟瞰图

入口景观创意设计

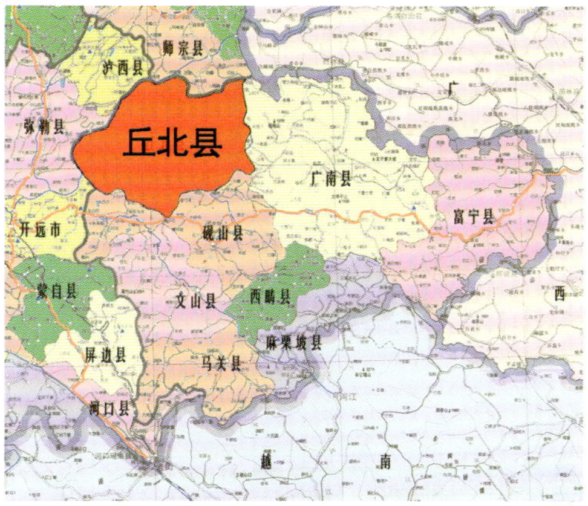

丘北县在文山州的位置

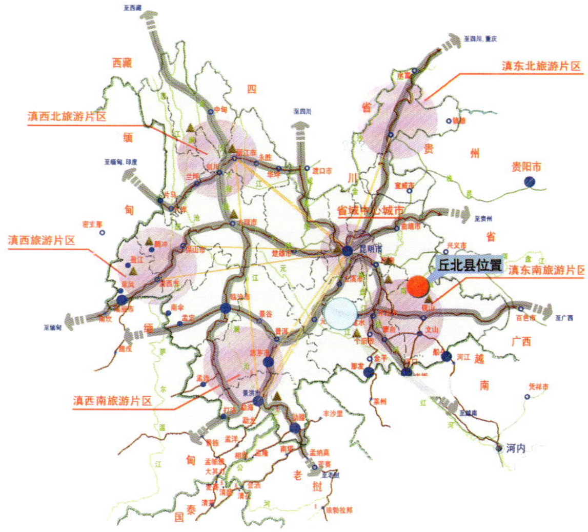

旅游区位示意

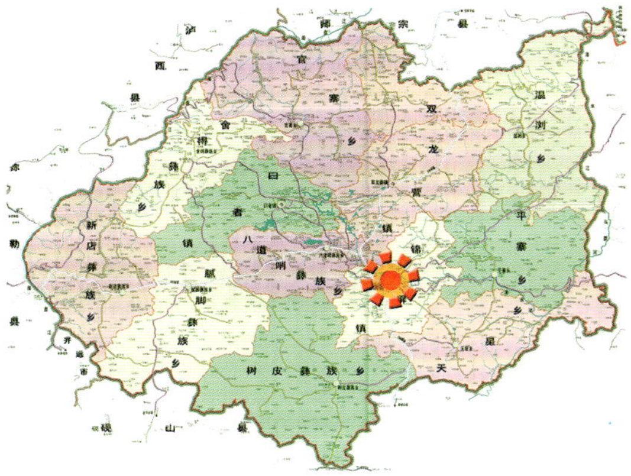

丘北县城在县域的位置

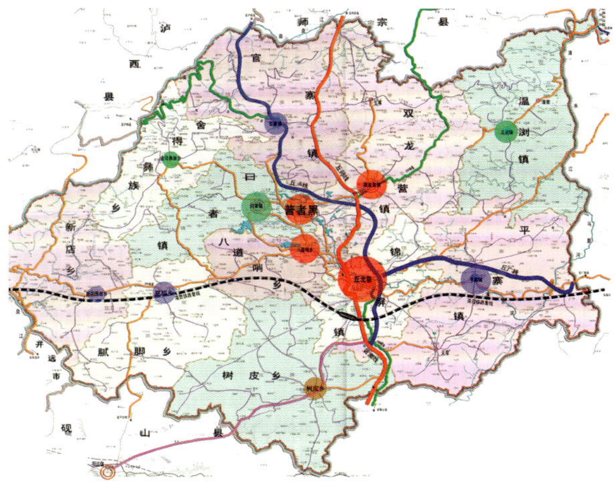

交通区位示意

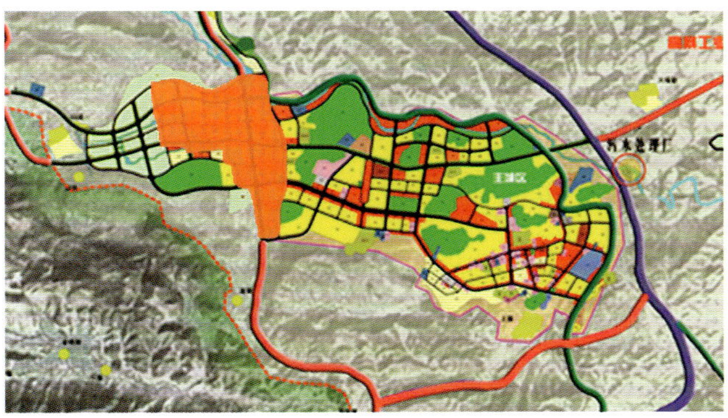

小新寨中心区位置

地理区位示意

1 地理区位优势

文山州位于中国—东盟自由贸易区和泛珠三角经济区两大国际国内区域合作的结合部,区位条件优越。丘北县距州府文山78km,西临南盘江,东隔清水江与广南县毗邻,南与砚山县、开远市接壤,北与师宗县,广西壮族自治区西林县衔接。清水河流域位于丘北中心片区和普者黑景区之间,地理区位优势明显。

2 交通区位分析

丘北县内现有东西向的305以及南北向的206省道,新建的丘广公路、丘泸公路、丘师公路、普炭公路等,构成了西至弥勒、泸西,北至师宗、曲靖,南至砚山、炭房、广西,东至广南的公路交通格局。即将开始建设的南昆铁路复线将使整个丘北的交通通达性大大提高。县城距离即将通航的文山普者黑机场85km,为丘北的经济发展带来了便捷的空中通道优势。清水河流域规规划区西面为普炭公路,东面为305省道,交通区位优势明显。

3 旅游区位优势

临近普者黑4A级国家旅游区和小新寨旅游服务组团,借普者黑的旅游资源和小新寨旅游服务功能,带动区域经济文化的发展。

Planning and Design of Xiaoxinzhai Area

小新寨片区规划设计

小新寨片区鸟瞰图

现状分析图

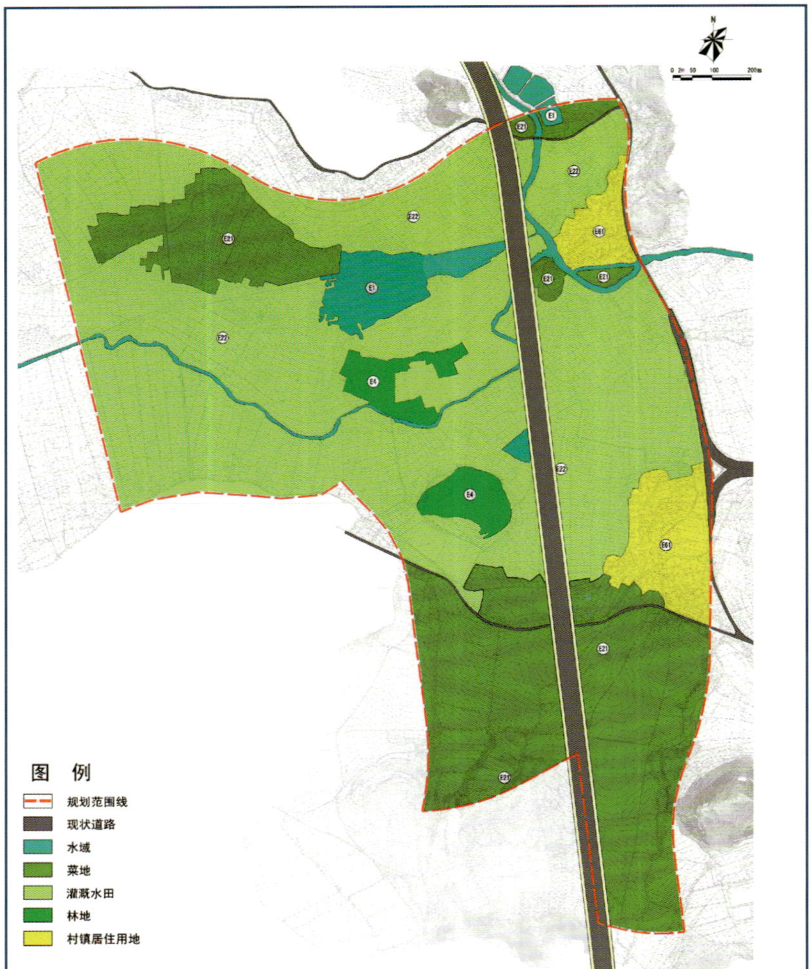

现状用地一览表

序号	用地性质	F1	F2	用地代号	面积（公顷）	比例（%）
1	水域和其他用地			E	204.67	99.24%
	其中	水域		E1	8.87	
		耕地	菜地	E21	62.08	
			灌溉水田	E22	117.10	
		林地		E4	5.72	
		村镇居住用地		E6	10.90	
2	其他				1.56	0.76%
3	总用地				206.23	100.00%

总平面图

功能结构分析图

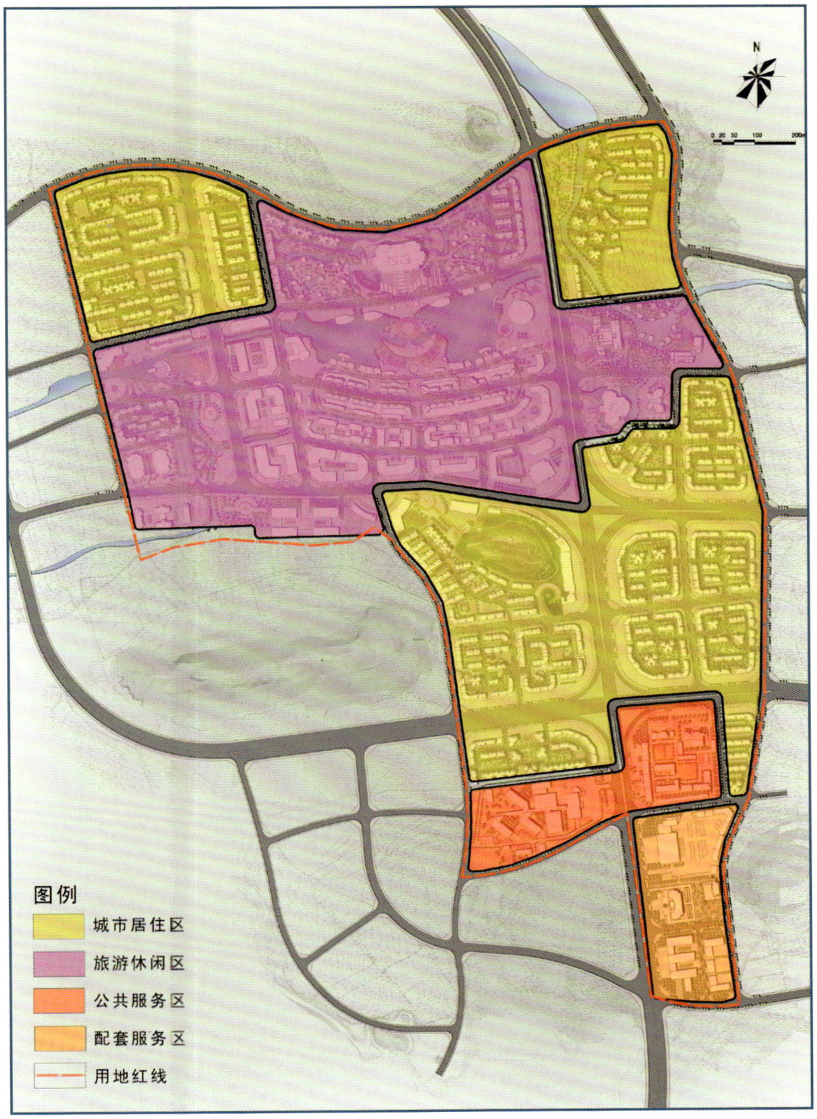

道路交通分析图

绿色景观系统规划图

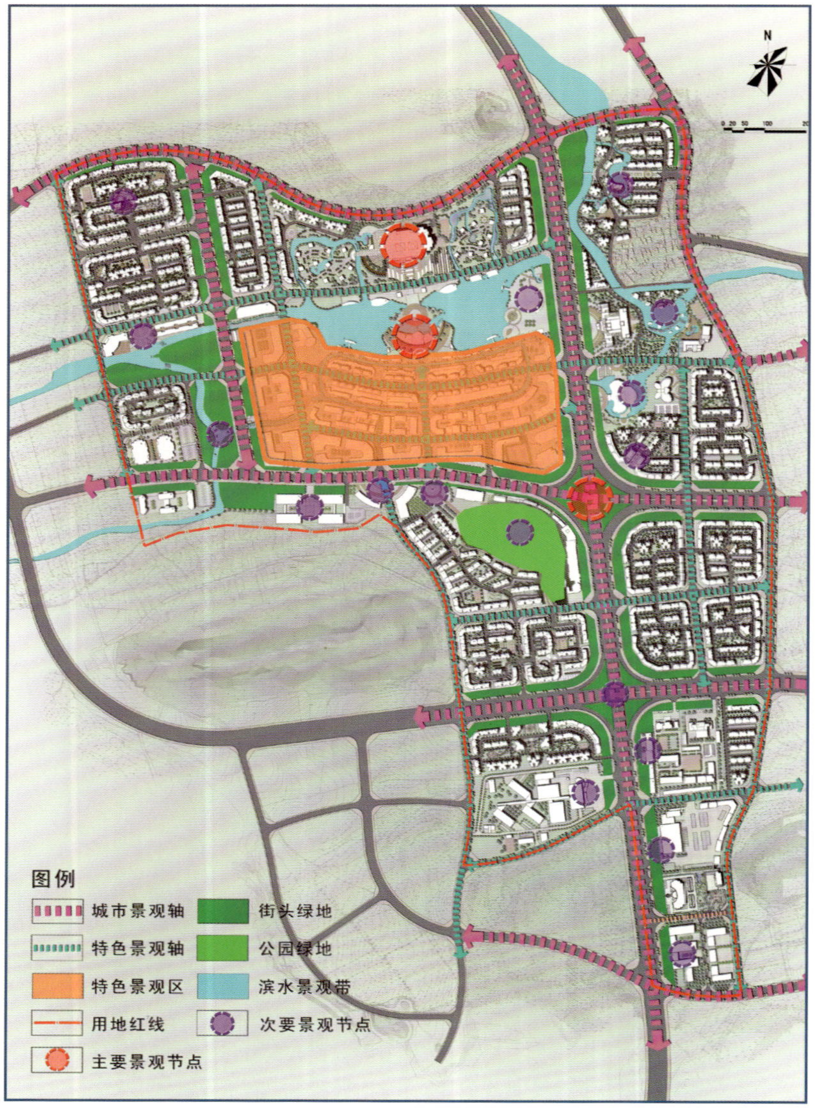

公共服务设施规划图

植物配置的原则：
1、适地适树，以云南当地乡土植物为主，尤以丘北地区植物为主充分考虑小气候影响以适应本地气候、水土、保证规划所运用树种能在本地区正常生长。并适当选用驯化的外来及野生植物，构成丰富的绿化空间。
2、选择抗逆性强的，对酸、碱、旱、涝、沙性及较恶劣的环境有较强的适应性并对病虫害、烟尘、污染有强抗性的树种，以保护当地的生态环境。
3、注意速生树种与慢生树种的结合，常绿树种与落叶树种的搭配。花卉点缀，草坪地被攀援植物的应用。
4、根据不同的功能特点选择不同植物种类。
N
0 20 50 100 200m
植物简图：
香樟
广玉兰
银杏
垂柳
四季桂
董棕
慈竹
杜鹃
油茶
紫荆
滇丁香
八角金盘
油茶
龟甲冬青
A：道路系统
行道树的选择以分枝点高，具有深根性、生命力强的植物。主要采用香樟、四季桂。并在行道树下配置草本植物细叶沿街草。主干道路布置形式为四板五带式，绿带以种植灌木或绿篱造型为主，选用铁树、金叶女贞、紫叶小檗等
停车场绿地以规则式周边绿化为主主要采用圆柏、紫叶李、红花檵木、女贞、杜鹃等。
B：城市居住区
居住区内采用的植物有常绿乔灌木：石楠、四季桂、云南山茶、短穗鱼尾葵、圆柏、云南含笑、滇丁香、苏铁、红花檵木。落叶乔灌木：鸡蛋花、紫荆、紫薇、小叶女贞。竹类：慈竹。宅旁栽植叶子花、八仙花、杜鹃等营造舒适轻松的环境。
居住区公园山地以保护原有的植被为主，并再原有基础上增加丘北云杉、丘北槭、滇藏木兰等。
C：度假休闲区
此区域利用各种乔灌木草本搭配营造优美的度假休憩环境。采用的植物有常绿乔灌木董棕、四季桂、海桐、滇丁香、油茶、云南黄素馨。落叶乔灌木：紫荆、紫薇、山玉兰、二乔玉兰、金叶女贞、紫叶小檗。并运用一二年生草本花卉、宿根花卉配置花坛：彩叶草、宿根福禄考、羽衣甘蓝等。沿河绿地采用垂柳、冬樱花、红枫、大花白杜鹃、八角金盘、杜鹃、菖蒲配置。水上种植大片荷花体现地方特色。
D：公共建筑周围环境绿地
医院附属绿地植物配置中需要强化凸显特殊种类植物对于人们心理和生理的调节作用建议采用松科、槭树科、木兰科、忍冬科的植物。如丘北云杉、日本花柏、圆柏、松、白玉兰等。
公共建筑周围采用香樟、四季桂、广玉兰。

植物配置示意图

道路竖向规划图

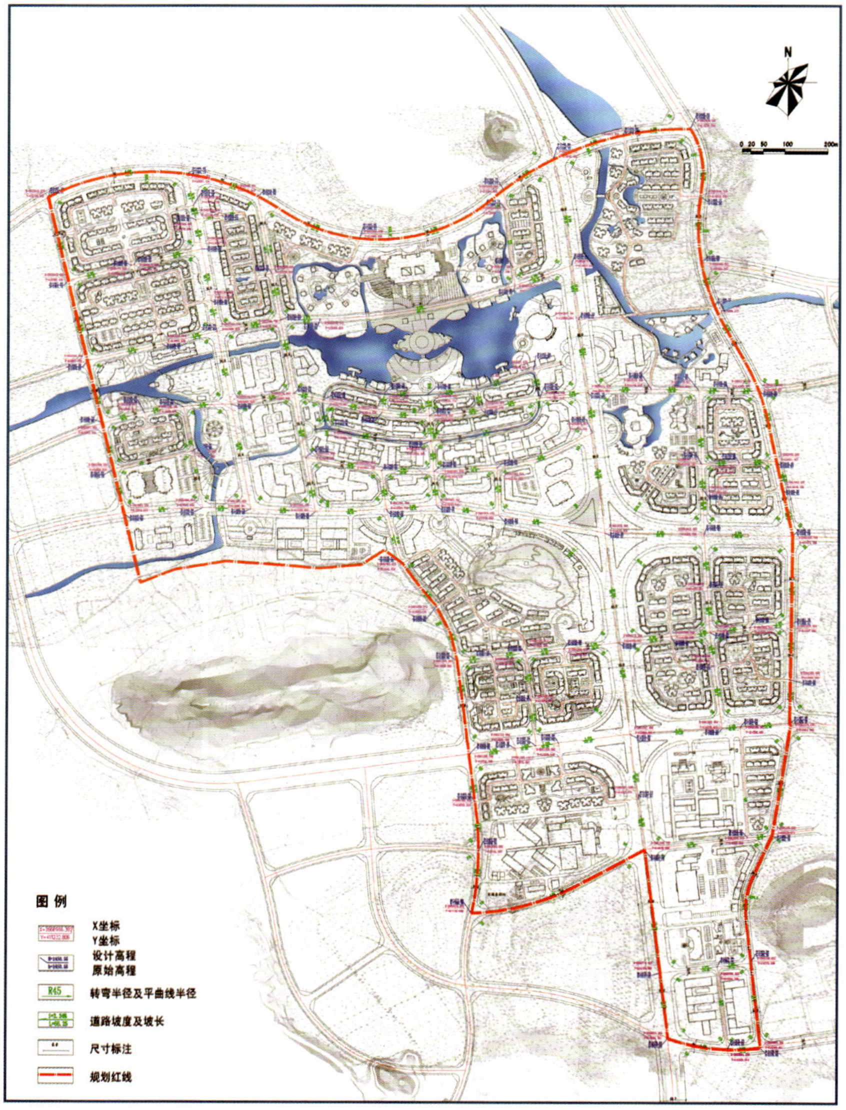

电力工程规划图

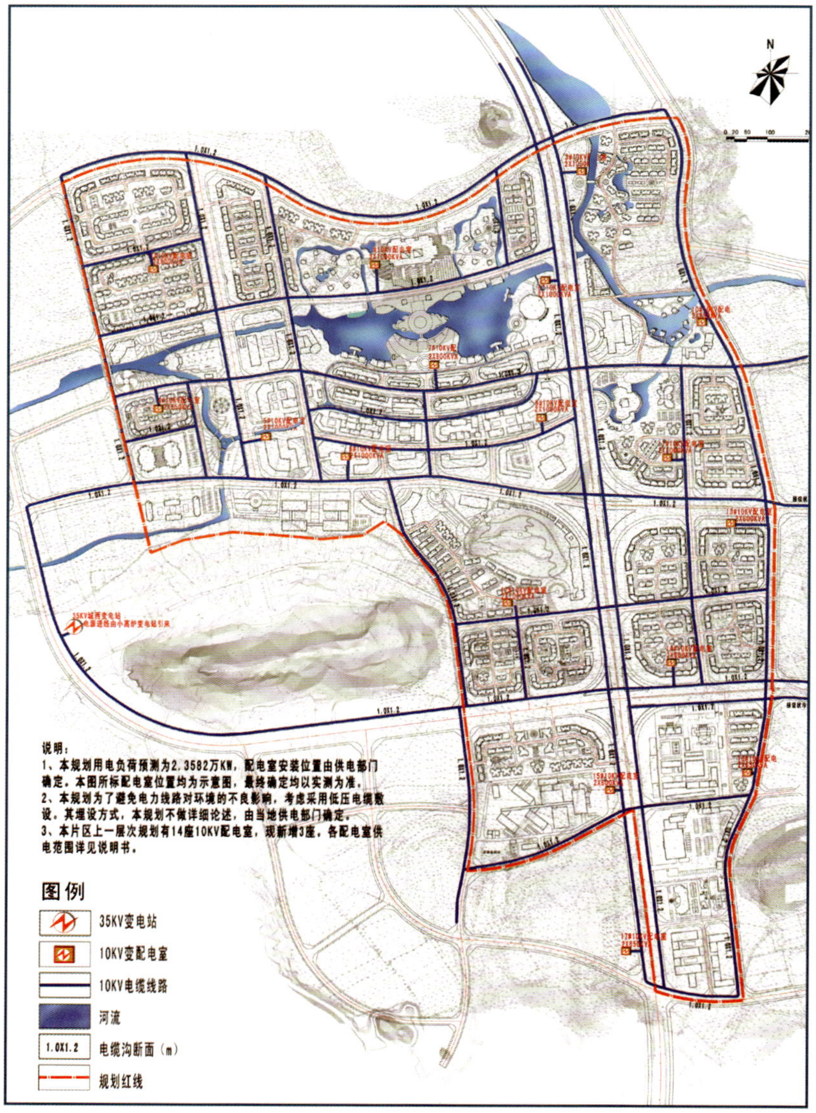

电信工程规划图

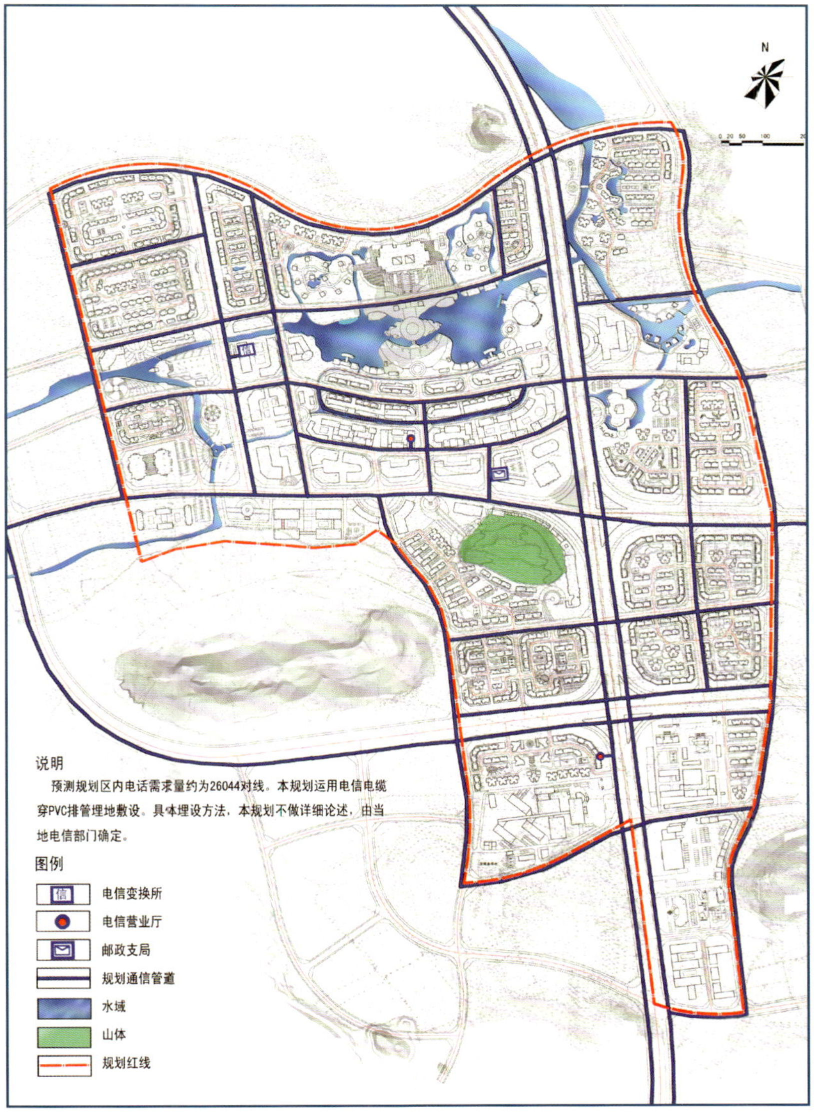

给水工程规划图

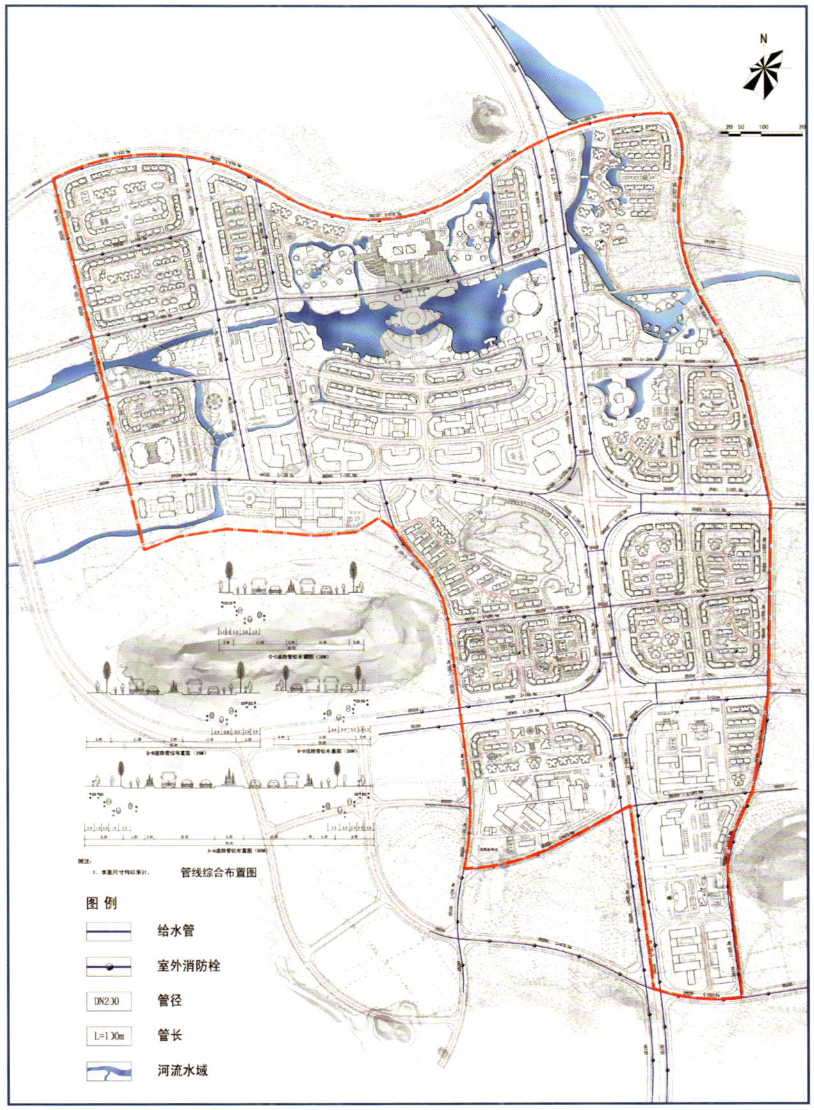

雨水工程规划图

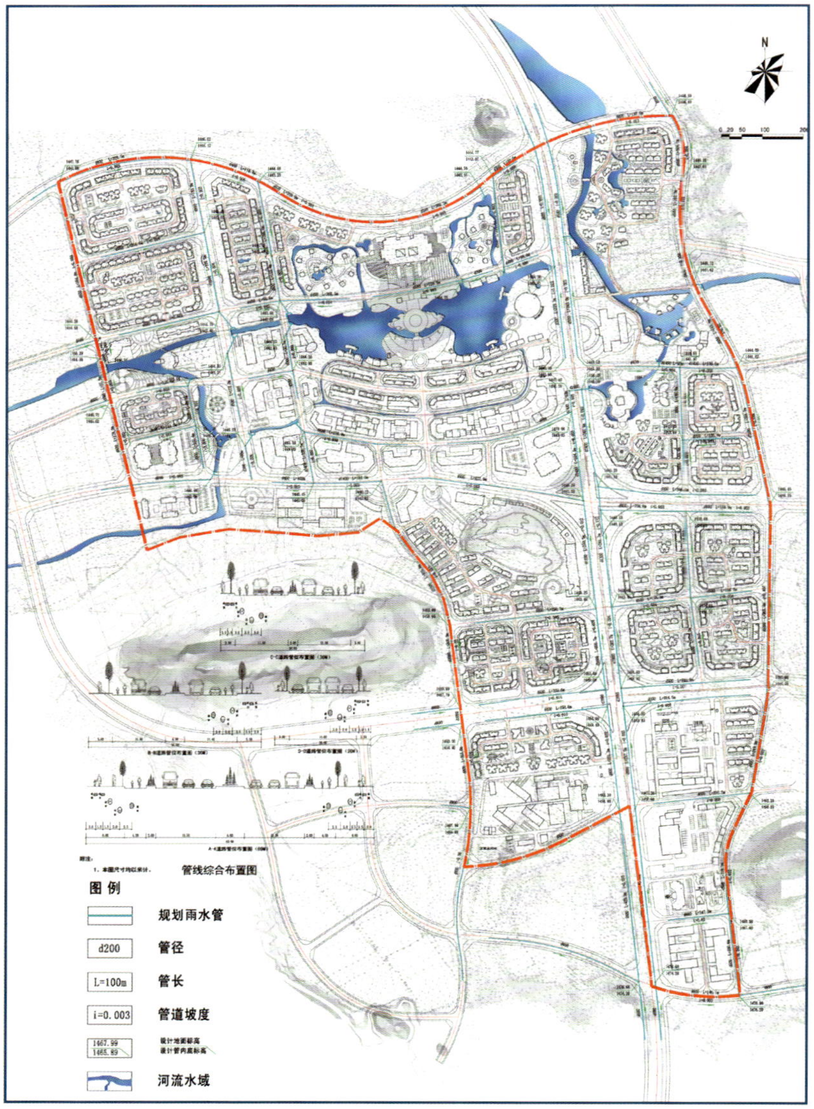

污水工程规划图

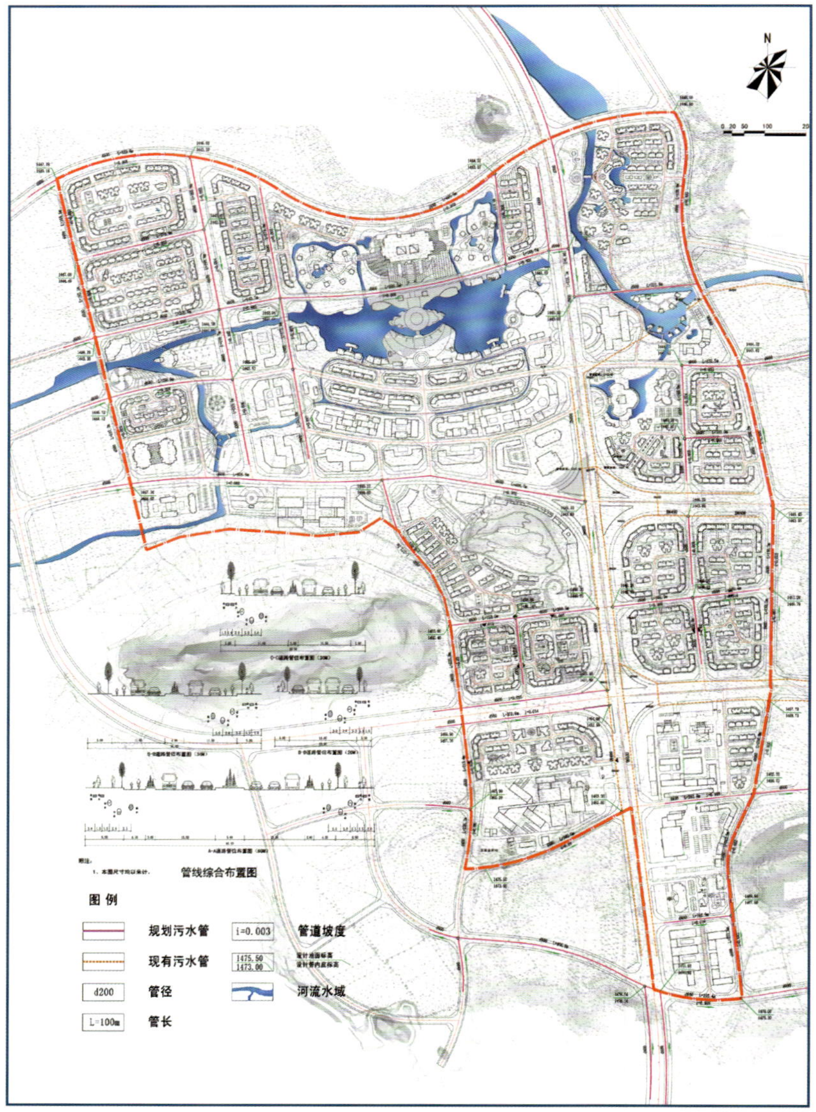

地块 3、4 鸟瞰效果图

荷塘月色酒店及表演广场鸟瞰图

①

①～②荷塘月色酒店效果图

②

①

②

①县歌剧院鸟瞰图②县歌剧院效果图③县歌剧院周边景观④县歌剧院周边景观⑤县歌剧院周边景观

③

④

⑤

风情街鸟瞰图

风情街效果图

①

②

①风情街入口牌坊 ②～④风情街效果图

风情街透视效果图

风情街透视效果图

HUGO
OPEN

①～②风情街透视效果图

③～④风情街夜景透视效果图

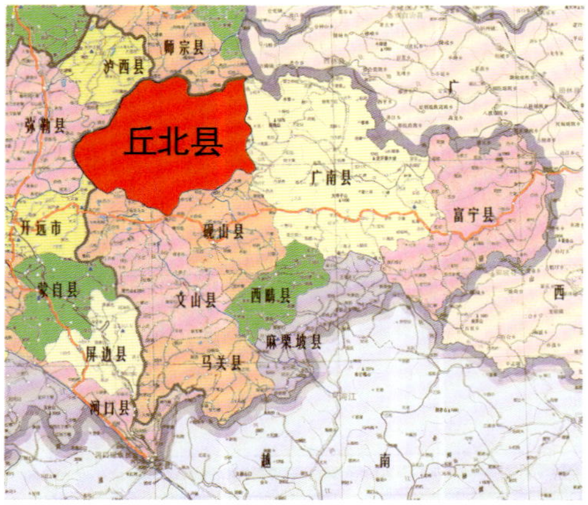

丘北县在文山州的位置

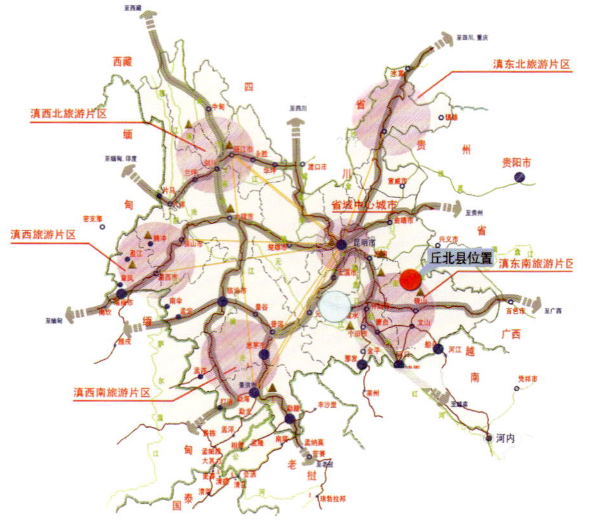

旅游区位示意

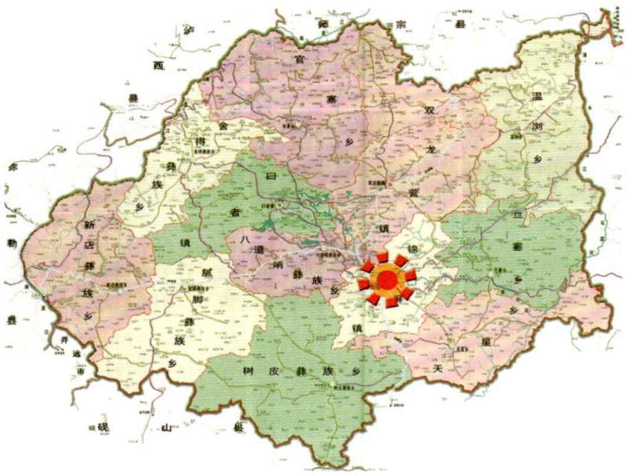

丘北县城在县域的位置

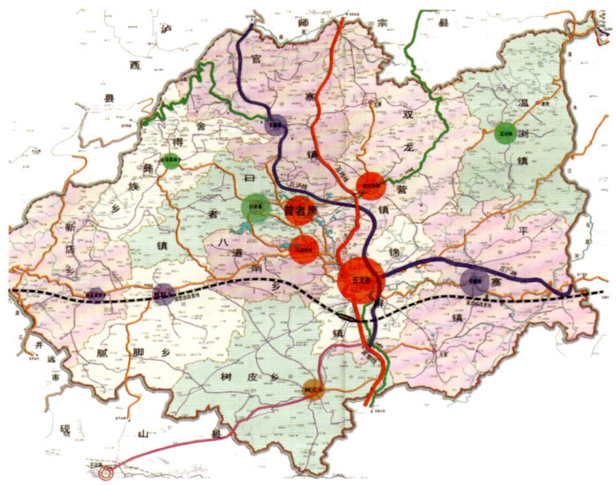

交通区位示意

清水河流域规划区在丘北城区的位置

地理区位示意

1 地理区位优势

文山州位于中国--东盟自由贸易区和泛珠三角经济区两大国际国内区域合作的结合部,区位条件优越。丘北县距州府文山78km,西临南盘江,东隔清水江与广南县毗邻,南与砚山县、开远市接壤,北与师宗县,广西壮族自治区西林县衔接。清水河流域位于丘北中心片区和普者黑景区之间,地理区位优势明显。

2 交通区位分析

丘北县内现有东西向的305以及南北向的206省道,新建的丘广公路、丘泸公路、丘师公路、普炭公路等,构成了西至弥勒、泸西,北至师宗、曲靖,南至砚山、炭房、广西,东至广南的公路交通格局。即将开始建设的南昆铁路复线将使整个丘北的交通通达性大大提高。县城距离即将通航的文山普者黑机场85km,为丘北的经济发展带来了便捷的空中通道优势。清水河流域规规划区西面为普炭公路,东面为305省道,交通区位优势明显。

3 旅游区位优势

临近普者黑4A级国家旅游区和小新寨旅游服务组团,借普者黑的旅游资源和小新寨旅游服务功能,带动区域经济文化的发展。

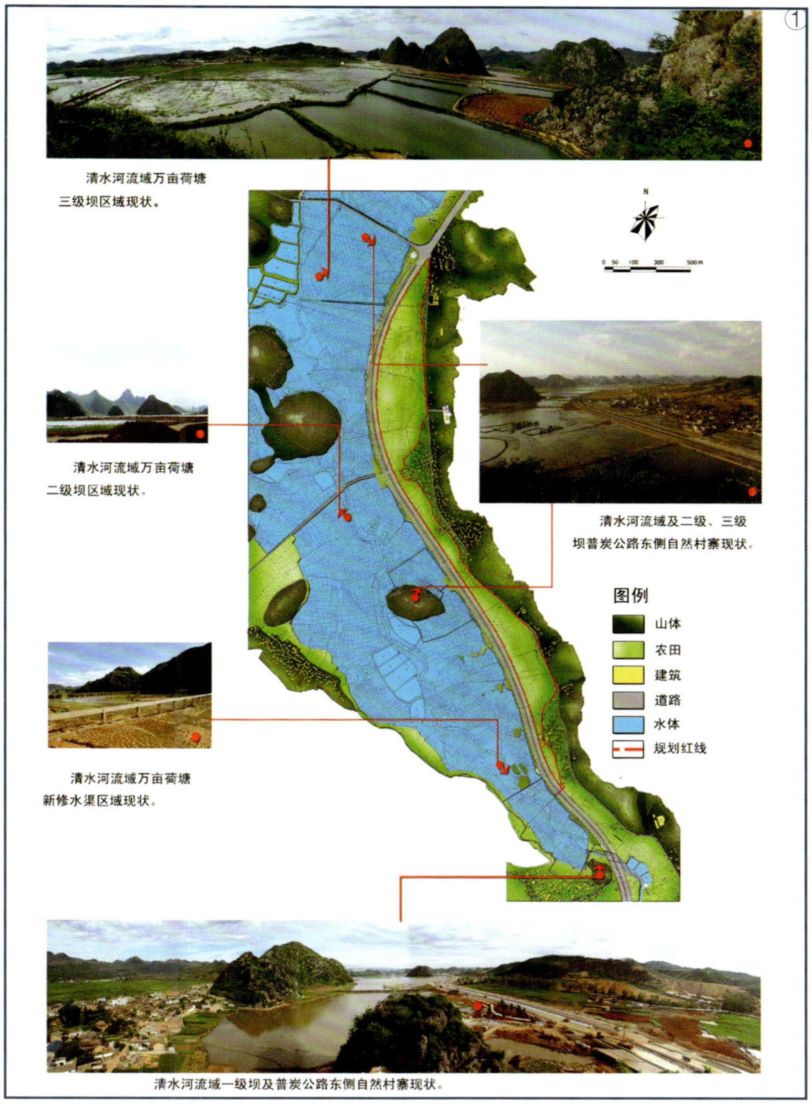

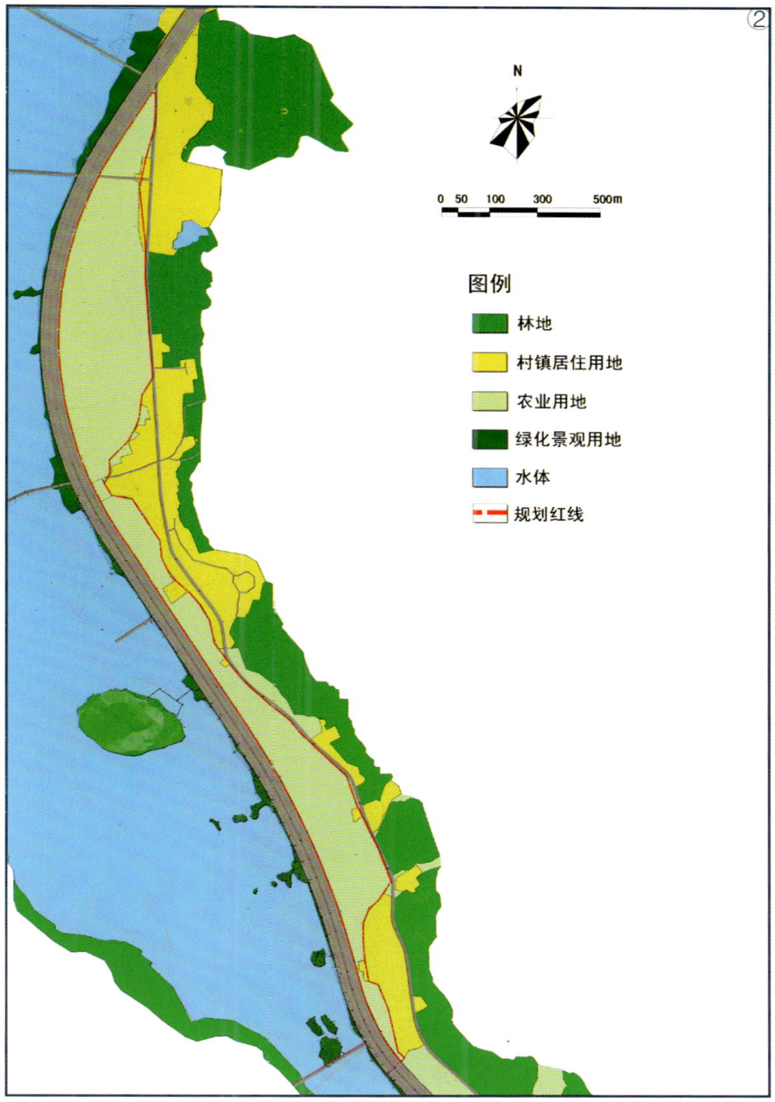

①现状实景图 ②用地现状图 ③入口人行天桥效果图

Planning and Design of Puzhehei Scenic Spot

普者黑景区规划设计

①入口景观山效果图 ②景区入口效果图 ③景区入口大门

Wanmu Lotus pond
万亩荷园

八角山总平面图

水上运动区总平面图

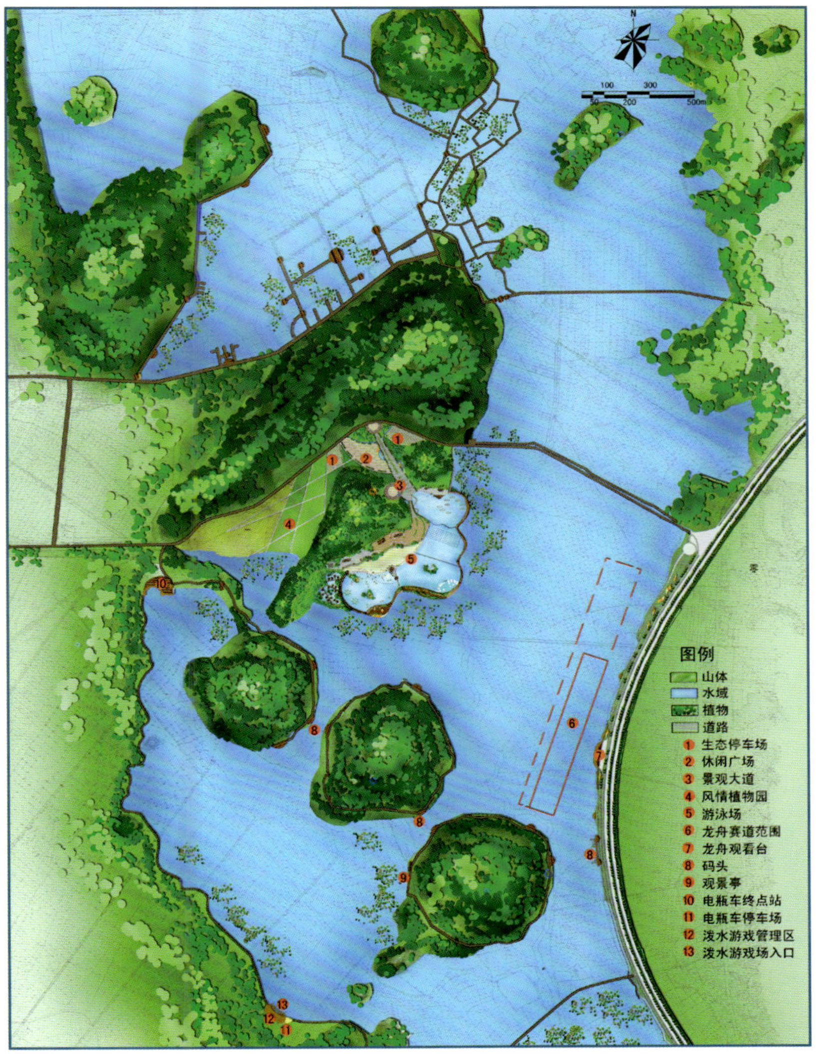

水上运动区功能结构分析图

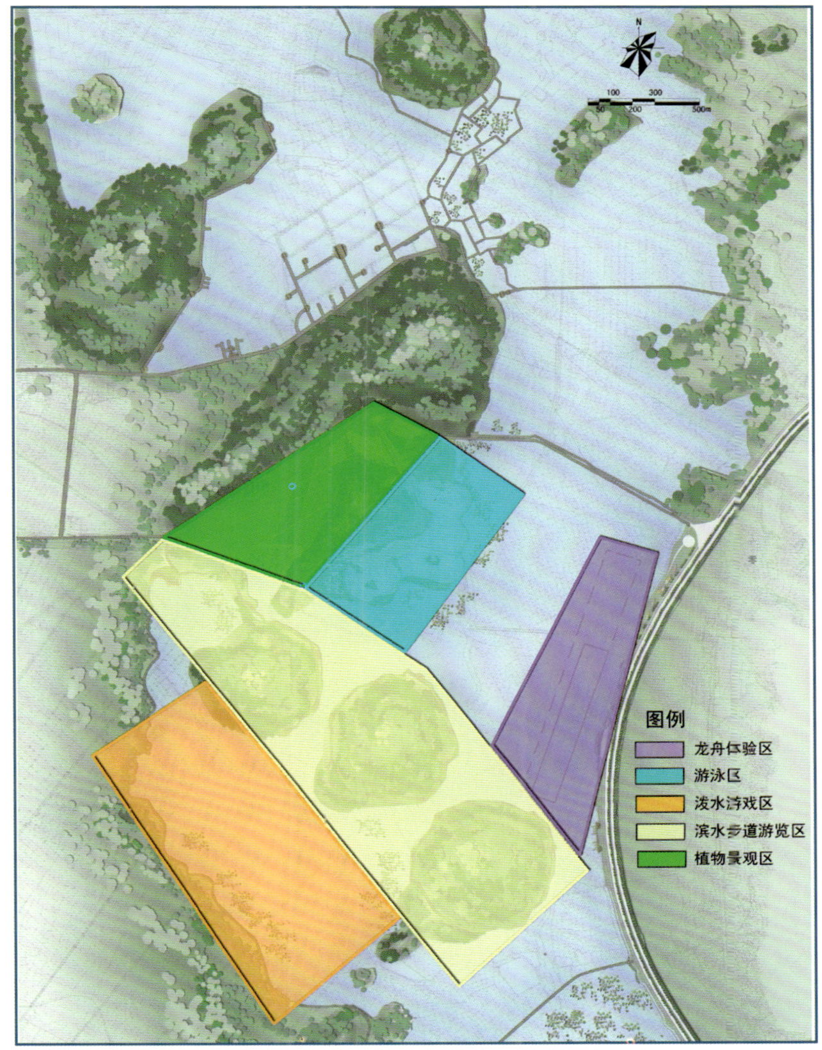

水上运动区综合交通分析图

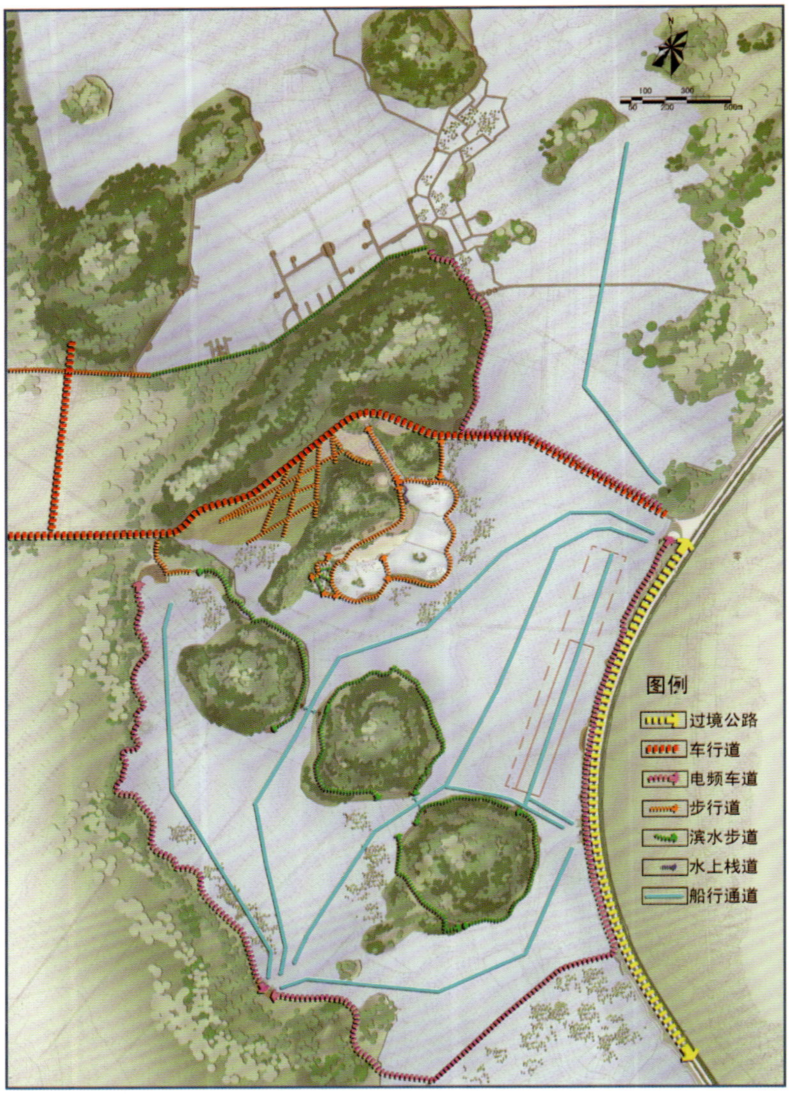

游泳场总平面图

N
0 20 50 100m
图例
1 景观大道
2 生态停车场
3 广场
4 综合管理区
5 特色餐饮区
6 无烟烧烤区
7 人造沙滩区
8 娱乐包房区
9 水球区
10 游泳区
11 家庭综合区
12 按摩池区
13 儿童设施区
14 幼儿戏水区
15 阳光浴场
16 休息看护区
17 风情植物园

游泳场竖向规划图

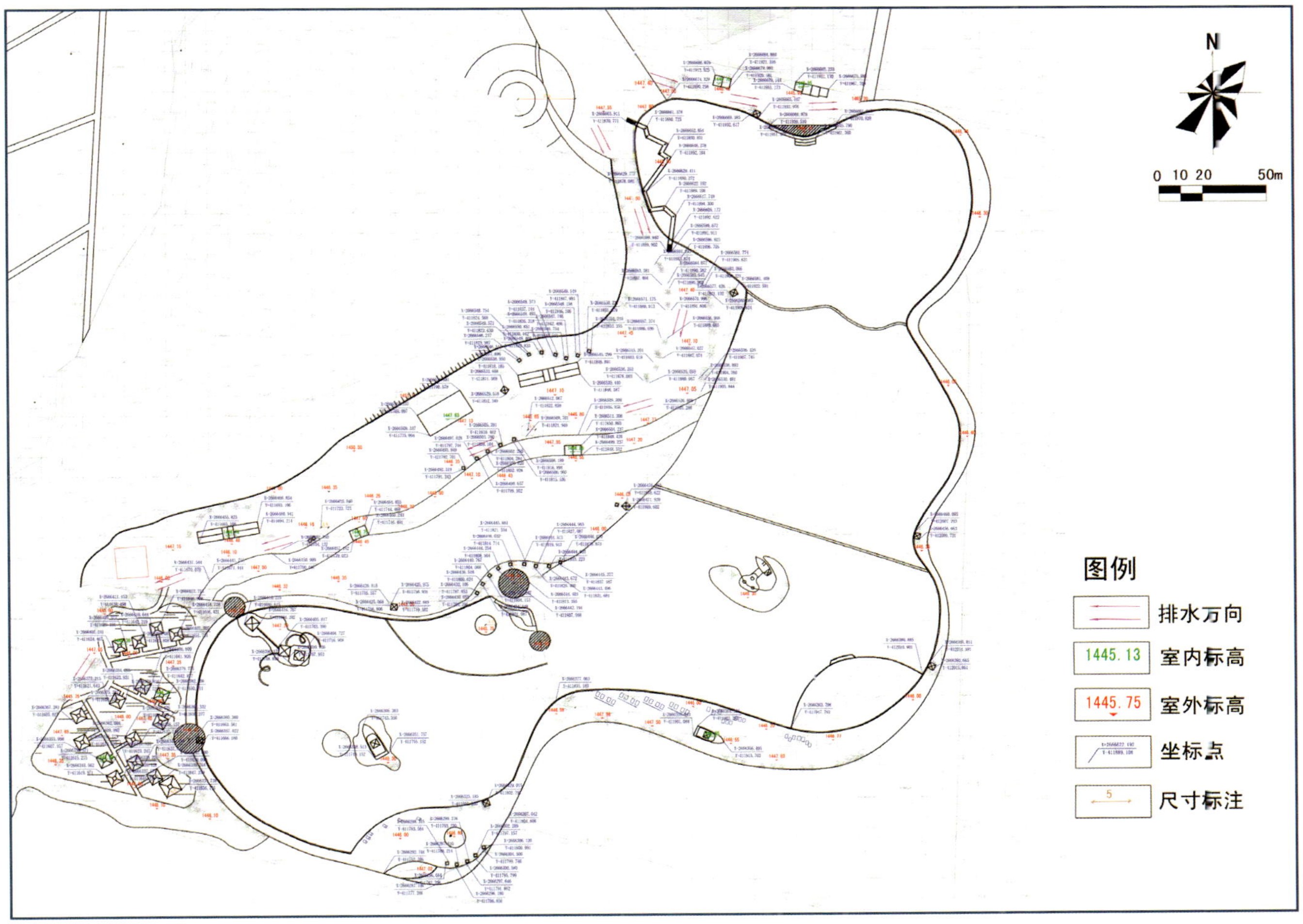
N
0 10 20 50m
图例
排水万向
1445.13 室内标高
1445.75 室外标高
坐标点
尺寸标注

①

①～②万亩荷田局部鸟瞰效果图

②

③～④景区风雨桥效果图

①八角山景区局部鸟瞰效果图 ②～③八角山局部景观透视效果图 ④八角山观瀑台 ⑤景观隔离带

游泳区鸟瞰效果图

①～②游泳区局部效果图

①

②

③

①～②泼水游戏区鸟瞰效果图 ③泼水游戏区透视效果图

Planning and Design of Folk Garden of Zhuang

壮族风情园规划设计

总平面图

功能结构规划图

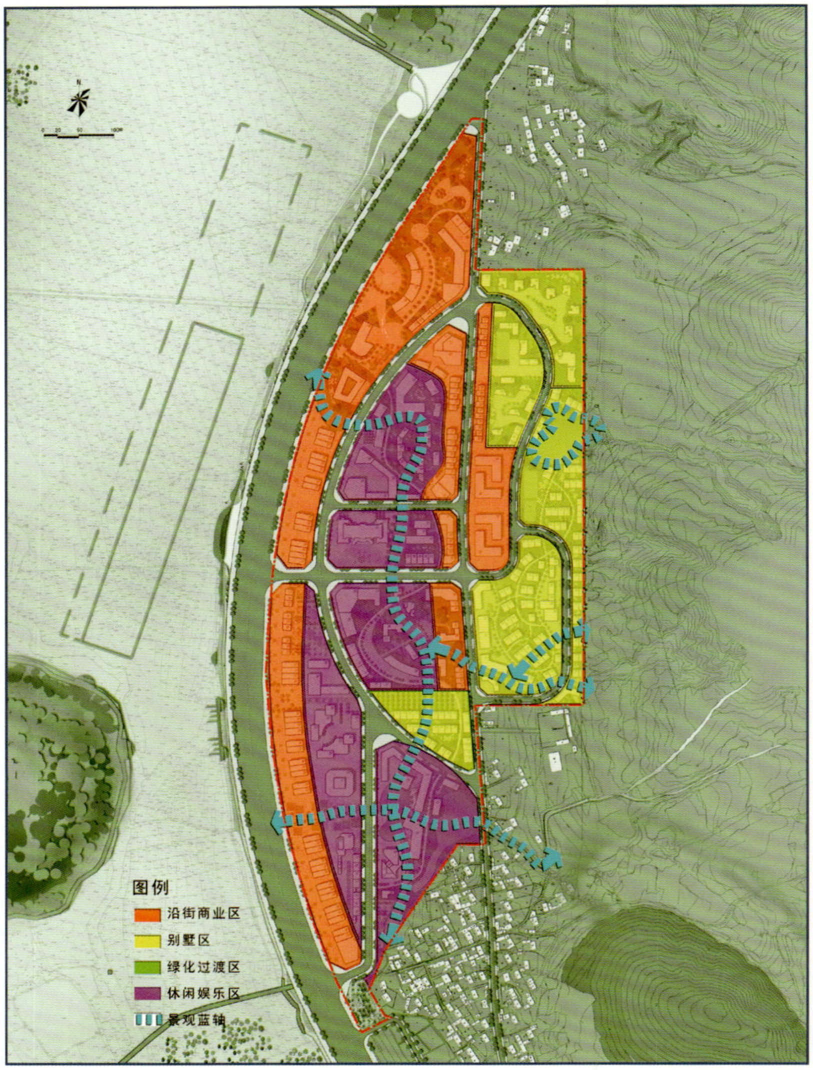

道路交通规划图

公共服务设施规划图

绿地景观规划图

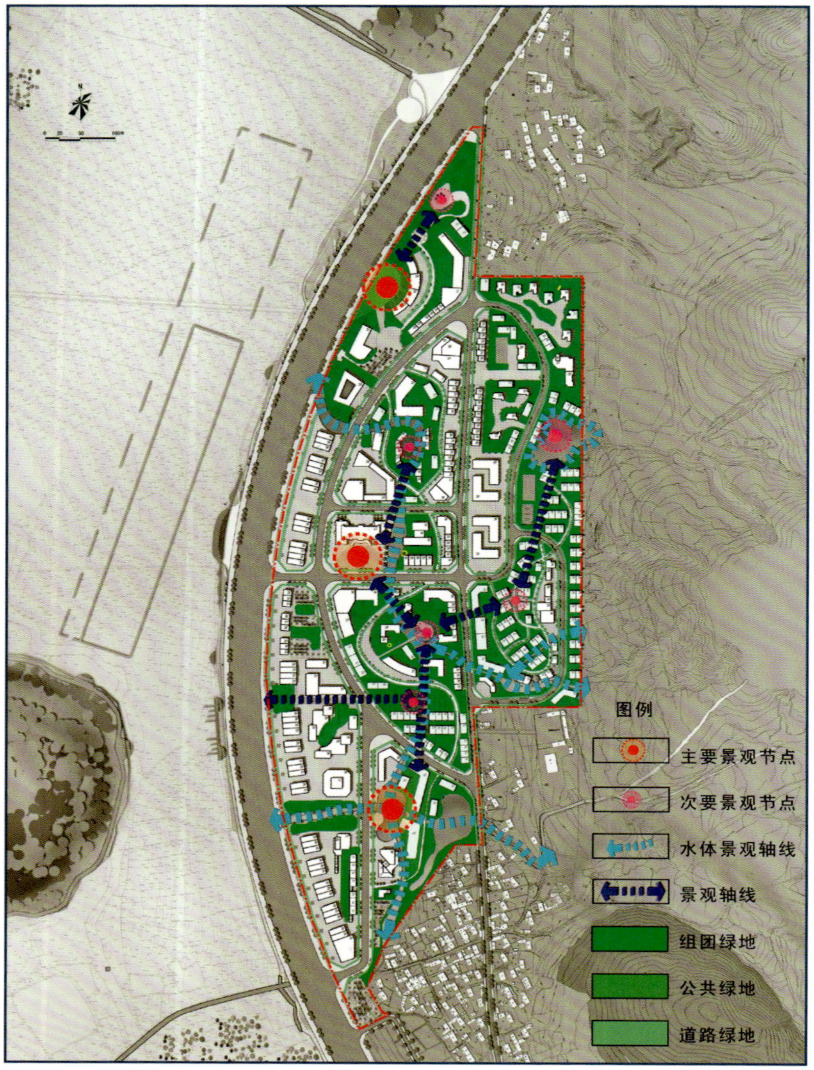

植物配置示意图

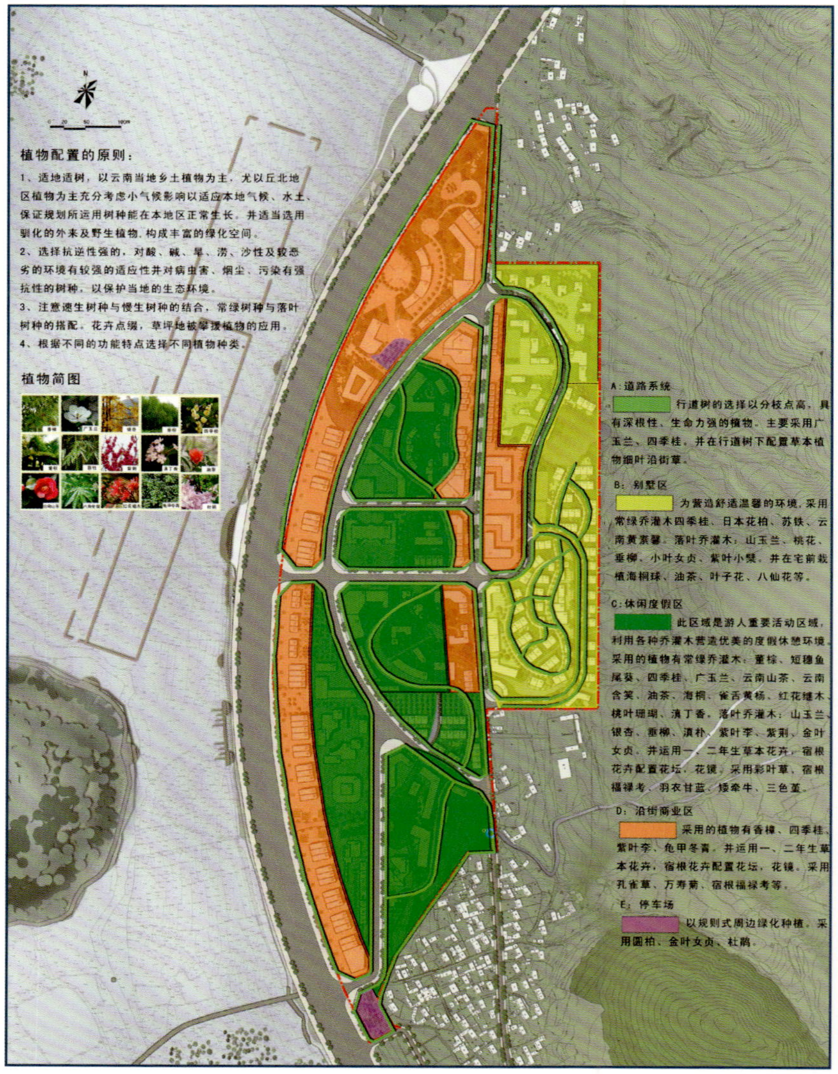

竖向规划图

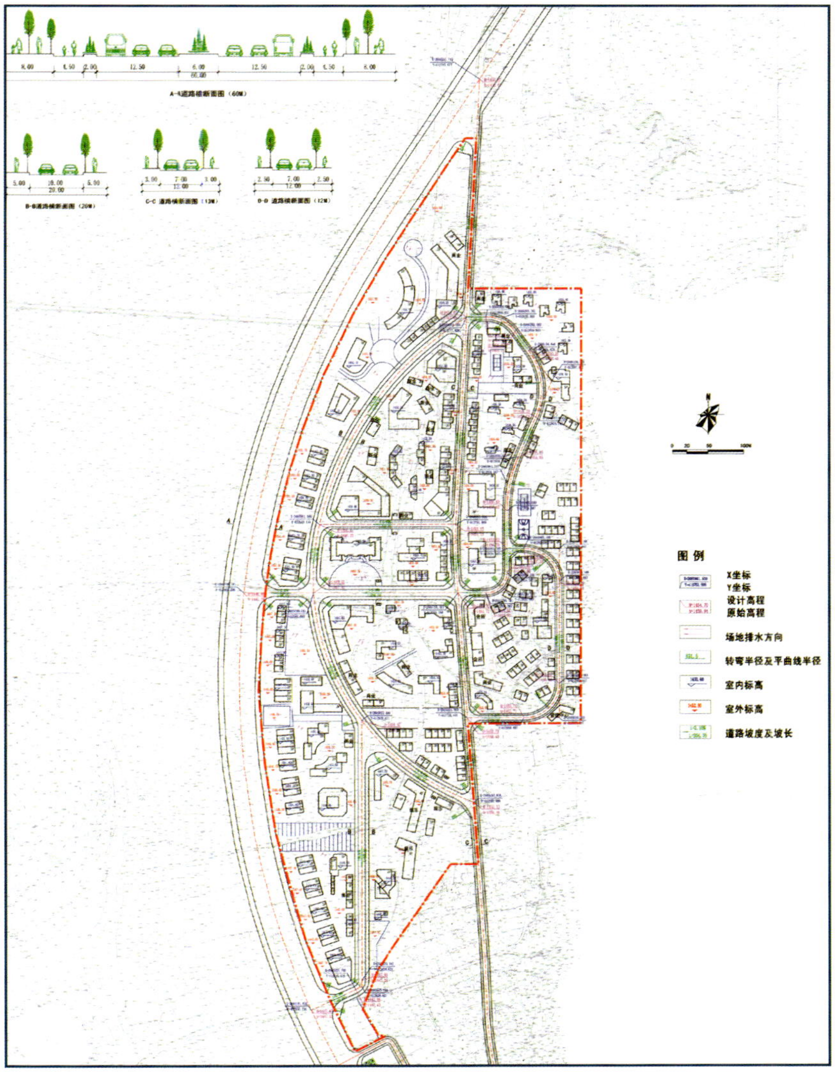
图 例
X坐标
Y坐标
设计高程
原始高程
场地排水方向
转弯半径及平曲线半径
室内标高
室外标高
道路坡度及坡长

建筑定位图

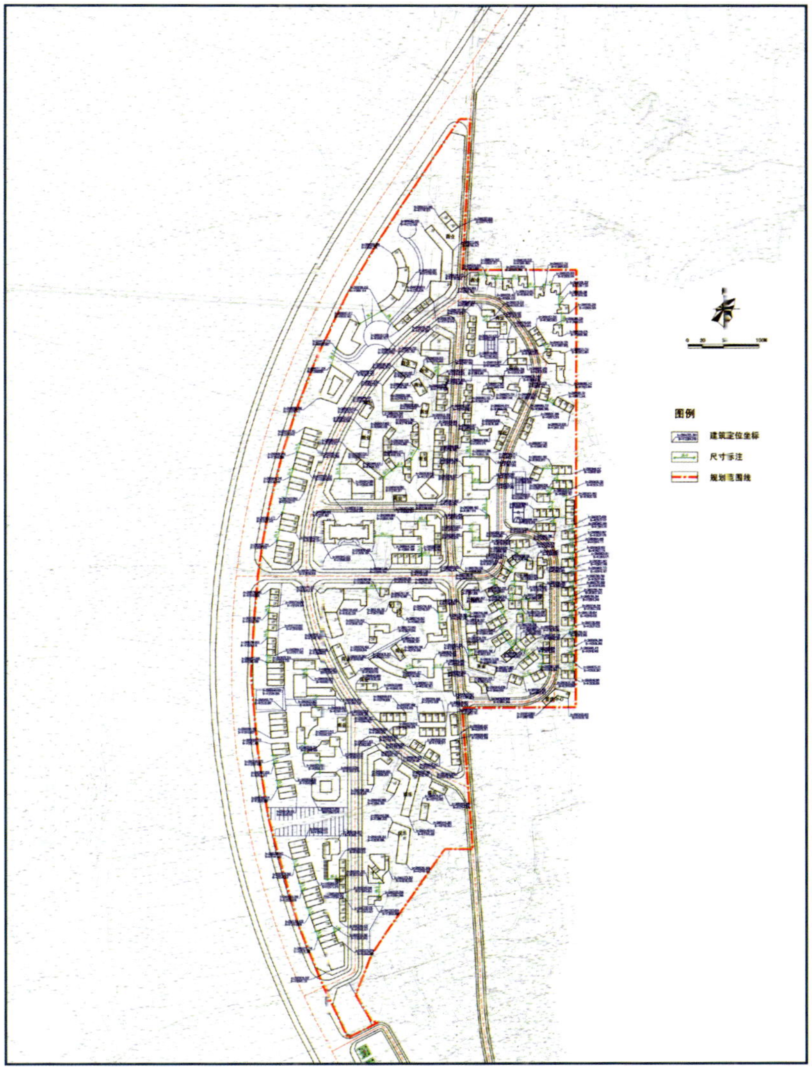
图例
建筑定位坐标
尺寸标注
规划范围线

给水工程规划图

雨水工程规划图

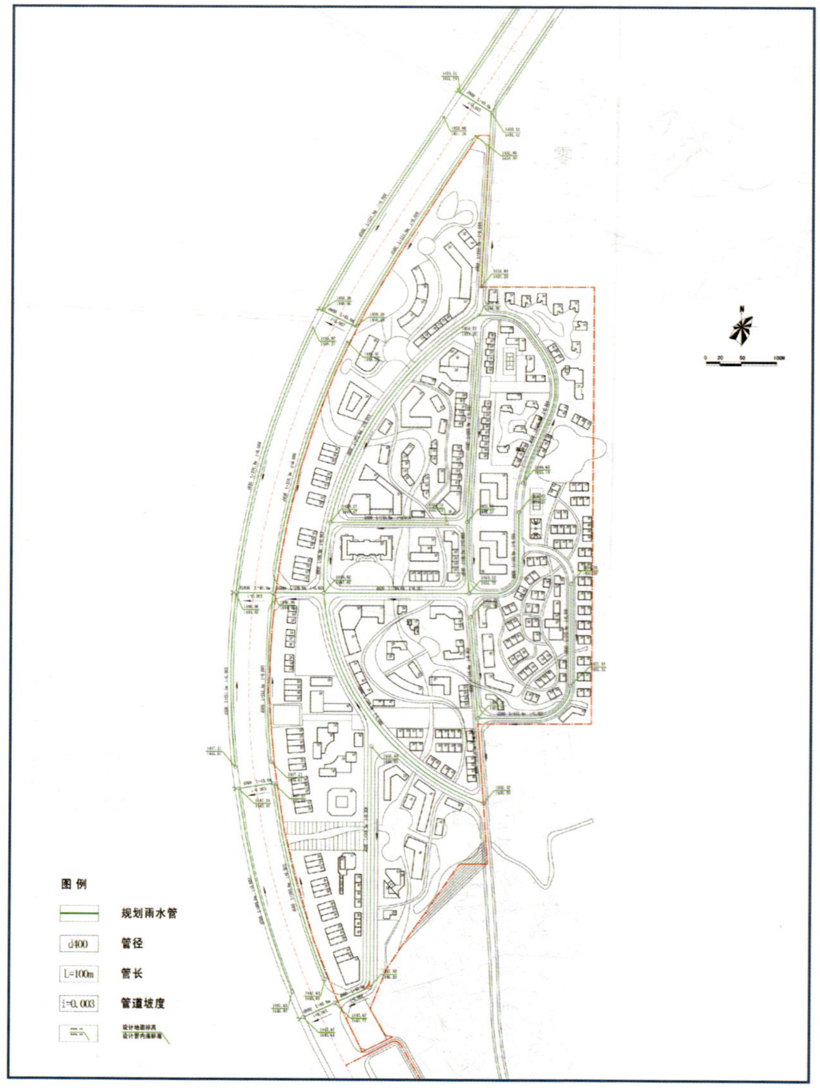

污水工程规划图

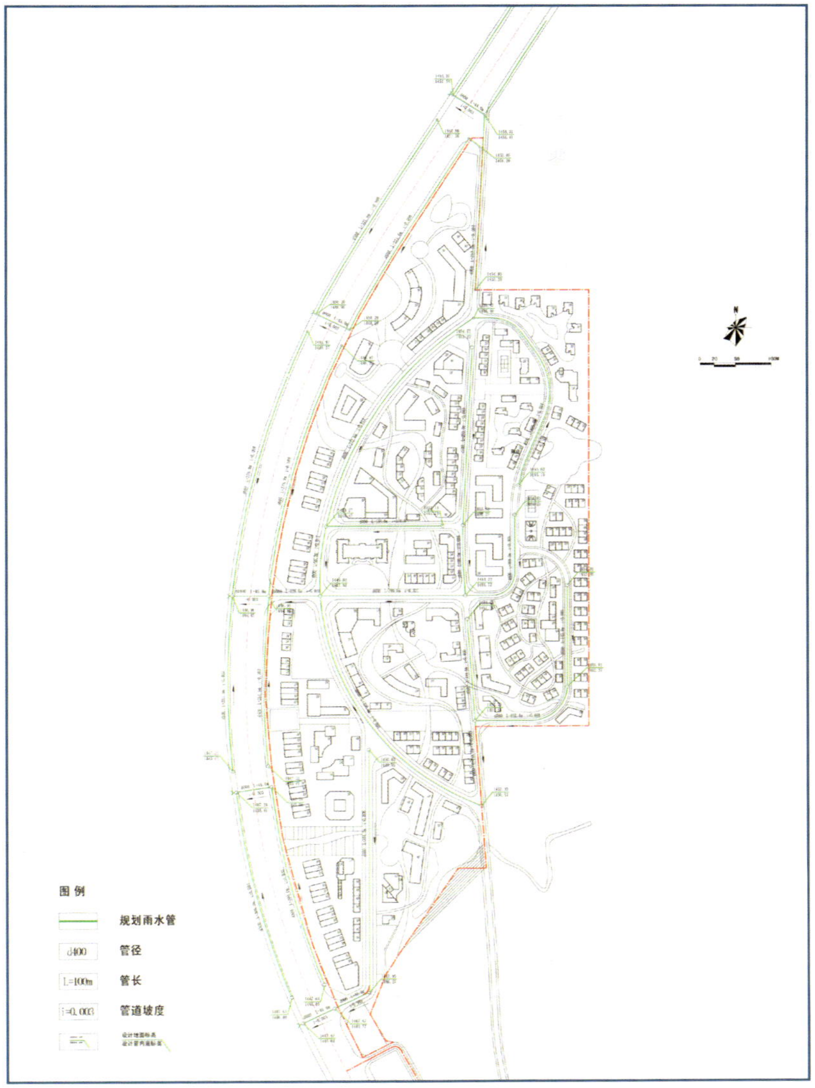

电力工程规划图

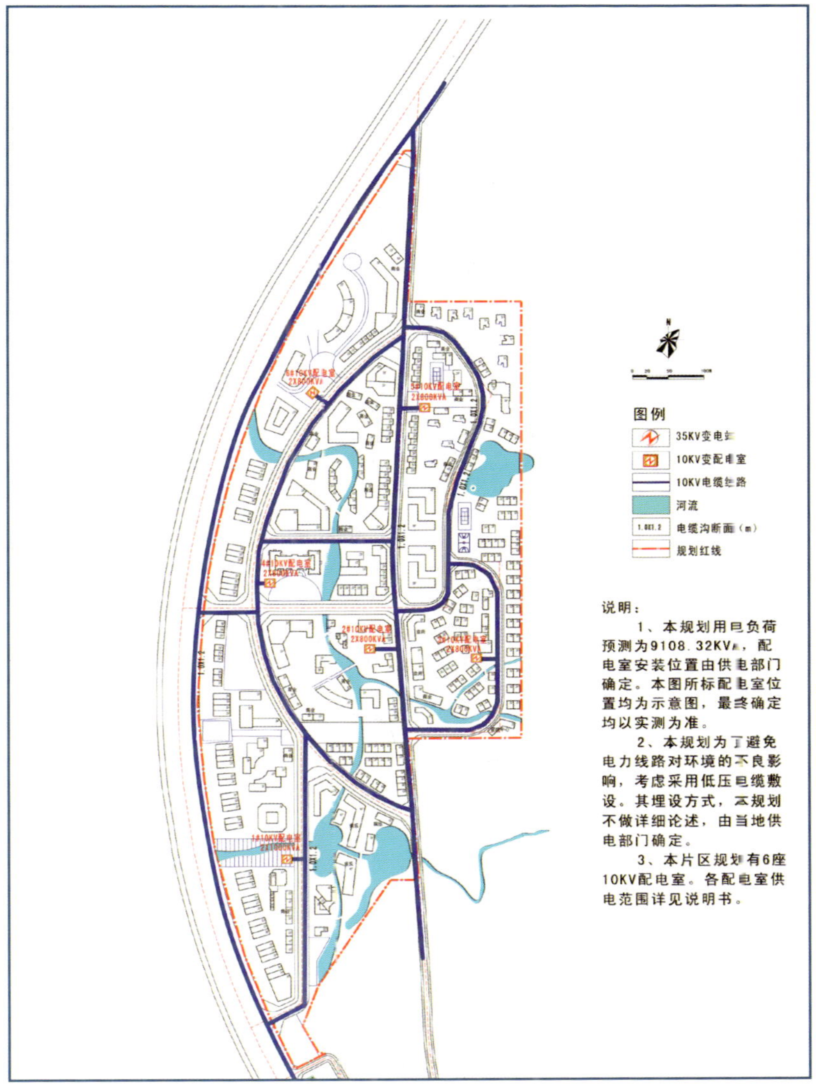

①

②

①食府鸟瞰效果图 ②食府透视效果图 ③食府大门效果图 ④食府透视效果图

①

②

③

①农家乐效果图 ②农家乐效果图 ③步行街效果图

①

②

①酒店效果图 ②酒店透视效果图 ③～④壮族风情园入口牌坊效果图

①

①～②沿街建筑透视效果图

②

Reference Documentation

参考文献：

[1] 杨永春．中国西部河谷型城市的发展和空间结构研究［D］．南京：南京大学，2003.
[2] 童乔慧．澳门城市环境与文脉研究［D］．南京：东南大学，2004.
[3] 邹耀勇．巴黎城市发展与保护史论［D］．上海：华东师范大学，2007.
[4] 蔡晓丰．城市风貌解析与控制［D］．上海：同济大学，2005.
[5] 李辉．城市公共空间的绿色建筑体系研究［D］．长春：东北师范大学，2004.
[6] 庄艳．城市化进程中历史地段的保护与更新——创造新地域建筑文化［D］．合肥：合肥大学，2001.
[7] 王建国．梁燕枫，刘海金．城市建筑文化问题浅析［J］．沈阳建筑大学学报（社会科学版）.2008,10：141–144.
[8] 陈宇．城市街道景观设计文化研究［D］．南京：东南大学，2006.
[9] 田国行．城市绿地景观规划的理论与方法［D］．北京：中国农业大学，2004.
[10] 陈李波．城市美学四题［D］．武汉：武汉大学，2006.
[11] 侯正华．城市特色危机与城市建筑风貌的自组织机制［D］．北京：清华大学，2003.
[12] 纪晓海．城市中心区历史建筑的保护［D］．大连：大连理工大学，2005.
[13] 胡仁茂．大空间建筑设计研究［D］．上海：同济大学，2006.
[14] 袁志准．当代中国城市建筑文化嬗变与社会转型［J］．艺术百家．2007,6：51–53.
[15] 胡群．对城市精神与建筑文化的思考［J］．中共福建省委党校学报．2008,7：94–96.
[16] 吴涛．梅洪元．关于城市更新与文脉和谐的思考［J］．华中建筑．2008,10：148–150.
[17] 秦军．贺忠超，李新家．贵州黎平港澳新区城市设计探讨［J］．规划师．2008,7：67–70.
[18] 陈洁行．建筑——城市文化的具象表现［J］．浙江树人大学学报．2008,8：83–86.
[19] 李蕾．建筑与城市的本土观［D］．上海：同济大学，2006.
[20] 高源．美国现代城市设计运作研究［D］．南京：东南大学，2005.
[21] 林玮．黔北风情一条街的设计探索［D］．重庆：重庆大学，2007.
[22] 康志忠．数字城市中街道景观主体三维可视化的快速实现研究［D］．武汉：武汉大学，2004.
[23] 黄麟．我国城市价值居所理论与发展探析［D］．成都：西南财经大学，2005.
[24] 李亮．中国城市规划变革背景下的城市设计研究［D］．北京：清华大学，2006.
[25] 梁梅．中国当代城市环境设计的美学分析与批判［D］．北京：中央美术学院，2005.
[26] 鲍宗豪等．城市的素质、风骨与灵魂［M］．上海：上海人民出版社，2005：16–22，41,48,320–354.
[27] 李德华．城市规划原理［M］．北京：中国建筑工业出版社，2001,2,193–200,479.
[28] 曹伟等．城市规划设计十二讲［M］．北京：机械工业出版社，2008：2–7，23–28,140–150.